Probability and Mathematical Statistics (Continu

SCHEFFE · The Analysis of Variance
SEBER · Linear Regression Analysis
WILKS · Mathematical Statistics
WILLIAMS · Diffusions, Markov Processes, and Martingales, Volume I: Foundations
ZACKS · The Theory of Statistical Inference

Applied Probability and Statistics

BAILEY · The Elements of Stochastic Processes with Applications to the Natural Sciences
BAILEY · Mathematics, Statistics and Systems for Health
BARNETT and LEWIS · Outliers in Statistical Data
BARTHOLOMEW · Stochastic Models for Social Processes, *Second Edition*
BARTHOLOMEW and FORBES · Statistical Techniques in Manpower Planning
BECK and ARNOLD · Parameter Estimation in Engineering and Science
BELSLEY, KUH, and WELSCH · Regression Diagnostics: Identifying Influential Data and Sources of Collinearity
BENNETT and FRANKLIN · Statistical Analysis in Chemistry and the Chemical Industry
BHAT · Elements of Applied Stochastic Processes
BLOOMFIELD · Fourier Analysis of Time Series: An Introduction
BOX · R. A. Fisher, The Life of a Scientist
BOX and DRAPER · Evolutionary Operation: A Statistical Method for Process Improvement
BOX, HUNTER, and HUNTER · Statistics for Experimenters: An Introduction to Design, Data Analysis, and Model Building
BROWN and HOLLANDER · Statistics: A Biomedical Introduction
BROWNLEE · Statistical Theory and Methodology in Science and Engineering, *Second Edition*
BURY · Statistical Models in Applied Science
CHAMBERS · Computational Methods for Data Analysis
CHATTERJEE and PRICE · Regression Analysis by Example
CHERNOFF and MOSES · Elementary Decision Theory
CHOW · Analysis and Control of Dynamic Economic Systems
CLELLAND, deCANI, BROWN, BURSK, and MURRAY · Basic Statistics with Business Applications, *Second Edition*
COCHRAN · Sampling Techniques, *Third Edition*
COCHRAN and COX · Experimental Designs, *Second Edition*
COX · Planning of Experiments
COX and MILLER · The Theory of Stochastic Processes, *Second Edition*
DANIEL · Biostatistics: A Foundation for Analysis in the Health Sciences, *Second Edition*
DANIEL · Applications of Statistics to Industrial Experimentation
DANIEL and Wood · Fitting Equations to Data: Computer Analysis of Multifactor Data, *Second Edition*
DAVID · Order Statistics
DEMING · Sample Design in Business Research
DODGE and ROMIG · Sampling Inspection Tables, *Second Edition*
DRAPER and SMITH · Applied Regression Analysis

continued on back

NORMAN L. JOHNSON

University of North Carolina, Chapel Hill

SAMUEL KOTZ

Temple University, Philadelphia

discrete

distributions in statistics

distributions

A WILEY-INTERSCIENCE PUBLICATION

JOHN WILEY & SONS,

New York • Chichester • Brisbane • Toronto • Singapore

ISBN 0-471-44360-3

PRINTED IN THE U.S.A.

10 9 8 7 6

General Preface

The purpose of this compendium is to give an account of the properties and uses of statistical distributions at the present time. The idea was originally suggested by one author (Samuel Kotz) while he was visiting the University of North Carolina at Chapel Hill as a Research Associate in 1962–1963. He felt that a collection of the more important facts about the commonly used distributions in statistical theory and practice would be useful to a wide variety of scientific workers.

While preparing the book, it became apparent that "important" and "commonly" needed to be rather broadly interpreted because of the differing needs and interests of possible users. We have, therefore, included a considerable amount of detailed information, often introducing less frequently used distributions related to those other which are better-known. However, we have tried to adhere to our original intention of excluding theoretical minutiae of no apparent practical importance. We do not claim our treatment is exhaustive, but we hope we have included references to most practical items of information.

This is not an introductory text-book on general statistical theory; it is, rather, a book for persons who wish to apply statistical methods, and who have some knowledge of standard statistical techniques.

For this reason no systematic development of statistical theory is provided beyond what is necessary to make the text intelligible. We hope that the book will also prove useful as a textbook and a source book for graduate courses and seminars in statistics, and possibly for discussion or seminar groups in industrial research laboratories.

Some arbitrary choices have been unavoidable. One such choice deserves special explanation: methods of estimation for parameters occurring in the specification of distributions have been included, but tests of significance have been excluded. We felt that the form of distribution is usually well-established in situations associated with estimation, but not in cases using tests of significance. At this point we would like to emphasize the usefulness of a rough classification of distributions as 'modelling' or 'sampling' distributions. These classes are neither mutually exclusive nor exhaustive. However, it is true that some distributions (e.g., Wishart, or correlation coefficient) are almost always encountered only as a consequence of application of sampling theory to well-established population distributions. Other distributions (e.g., Weibull or logistic) are mostly used in the construction of models of variations in real populations.

We have tried to arrange the order of discussion to produce a text which develops as naturally as possible. Although some inversions have appeared desirable in the discussion of some distributions, there are broad similarities in all the treatments and the same kind of information is provided in each case. The "historical remarks" found in the discussions are not intended to be full accounts, but are merely to help assess the part played by the relevant distribution in the development of applied statistical methods.

We have included at the end of each chapter a fairly comprehensive bibliography including papers and books mentioned in the text. This inevitably produces some repetition of titles, but is, in our opinion, important to a reader who would like a more extended treatment of a particular topic.

References are indicated by numbers in square brackets in the text e.g. [4]. References and equations are numbered separately in each chapter.

The production of a work such as the present necessarily entails the assembly of substantial amounts of information, much of which is only available from published papers, and even less readily accessible sources. We would like to thank Miss Judy Allen, Miss Lorna Lansitie, Mr. K. L. Weldon, Mr. H. Spring, Mr. B. G. A. Kelly, Mr. B. Zemmel, Mr. P. Cehn, Mr. W. N. Selander and Dr. T. Sugiyama for much help in this respect.

We are particularly indebted to Professor Herman Chernoff who read our manuscript, for valuable comments and suggestions on matters of basic importance; and to Dr. D. W. Boyd and Mr. E. E. Pickett for their assistance in compilation of the preliminary drafts of the chapters on non-central χ^2 distribution, quadratic forms, and Pareto and inverse Gaussian distributions.

We would also like to record our appreciation of Miss D. Coles, Miss V. Lewis, Mrs. G. Bem, Miss C. M. Sawyer and Miss S. Lavender for their typing work and for other technical assistance.

The assistance of our wives, especially in proofreading, cannot be overestimated.

The financial support of the U. S. Air Force Office of Scientific Research and the stimulating assistance of Mrs. Rowena Swanson of the Information Sciences Division are sincerely appreciated.

The hospitality of the Department of Statistics at Chapel Hill during the summer of 1966 and Cornell University during the summer of 1968 aided in the solution of organizational difficulties connected with this project. The help of the Librarians at the University of Toronto, the University of North Carolina at Chapel Hill, Temple University, and Cornell University is also much appreciated.

In the course of the preparation of these volumes, we were greatly assisted by existing bibliographical literature on Statistical Distributions. Among them are F. A. Haight's *Index to the Distributions of Mathematical Statistics*, J. A. Greenwood and H. O. Hartley's *Guide to Tables in Mathematical Statistics*, G. P. Patil and S. W. Joshi's *Bibliography of Classical and Contagious Discrete Distributions*, W. Buckland's *Statistical Assessment of the Life Characteristic* and L. S. Deming's *Selected Bibliographies of Statistical Literature 1930 to 1957*.

Last but not least, we would like to thank the authors from many parts of the globe, too numerous to be mentioned individually, who so generously supplied us with the reprints of their publications and thus facilitated the compilation of these volumes.

Norman L. Johnson; Samuel Kotz

Preface

In order to avoid a long delay between the completion of the text and publication, and because of substantial additions to the literature in the recent years, it was decided to publish this work in three volumes.

A natural division is formed by the classes of discrete and continuous distributions. (Mixed distributions are not discussed explicitly, but only incidentally.) The present volume is concerned only with discrete distributions. However, reference to continuous distributions must inevitably occur — for example, in connection with limiting forms of a distribution. There will also be occasional references to topics (such as the Cornish-Fisher expansion) which will be more fully discussed in the succeeding volumes dealing with continuous distributions. Chapter 1 is an introduction *to the entire compendium*, not only to this volume. It need not be mastered in detail before proceeding to later chapters. It is a set of references to methods used at various points in the book and may be consulted as need arises.

Contents

3

Binomial Distribution 50

4

Poisson Distribution 87

5

Negative Binomial Distribution 122

6

Hypergeometric Distribution 143

7

Logarithmic Series Distribution

8

Some Compound, Generalized and Modified Discrete Distributions

9

Contagious Distributions

1

Preliminary Information

1. Introduction

This work contains descriptions of the many different distributions used in statistical theory and practice, each with its own peculiarities distinguishing it from others. However, as we will have occasion to use the same general ideas repeatedly, it is convenient to collect the appropriate definitions and methods in one place. This chapter does just that. The collection will incidentally serve the additional purpose of allowing us to explain the sense in which we shall use various terms throughout the books.

A further feature of the chapter is a collection of mathematical formulas which we shall use occasionally. In subsequent chapters, though there will be no reference to this chapter, refer to it when an unfamiliar and apparently undefined symbol is encountered. Give special notice to the symbols used in the calculus of finite differences, described in Section 2.

This work is intended mainly for reference. However, although detailed proofs will not generally be given, outlines for proofs are given in some places, and this may facilitate the use of the text as a textbook for a formal course, an informal seminar, or self-study.

2. Finite Difference Calculus

We shall occasionally use operators from the calculus of finite differences. These operators are defined as follows.

The *displacement operator* E increases the argument of a function by 1, i.e.

(1) $$Ef(x) = f(x + 1).$$

Evidently

$$E[Ef(x)] = Ef(x + 1) = f(x + 2).$$

This equation may be summarized as

(2) $$E^2 f(x) = f(x + 2).$$

More generally

(2)′ $$E^n f(x) = f(x + n)$$

for any positive integer n, and we interpret $E^h f(x)$ as $f(x + h)$ for any real h.

The *forward difference operator* Δ is defined by

(3) $$\Delta f(x) = f(x + 1) - f(x).$$

Noting that

(4) $$f(x + 1) - f(x) = E f(x) - f(x) = (E - 1) f(x),$$

we have the *symbolic* (or *operational*) relation

(5) $$\Delta \equiv E - 1.$$

If n is an integer then

(6) $$\begin{aligned} \Delta^n f(x) &= (E - 1)^n f(x) \\ &= \sum_{j=0}^{n} \binom{n}{j} (-1)^j E^{n-j} f(x) \\ &= \sum_{j=0}^{n} \binom{n}{j} (-1)^j f(x + n - j). \end{aligned}$$

The quantity $\Delta^n f(x)$ is called the *n-th forward difference* of $f(x)$. Also rewriting (5) as

$$E \equiv 1 + \Delta$$

we can use this to recast (2)′ as

(7) $$\begin{aligned} f(x + n) &= (1 + \Delta)^n f(x) \\ &= \sum_{j=0}^{n} \binom{n}{j} \Delta^j f(x). \end{aligned}$$

Replacing n by h, which may be any real number, and using the interpretation of $E^h f(x)$ as $f(x + h)$, we have

(8) $$f(x + h) = (1 + \Delta)^h f(x) = f(x) + h \Delta f(x) + \frac{h(h - 1)}{2!} \Delta^2 f(x) + \cdots .$$

The series on the right hand side need not terminate. However, if h is small and $\Delta^n f(x)$ decreases rapidly enough as n increases, then a good approximation to $f(x + h)$ may be obtained with but few terms of the expansion. This expansion may then be used to interpolate values of $f(x + h)$, given values $f(x)$, $f(x + 1), \ldots$ at unit intervals. (A similar formula applies whenever the values are at *equal* intervals of the argument. The common interval can be transformed to 1 by an appropriate scale transformation). Formula (8), used in this way, is called *Newton's forward difference (interpolation) formula.*

The *backward difference operator*, ∇, is defined, similarly, by the equation

$$\nabla f(x) = f(x) - f(x-1) = (1 - E^{-1})f(x). \tag{9}$$

Note that $\nabla \equiv \Delta E^{-1} \equiv E^{-1}\Delta$. There is a backward difference interpolation formula analogous to Newton's forward difference formula.

The *central difference operator*, δ, is defined by

$$\begin{aligned} \delta f(x) &= f(x + \tfrac{1}{2}) - f(x - \tfrac{1}{2}) \\ &= (E^{\frac{1}{2}} - E^{-\frac{1}{2}})f(x). \end{aligned} \tag{10}$$

Note that $\delta \equiv \Delta E^{-\frac{1}{2}} = E^{-\frac{1}{2}}\Delta$. *Everett's central difference* interpolation formula

$$\begin{aligned} f(x+h) = (1-h)f(x) + hf(x+1) &- \tfrac{1}{6}(1-h)\{1-(1-h)^2\}\,\delta^2 f(x) \\ &- \tfrac{1}{6}h(1-h^2)\,\delta^2 f(x+1) + \cdots \end{aligned}$$

is especially useful for computation.

Finally, we introduce (from the *differential* calculus) the differentiation operator D, defined by

$$Df(x) = f'(x) = df(x)/dx. \tag{11}$$

If the function $f(x)$ can be expressed in terms of a Taylor series then

$$f(x+h) = \sum_{j=0}^{\infty} (h^j/j!)D^j f(x). \tag{12}$$

The operator acting on $f(x)$ can be written formally as

$$\sum_{j=0}^{\infty} (hD)^j/j! \equiv e^{hD}. \tag{13}$$

Comparing (8) with (12) we have (again formally)

$$e^{hD} \equiv (1+\Delta)^h \qquad \text{or} \qquad e^D \equiv 1 + \Delta. \tag{14}$$

Although this is only a *formal* relation among operators, it gives exact results when $f(x)$ is a polynomial of finite order, and useful approximations in other cases, especially when $D^j f(x)$ and $\Delta^j f(x)$ decrease rapidly as j increases.

Re-writing $\quad e^D \equiv 1 + \Delta \quad$ as $\quad D \equiv \log(1 + \Delta)$

we obtain a *numerical differentiation** formula

$$f'(x) = Df(x) = \Delta f(x) - \tfrac{1}{2}\Delta^2 f(x) + \tfrac{1}{3}\Delta^3 f(x) - \cdots. \tag{15}$$

*This is not the only numerical differentiation formula. There are others which are sometimes more accurate. This one is quoted only as an example.

There is an expansion in terms of forward differences $\Delta^j f(x)$ analogous to the Taylor series expansion (12) in terms of $D^j f(x)$. The powers h^j have to be replaced by *descending factorials*

$$h^{(j)} = h(h-1)\cdots(h-j+1),^* \tag{16}$$

(note that there are j terms in the product) and we can rewrite the expansion (8) as

$$f(x+h) = \sum_{j=0}^{\infty} (h^{(j)}/j!)\,\Delta^j f(x). \tag{17}$$

If $f(x)$ is a polynomial of degree N, the expansion of (17) ends with the term containing $\Delta^N f(x)$, just as (12) would end with the term containing $D^N f(x)$.

In particular (putting $x = 0$, $h = x$, and $f(x) = x^N$)

$$x^N = \sum_{j=0}^{N} (x^{(j)}/j!)\,\Delta^j 0^N \tag{18}$$

where $\Delta^j 0^N$ means $\Delta^j x^N|_{x=0}$, and is called a "*difference of zero.*" The multiplier $\Delta^j 0^N/j!$ of $x^{(j)}$ in (18) is called a *Stirling number of the second kind.* The notation $\mathcal{S}_N^{(j)} = \Delta^j 0^N/j!$ is sometimes used. Such numbers have been tabulated in a number of places (e.g. Abramowitz and Stegun [1]).

Equation (18) can be inverted to give expansions for $x^{(N)}$ as polynomials of x. The coefficients are called *Stirling numbers of the first kind*, and written $S_N^{(j)}$,† so that $x^{(N)} = \sum_{j=0}^{n} S_N^{(j)} x^j$.

The Stirling numbers of first kind $|S_m^{(p)}|$ for $m = 1(1)20$‡ and $p = 1(1)m$ and the Stirling numbers of second kind $\Delta^j 0^N/j!$ ($N = 1(1)25$ and $j = 1(1)N$) are tabulated in David et al. [3].

Applying the difference operator Δ to the descending factorial $x^{(N)}$ we have

(19)

$$\begin{aligned}\Delta x^{(N)} &= (x+1)^{(N)} - x^{(N)}\\ &= (x+1)x(x-1)\cdots(x-N+2) - x(x-1)(x-2)\cdots(x-N+1)\\ &= \{x+1-(x-N+1)\}x(x-1)\cdots(x-N+2)\\ &= Nx^{(N-1)}.\end{aligned}$$

Repeating the operation we find

$$\Delta^j x^{(N)} = N^{(j)} x^{(N-j)}; \qquad (j \le N) \tag{20}$$

*If h is a positive integer then $h^{(j)} = 0$ for $j > h$.

†The superfix (j) on the Stirling number does *not* indicate a descending factorial.

‡The notation $a(\delta)b$ is used throughout the book. It refers to all the values between a and b at δ intervals.

the analogy with the differential calculus formula

(21) $$D^j x^N = N^{(j)} x^{N-j} \qquad (j \leq N)$$

is apparent, and just as $D^j x^N = 0$ for $j > N$, so $\Delta^j x^{(N)} = 0$ for $j > N$.

Since any polynomial of degree N can be expressed as a linear function of $x^{(N)}, x^{(N-1)}, \ldots x^{(1)}$ and 1, it follows that the jth difference of such a polynomial is zero if $j > N$. That is why if $f(x)$ is a polynomial of degree N, the expansion in (17) ends with the term containing $x^{(N)}$.

The *ascending factorial* defined by

(22) $$h^{[j]} = h(h+1)\cdots(h+j-1),^*$$

is related to the operator ∇ the same way that $x^{(N)}$ is related to Δ. Thus

$$f(x+h) = \sum_{j=0}^{\infty} (h^{[j]}/j!)\nabla^j f(x)$$

and

$$\begin{aligned} \nabla x^{[N]} &= x^{[N]} - (x-1)^{[N]} \\ &= N x^{[N-1]}. \end{aligned}$$

The binomial coefficient $\binom{n}{r} = n!\,[(n-r)!r]^{-1}$ may be written $n^{(r)}/r!$. Similarly the multinomial coefficient

(23) $$\binom{n}{r_1, r_2, \ldots, r_k} = \frac{n!}{r_1! r_2! \ldots r_k!}$$

may be written

$$n^{(n-r_k)}[r_1! r_2! \ldots r_{k-1}!]^{-1}.$$

The notation $\binom{n}{r}$ can also be used to represent the quantity

$$\Gamma(n+1)/[\Gamma(r+1)\Gamma(n-r+1)]$$

for n, r *any* real numbers. $\Gamma(\cdot)$, the gamma function, will be defined in Section 3. Similarly

$$\binom{n}{r_1, r_2, \ldots, r_k} = \Gamma(n+1)\left[\prod_{j=1}^{k} \Gamma(r_j+1)\right]^{-1}.$$

3. Special Mathematical Functions: Definitions and Formulas

Here we define functions which we shall use in the discussion of statistical distributions. Only those properties likely to be useful in the discussion will be described. A full presentation of the functions can be found in mathematical textbooks. One purpose of this section is establishing the notation that we

*If x is a negative integer then $x^{[j]} = 0$ for $j > |x|$.

shall use. Definitions will not always be repeated in subsequent chapters. Definitions of the factorial, polynomial, exponential, logarithmic, and trigonometric functions are not given; we assume that the reader is conversant with these. Except where stated otherwise, we are using real (not complex) variables, and 'log' means natural logarithm (i.e. to base e).

The *gamma function* with argument α is defined by

$$\Gamma(\alpha) = \int_0^\infty t^{\alpha-1}e^{-t}\,dt \qquad (\alpha > 0). \tag{24}$$

By combining the recurrence formula,

$$\Gamma(\alpha + 1) = \alpha\Gamma(\alpha), \tag{25}$$

with the particular values,

$$\Gamma(\tfrac{1}{2}) = \sqrt{\pi};\ \Gamma(1) = 1, \tag{26}$$

it is possible to evaluate the gamma function for most arguments needed in the course of our work. Note, that if α is an integer, then form (25) and (26):

$$\Gamma(\alpha + 1) = \alpha!. \tag{27}$$

In statistical work we often use the *incomplete gamma function**

$$\Gamma_T(\alpha) = \int_0^T t^{\alpha-1}e^{-t}\,dt. \tag{28}$$

We use the ratio

$$\Gamma_T(\alpha)/\Gamma(\alpha) \tag{29}$$

more than (28) itself. This ratio may be called the *incomplete gamma function ratio* (the word 'ratio' is often omitted). In the tables of Pearson [11] the function tabulated is:

$$I(u,p) = \Gamma_{u\sqrt{p+1}}(p + 1)/\Gamma(p + 1) \tag{30}$$

to seven decimal places for $p = -1(0.05)0(0.1)5(0.2)50$ with u at intervals of 0.1. Harter [7] gives $I(u,p)$ to nine decimal places for $p = -0.5(0.5)74(1)164$ also with u at intervals of 0.1. There are also a number of tables giving solutions (for u) of the equation $I(u,p) = P$, with which we are not concerned here.

Stirling's formula, in the form

$$\Gamma(\alpha + 1) = (2\pi)^{\frac{1}{2}}\alpha^{\alpha+\frac{1}{2}} \exp[-\alpha + \theta_\alpha/(12\alpha)] \tag{31}$$

with $0 < \theta_\alpha < 1$, is useful, especially in approximation of binomial and multinomial coefficients.

*$\Gamma(\alpha)$ may be called a *complete* gamma function to distinguish it from $\Gamma_T(\alpha)$.

The derivatives of the logarithm of $\Gamma(\alpha)$ are also useful, though not needed as often as the gamma function itself. The function

$$\psi(\alpha) = \frac{d}{d\alpha}\{\log \Gamma(\alpha)\} = \Gamma'(\alpha)/\Gamma(\alpha) \tag{32}$$

is called the *digamma function* (with argument α) or the *psi function*.

Similarly

$$\psi'(\alpha) = \frac{d}{d\alpha}\{\psi(\alpha)\} = \frac{d^2}{d\alpha^2}\{\log \Gamma(\alpha)\} \tag{33}$$

is called the *trigamma function*, and generally

$$\psi^{(s)}(\alpha) = \frac{d^s}{d\alpha^s}\{\psi(\alpha)\} = \frac{d^{s+1}}{d\alpha^{s+1}}\{\log \Gamma(\alpha)\} \tag{34}$$

is called the $(s + 2) -$ *gamma function.*

Extensive tables of the digamma, trigamma, tetragamma, pentagamma and hexagamma functions are contained in Davis [4]. Shorter, but useful, tables are contained in Abramowitz and Stegun [1].

From the recurrence formula (25) for the gamma function, the recurrence formula

$$\psi(\alpha + 1) = \psi(\alpha) + \alpha^{-1} \tag{35}$$

for the digamma function can be derived.

Particular values are

$$\psi(1) = -\gamma = -0.577216; \ \psi(\tfrac{1}{2}) = -\gamma - 2\log_e 2 = -1.963510 \tag{36}$$

where γ is Euler's constant,

$$\gamma = \lim_{N\to\infty}\left\{\sum_{j=1}^{N} j^{-1} - \log N\right\}.$$

A very good approximate formula for $\psi(\alpha)$ is

$$\psi(\alpha) \doteqdot \log(\alpha - \tfrac{1}{2}) \tag{37}$$

provided $\alpha \geq 2$.

The derived formulas $\psi'(\alpha) = (\alpha - \frac{1}{2})^{-1}$; $\psi''(\alpha) = -(\alpha - \frac{1}{2})^2$, etc., are also quite accurate unless α is small. For example:

$$\psi'(2) = 0.64493; \qquad (2 - \tfrac{1}{2})^{-1} = 0.66667$$

and

$$\psi''(2) = -0.40411; \ -(2 - \tfrac{1}{2})^{-2} = -0.44444.$$

The *beta function* with arguments α and β is defined by

$$B(\alpha,\beta) = \int_0^1 t^{\alpha-1}(1 - t)^{\beta-1}\,dt \qquad (\alpha > 0; \beta > 0). \tag{38}$$

There is a relationship between beta and gamma functions. It is

(39) $$B(\alpha,\beta) = \Gamma(\alpha)\Gamma(\beta)/\Gamma(\alpha + \beta).$$

Just as we often need to calculate the incomplete gamma function, so we need also values of the *incomplete beta function*

(40) $$B_T(\alpha,\beta) = \int_0^T t^{\alpha-1}(1 - t)^{\beta-1}\,dt \qquad (0 < T < 1),$$

or the *incomplete beta function ratio*

(41) $$I_T(\alpha,\beta) = B_T(\alpha,\beta)/B(\alpha,\beta).$$

Again the word 'ratio' is often omitted.

Extensive tables of $I_T(\alpha,\beta)$ to seven decimal places are contained in Pearson [12], for $T = 0.01(0.01)1.00$, α, $\beta = 0.5(0.5)11(1)50$; $\alpha \geq \beta$. Since $I_T(\alpha,\beta) = 1 - I_{1-T}(\beta,\alpha)$, the same tables can be used when $\beta > \alpha$. These may be supplemented, for small values of α, by the tables of Vogler [18], which give (*inter alia*) solutions of the equation $I_T(\alpha,\beta) = P$ to six significant figures for $\alpha = 0.50(0.05)1.00$, 1.1, 1.25, 1.75 and 2.25; $\beta = 0.5(0.5)5$, 7.5, 10, 12, 15, 20, 30, and 60, and $P = 0.0001$, 0.001, 0.005, 0.01, 0.025, 0.05, 0.1, 0.25, 0.5. (They do not give values of $I_T(\alpha,\beta)$ at regular intervals for T.) Both Pearson [12] and Vogler [18] give values for the complete beta function $B(\alpha,\beta)$.

The *hypergeometric function* with arguments α, β, γ, and x is

(42) $$F(\alpha,\beta;\gamma;x) = 1 + \frac{\alpha\beta}{\gamma}\cdot\frac{x}{1!} + \frac{\alpha(\alpha+1)\beta(\beta+1)}{\gamma(\gamma+1)}\cdot\frac{x^2}{2!} + \cdots$$
$$= \sum_{j=0}^{\infty} \frac{\alpha^{[j]}\beta^{[j]}}{\gamma^{[j]}}\cdot\frac{x^j}{j!} \qquad (\gamma > 0).$$

This series converges for $|x| < 1$.

The *confluent hypergeometric function* with arguments α, γ and x is

(43) $$M(\alpha;\gamma;x) = 1 + \frac{\alpha}{\gamma}\,\frac{x}{1!} + \frac{\alpha(\alpha+1)}{\gamma(\gamma+1)}\cdot\frac{x^2}{2!} + \cdots$$
$$= \sum_{j=0}^{\infty} \frac{\alpha^{[j]}}{\gamma^{[j]}}\cdot\frac{x^j}{j!} \qquad (\gamma > 0).$$

This series also converges for $|x| < 1$. Note that if α or β is a non-positive integer, then $\alpha^{[j]}$ or $\beta^{[j]} = 0$ for $j > |\alpha|$ or $|\beta|$, therefore $F(\alpha, \beta;\gamma;x)$ is a polynomial in x of degree $|\alpha|$ or $|\beta|$. Similarly, if α is a non-positive integer, $M(\alpha;\gamma;x)$ is a polynomial in x of degree $|\alpha|$.

The integral representations

(44) $$F(\alpha,\beta;\gamma;x) = \frac{\Gamma(\gamma)}{\Gamma(\beta)\Gamma(\gamma-\beta)}\int_0^1 t^{\beta-1}(1 - t)^{\gamma-\beta-1}(1 - tx)^{-\alpha}\,dt$$
$$(\beta > 0)$$

and

$$M(\alpha;\gamma;x) = \frac{\Gamma(\gamma)}{\Gamma(\alpha)\Gamma(\gamma-\alpha)} \int_0^1 t^{\alpha-1}(1-t)^{\gamma-\alpha-1} e^{tx}\, dt \qquad (\alpha > 0) \tag{45}$$

are sometimes useful.

For both $F(\cdot)$ and $M(\cdot)$ there are recurrence relations between values of the function for different values of the arguments. We mention here only a few such relations.

$$\begin{aligned} F(\alpha,\beta;\gamma;x) &= (1-x)^{\gamma-\alpha-\beta}F(\gamma-\alpha,\gamma-\beta;\gamma;x) \\ &= (1-x)^{-\alpha}F(\alpha,\gamma-\beta;\gamma;x(x-1)^{-1}) \end{aligned} \tag{46}$$

$$\frac{d}{dx}F(\alpha,\beta;\gamma;x) = (\alpha\beta/\gamma)F(\alpha+1,\beta+1;\gamma+1;x) \tag{47}$$

$$D^nF(\alpha,\beta;\gamma;x) = (\alpha^{(n)}\beta^{(n)}/\gamma^{(n)})F(\alpha+n,\beta+n;\gamma+n;x) \tag{48}$$

$$M(\alpha;\gamma;x) = e^xM(\gamma-\alpha;\gamma;-x) \tag{49*}$$

$$\frac{d}{dx}M(\alpha;\gamma;x) = (\alpha/\gamma)M(\alpha+1;\gamma+1;x) \tag{50}$$

$$D^nM(\alpha;\gamma;x) = (\alpha^{(n)}/\gamma^{(n)})M(\alpha+n;\gamma+n;x). \tag{51}$$

Rushton and Lang [14] give tables of the confluent hypergeometric function, in a form particularly convenient for statistical applications. The tables in Abramowitz and Stegun [1] are also useful.

An approximate formula for the confluent hypergeometric function is:

$$M(\alpha;\gamma;x) = \Gamma(\gamma)e^{\frac{1}{2}x}\{x(\tfrac{1}{2}\gamma-\alpha)\}^{\frac{1}{4}-\frac{1}{2}\gamma}\cos\{2\{x(\tfrac{1}{2}\gamma-\alpha)\}^{\frac{1}{2}}-(\tfrac{1}{2}\gamma-\tfrac{1}{4})\pi\} \tag{52}$$

for α large (γ and x fixed).

The *Bessel function of the first kind*, $J_\nu(x)$, can be defined by the following equation

$$J_\nu(x) = (\tfrac{1}{2}x)^\nu \sum_{j=0}^{\infty} \frac{(-\frac{1}{4}x^2)^j}{j!\,\Gamma(\nu+j+1)}. \tag{53}$$

ν is the order of the function. In this book we shall use the *modified Bessel function of the first kind*

$$I_\nu(x) = -i^\nu J_\nu(ix) = (\tfrac{1}{2}x)^\nu \sum_{j=0}^{\infty} \frac{(\frac{1}{4}x^2)^j}{j!\,\Gamma(\nu+j+1)}.^\dagger \tag{54}$$

When $|x|$ is small (ν being fixed)

$$I_\nu(x) \doteqdot (\tfrac{1}{2}x)^\nu/\Gamma(\nu+1). \tag{55}$$

*(49) is one of the *Kummer transformations*.

†In (54) i stands for $\sqrt{-1}$.

When $|x|$ is large (ν being fixed)

$$I_\nu(x) \doteqdot \frac{e^x}{\sqrt{2\pi x}}\left\{1 - \frac{4\nu^2 - 1^2}{8x} + \frac{(4\nu^2 - 1^2)(4\nu^2 - 3^2)}{2!(8x)^2} - \cdots\right\}. \tag{56}$$

When ν is large

$$I_\nu(x) \doteqdot \frac{1}{\sqrt{2\pi\nu}} e^{\nu\eta}(1 + x^2)^{-\frac{1}{4}}, \tag{57}$$

where $\eta = (1 + x^2)^{\frac{1}{2}} - \log[x^{-1}\{1 + (1 + x^2)^{\frac{1}{2}}\}]$. Abramowitz and Stegun [1] have a more detailed discussion of these formulas.

The *Riemann zeta function* is defined by the equation

$$\zeta(x) = \sum_{j=1}^{\infty} j^{-x}. \tag{58}$$

The series is convergent for $x > 1$, and it is only for these values of x that we use the function. A generalized form of the Riemann zeta function is defined by

$$\zeta(x,a) = \sum_{j=1}^{\infty} (j + a)^{-x} \qquad (x > 1; a > -1).$$

An approximate formula for $\zeta(x)$ is

$$\zeta(x) \doteqdot 1 + 2^{-(x+1)}\frac{2x^2 + 8.4x + 21.6}{(x - 1)(x + 7)}. \tag{59}$$

Particular values are:

$$\zeta(2) = \pi^2/6; \ \zeta(4) = \pi^4/90.$$

Values of $\zeta(n)$ for $n = 2(1)42$, to 20 decimal places, are given in Abramowitz and Stegun [1] and in Davis [4].

A general formula, for even values of the argument, is

$$\zeta(2r) = \frac{(2\pi)^{2r}}{2\{(2r)!\}}|B_2|_r \tag{60}$$

where $B_0, B_1, \ldots B_r, \ldots$ denote the *Bernoulli numbers*. These are the coefficients of $1, t, \ldots t^r/r!, \ldots$ in the expansion of $t(e^t - 1)^{-1}$. Numerical values are:

$$B_0 = 1; \ B_1 = -\tfrac{1}{2}; \ B_2 = \tfrac{1}{6}; \ B_4 = -\tfrac{1}{30}; \ B_6 = \tfrac{1}{42}; \ B_8 = -\tfrac{1}{30}$$
$$(B_{2r+1} = 0 \text{ for } r > 0).$$

The *Bernoulli polynomials*, $B_0(x), B_1(x), \ldots B_r(x), \ldots$ are defined as the coefficients of $1, t, \ldots t^r/r!, \ldots$ in the expansion of $te^{tx}(e^t - 1)^{-1}$, so that $B_r(0) = B_r$. A useful formula is

$$\sum_{j=1}^{N} j^r = (r + 1)^{-1}\{B_{r+1}(N + 1) - B_{r+1}\}. \tag{61}$$

The first seven Bernoulli polynomials are:

$$B_0(x) = 1;\ B_1(x) = x - \tfrac{1}{2};\ B_2(x) = x^2 - x + \tfrac{1}{6};$$
$$B_3(x) = x^3 - \tfrac{3}{2}x^2 + \tfrac{1}{2}x;\ B_4(x) = x^4 - 2x^3 + x^2 - \tfrac{1}{30};$$
$$B_5(x) = x^5 - \tfrac{5}{2}x^2 + \tfrac{5}{3}x^3 - \tfrac{1}{6}x;\ B_6(x) = x^6 - 3x^5 + \tfrac{5}{2}x^4 - \tfrac{1}{2}x^2 + \tfrac{1}{42}.$$

These polynomials satisfy the recurrence formulas

(62) $$B_r(x + h) = \sum_{j=0}^{r} \binom{r}{j} B_j(x)h^{n-j}*$$

and

(63) $$dB_r(x)/dx = rB_{r-1}(x) \qquad (r > 0).$$

Bernoulli polynomials $B_n(x)$ for $n = 0(1)12$, and Bernoulli numbers B_n for $n = 1(1)12$, are tabulated in David et al. [3].

The *Euler polynomials*, $E_r(x)$, are defined by the identity

(64) $$2e^{tx}(e^t + 1)^{-1} \equiv \sum_{j=0}^{\infty} E_j(x)(t^j/j!).$$

The *Euler numbers*, E_r, are defined by

(65) $$E_r = 2^r E_r(\tfrac{1}{2}).$$

These numbers are all integers for r even, and zero for r odd.

$$E_0 = 1;\ E_2 = -1;\ E_4 = 5;\ E_6 = -61;\ E_8 = 1385.$$

Further values are given in Davis [4].

If $(m + n)$ is odd then

(66) $$\int_0^1 B_m(x)B_n(x)\,dx = 0 = \int_0^1 E_m(x)E_n(x)\,dx.$$

Both the polynomials $B_m(x)$, $B_n(x)$ and the polynomials $E_m(x)$, $E_n(x)$ are *orthogonal* over the interval (0,1) with uniform weight function. For a further discussion of Bernoulli and Euler polynomials, see Milne-Thomson [10].

If, for a family of polynomials in x, $\{P_r(x)\}$, with $P_r(x)$ of degree r the equation

(67) $$\int_{-\infty}^{\infty} w(x)P_m(x)P_n(x)\,dx = 0$$

is satisfied whenever $m \neq n$, the family is said to be *orthogonal* with respect to

*Symbolically $B_r(x + h) = (E + h)^r B_0(x)$ with the displacement operator E applying to the subscript.

the weight function $w(x)$. In particular cases, $w(x)$ may be zero outside certain intervals (as, for example, in (66)).

We make especial use of two families of orthogonal polynomials, with weight functions

$$w(x) = e^{-\frac{1}{2}x^2} \quad \text{and}$$

$$w(x) = \begin{cases} x^a e^{-x} & (x \geq 0;\ a > 0) \\ 0 & (x < 0) \end{cases}$$

respectively. The first of these families we called the *Hermite polynomials*, the second the *generalized Laguerre polynomials.*

The rth *Hermite polynomial* is defined by

(68) $$H_r(x) = (-1)^r e^{\frac{1}{2}x^2} D^r e^{-\frac{1}{2}x^2} \qquad (r = 0,1,\ldots).$$

From (68) it follows that

$$H_0(x) = 1;\ H_1(x) = x;\ H_2(x) = x^2 - 1;$$

$$H_3(x) = x^3 - 3x;\ H_4(x) = x^4 - 6x + 3;\ H_5(x) = x^5 - 10x^3 + 15x$$

and generally

(69) $$H_r(x) = x^r - \frac{r^{(2)}}{2 \cdot 1!} x^{r-2} + \frac{r^{(4)}}{2^2 \cdot 2!} x^{r-4} - \cdots.$$

The general term is $(-1)^j r^{(2j)} x^{r-2j}/(2^j \cdot j!)$. The series terminates automatically after $j = [(r-1)/2]$.*

The rth *generalized Laguerre polynomial* of order a, $L_r^{(a)}(x)$ is

(70) $$\sum_{j=0}^{r} (-1)^j \binom{r+a}{r-j} \frac{x^j}{j!} = \binom{r+a}{r} M(-r;a+1;x).$$

The recurrence formula

(71) $$L_r^{(a+1)}(x) = x^{-1}[(x-r)L_r^{(a)}(x) + (a+r)L_{r-1}^{(a)}(x)]$$

is useful in computation.

Other families of orthogonal polynomials occasionally used in statistical theory are Jacobi, Chebyshev, Krawtchouk, and Charlier. The weight function corresponding to the Jacobi polynomial is of the form

$$w(x) = \begin{cases} x^{\alpha-1}(1-x)^{\beta-1} & (0 \leq x \leq 1) \\ 0 & \text{for other values of } x. \end{cases}$$

The last three families have the following point weight functions:

$$\text{Chebyshev polynomials } w(x) = \begin{cases} 1 & (x = x_1, x_2, \ldots x_n) \\ 0 & \text{for other values of } x \end{cases}$$

*Here and in some other places $[r]$ = the largest integer $\leq r$.

$$\text{Krawtchouk polynomials } w(x) = \begin{cases} \binom{N}{x} p^x(1-p)^{N-x} & (x = 0,1,2,\ldots N) \\ 0 & \text{for other values of } x \end{cases}$$

$$\text{Charlier polynomials } w(x) = \begin{cases} e^{-\theta}\theta^x/x! & (x = 0,1,2,3,4,\ldots) \\ 0 & \text{for other values of } x \end{cases}$$

4. Calculus of Probability

The *probability of an event E*, $\Pr[E]$, is a real number, $0 \leq \Pr[E] \leq 1$, which should be defined so as to correspond with our intuitive notion that the probability is the proportion (*relative frequency*) of times that E might be expected to occur in repeated observation under approximately specified conditions.

Apart from simple events, we consider compound events composed of combinations of simple events. The compound event defined as "either E_1 or E_2, or both" is called the *logical sum* or *union* of E_1 and E_2, and written symbolically $E_1 + E_2$ or $E_1 \cup E_2$. (The two names and symbols mean the same thing.) The compound event defined as "both E_1 and E_2" is called the *logical product* or *intersection* of E_1 and E_2, and written symbolically E_1E_2 or $E_1 \cap E_2$. (Again the two names and symbols refer to the same concept.) These definitions can be extended to combinations of any number of events. Thus $E_1 + E_2 + \cdots + E_k$ or $\bigcup_{j=1}^{k} E_j$, means "at least one of $E_1, E_2, \ldots, E_k$, while $E_1E_2 \cdots E_k$ or $\bigcap_{j=1}^{k} E_j$ means "every one of $E_1, E_2, \ldots, E_k$." By further extension we have such compound events as $(E_1 \cup E_2) \cap E_3$, meaning "both E_3, and at least one of E_1 and E_2." By a natural extension we can form compounds of enumerable infinities of events such as $\bigcup_{j=1}^{\infty} E_j$ and $\bigcap_{j=1}^{\infty} E_j$. It is even possible to conceive of compounds of nonenumerable infinities of events.

An important formula connecting probabilities of different, but related events is

$$\Pr[E_1 \cup E_2] = \Pr[E_1] + \Pr[E_2] - \Pr[E_1 \cap E_2]. \tag{72}$$

This formula can be extended to give *Boole's formula*

$$\Pr\left[\bigcup_{j=1}^{k} E_j\right] = \sum_{j=1}^{k} \Pr[E_j] - \sum\sum \Pr[E_{j_1} \cap E_{j_2}] + \sum\sum\sum \Pr[E_{j_1} \cap E_{j_2} \cap E_{j_3}] - \cdots + (-1)^{k+1} \Pr\left[\bigcap_{j=1}^{k} E_j\right] \tag{73}$$

where the repeated summation signs $\sum \overset{m \text{ terms}}{\cdots\cdots} \sum$ mean "summation over all integers $j_1, j_2, \ldots, j_m$ subject to $1 \leq j_i \leq m$, $j_1 < j_2 < \cdots < j_m$." It is a useful feature of formula (73) that bounds for $\Pr\left[\bigcup_{j=1}^{k} E_j\right]$ may be obtained by stopping at any two consecutive sets of terms. For example:

$$(74) \qquad \sum_{j=1}^{k} \Pr[E_j] - \sum\sum \Pr[E_{j_1} \cap E_{j_2}] \le \Pr\left[\bigcup_{j=1}^{k} E_j\right] \le \sum_{j=1}^{k} \Pr[E_j].$$

Another important formula is

$$(75) \qquad \Pr[E_1 \cap E_2] = \Pr[E_1]\Pr[E_2 \mid E_1] = \Pr[E_2]\Pr[E_1 \mid E_2]$$

where $\Pr[E_2 \mid E_1]$ is the *conditional* probability of E_2, *given* E_1. A conditional probability applies under the restricted set of circumstances specified by the symbols *following* the vertical stroke.

If $\Pr[E_2 \mid E_1] = \Pr[E_2]$ the event E_2 is *independent* of the event E_1. Provided $\Pr[E_1]$ and $\Pr[E_2]$ are not both zero, if E_2 is independent of E_1, then E_1 is independent of E_2, and E_1 and E_2 are said to be *mutually independent.*

Equation (75) may be extended to

$$(76) \qquad \Pr\left[\bigcap_{j=1}^{k} E_j\right] = \Pr[E_1]\Pr[E_2 \mid E_1]\Pr[E_3 \mid E_1 \cap E_2] \cdots \Pr\left[E_k \,\middle|\, \bigcap_{j=1}^{k-1} E_j\right].$$

If E_j is independent of $\bigcap_{i=1}^{j-1} E_i$ for all j then (76) takes the simpler form

$$(77) \qquad \Pr\left[\bigcap_{j=1}^{k} E_j\right] = \prod_{j=1}^{k} \Pr[E_j].$$

This will be so, in particular, if $E_1, E_2, \ldots E_k$ are a *mutually independent set* of events, i.e. if for any two disjoint subsets $\{E_{a_1}, E_{a_2}, \ldots E_{a_r}\}$ and $\{E_{b_1}, \ldots E_{b_s}\}$ the events $\bigcap_{j=1}^{r} E_{a_j}$ and $\bigcap_{j=1}^{s} E_{b_j}$ are mutually independent.

The event "negation of E" is called the *complement* of E, and often denoted by $\overline{E}$. Useful formulas involving complements are

$$(78) \qquad \Pr[\overline{E}] = 1 - \Pr[E]$$

$$(79) \qquad \overline{E_1 \cup E_2} = \overline{E_1} \cap \overline{E_2}$$

and

$$(80) \qquad \overline{E_1 \cap E_2} = \overline{E_1} \cup \overline{E_2}.$$

If $\Pr[E_1 \cap E_2] = 0$ the events E_1 and E_2 are called *mutually exclusive* (or *mutually incompatible*).

If every pair of the events $E_1, E_2, \ldots E_k$ is mutually exclusive then formula (73) becomes

$$(81) \qquad \Pr\left[\bigcup_{j=1}^{k} E_j\right] = \sum_{j=1}^{k} \Pr[E_j].$$

A (real) *random variable* **X** is a quantity taking real values in each of a series of observations such that $\Pr[\mathbf{X} \leq x]$ exists for all real values of x.*

The quantity $\Pr[\mathbf{X} \leq x]$, regarded as a function of x, is the *cumulative distribution function* of **X**. It is customarily denoted by $F_{\mathbf{X}}(x)$.

Clearly $F_{\mathbf{X}}(x)$ is a non-decreasing function of x, and $0 \leq F_{\mathbf{X}}(x) \leq 1$. If $\lim_{x\to-\infty} F_{\mathbf{X}}(x) = 0$ and $\lim_{x\to\infty} F_{\mathbf{X}}(x) = 1$ then the distribution is *proper*. We shall be concerned only with proper distributions.

The study of distributions is essentially a study of cumulative distribution functions. In all cases in this book the cumulative distribution function belongs to one of two important classes, or can be constructed by mixing elements from the two classes. In the first class $F_{\mathbf{X}}(x)$ is a step function with only an enumerable number of steps. If the height of step at x_j is p_j then

$$\Pr[\mathbf{X} = x_j] = p_j$$

and, if the distribution is proper, $\sum_j p_j = 1$. For all other values, or sets of values, the probability is zero. Random variables belonging to this class are called *discrete*; their distributions are called *discrete distributions*.

In the other major class, $F_{\mathbf{X}}(x)$ is absolutely continuous, and $F_{\mathbf{X}}(x)$ can be expressed as an integral.

$$F_{\mathbf{X}}(x) = \int_{-\infty}^{x} p_{\mathbf{X}}(x)\, dx. \tag{82}$$

Any function $p_{\mathbf{X}}(x)$ for which (82) is valid for any x is a *probability density function* of **X**. For convenience the subscript **X** may be dropped, when this does not cause confusion. Random variables in this class are called *continuous*; their distributions are called *continuous distributions*.

The above concepts can both be extended to the *joint distribution* of a finite number of random variables $\mathbf{X}_1, \mathbf{X}_2, \ldots \mathbf{X}_n$. The *joint cumulative distribution function* is

$$\Pr\left[\bigcap_{j=1}^{n} (\mathbf{X}_j \leq x_j)\right] = F_{\mathbf{X}_1\mathbf{X}_2,\ldots\mathbf{X}_n}(x_1,x_2,\ldots x_n) \tag{83}$$

(or simply $F(x_1,x_2,\ldots x_n)$).

If $\Pr\left[\bigcap_{j=1}^{n} (\mathbf{X}_j = x_j)\right]$ is zero except for an enumerable number of sets of values, $\{x_1^{(i)},x_2^{(i)},x_3^{(i)},\ldots x_n^{(i)}\}$, and $\sum_i \Pr\left[\bigcap_{j=1}^{n} (\mathbf{X}_j = x_j^{(i)})\right] = 1$ then we have a *discrete* distribution.

*We shall generally use lower case letters for random variables with discrete distributions and capital letters for random variables with continuous distributions. There are exceptions to this; we use $\boldsymbol{\theta}$ not Θ to represent an estimator of θ even though it has continuous distribution. In this volume, (but *not* in later volumes) bold face letters will be used to denote random variables.

If $F(x_1,x_2,\ldots x_n)$ is absolutely continuous then

$$(84)\qquad F(x_1,x_2,\ldots x_n) = \int_{-\infty}^{x_n}\int_{-\infty}^{x_{n-1}}\cdots\int_{-\infty}^{x_1} p(x_1,x_2,\ldots x_n)\,dx_1\ldots dx_n$$

where $p(x_1,x_2\ldots x_n)$ (or strictly $p_{\mathbf{X}_1,\ldots,\mathbf{X}_n}(x_1,\ldots,x_n)$) is a *joint probability density function* of $\mathbf{X}_1, \mathbf{X}_2, \ldots \mathbf{X}_n$.

For a discrete joint distribution

$$(85)\qquad \sum_i \Pr\left[\bigcap_{j=1}^{n-1}(\mathbf{X}_j = x_j)\cap(\mathbf{X}_n = x_n^{(i)})\right] = \Pr\left[\bigcap_{j=1}^{n-1}(\mathbf{X}_j = x_j)\right]$$

The summation is over all values of $x_n^{(i)}$ for which the probability is not zero. Similarly for a continuous joint distribution

$$(86)\qquad \int_{-\infty}^{\infty} p(x_1,x_2,\ldots x_n)\,dx_n = p(x_1,x_2,\ldots x_{n-1}).$$

By repeated summing or integration, it is possible to obtain the joint distribution of any subset of $\mathbf{X}_1, \mathbf{X}_2, \ldots \mathbf{X}_n$, in particular the distributions of each separate $\mathbf{X}_j$ can be obtained. These are called *marginal* distributions.

The *conditional joint distribution of* $\mathbf{X}_1, \mathbf{X}_2, \ldots, \mathbf{X}_r$ *given* $\mathbf{X}_{r+1}, \mathbf{X}_{r+2}, \ldots, \mathbf{X}_n$ (that is, the joint distribution of the subset of the first r random variables in the case where particular values have been given for the remaining $n - r$ variables) is defined by the cumulative distribution function.

$$(87)\qquad F(x_1,\ldots x_r \mid x_{r+1},\ldots,x_n) = \Pr\left[\bigcap_{j=1}^{r}(\mathbf{X}_j \le x_j)\,\middle|\,\bigcap_{j=r+1}^{n}(\mathbf{X}_j = x_j)\right]$$

for discrete distributions, and by the probability density function

$$(88)\qquad \begin{aligned} p(x_1,x_2,\ldots,x_r \mid x_{r+1},\ldots x_n) &= p(x_1,x_2,\ldots x_n)/p(x_{r+1},\ldots x_n) && (\text{for } p(x_{r+1},\ldots x_n) > 0)\\ &= 0 && (\text{for } p(x_{r+1},\ldots x_n) = 0)\end{aligned}$$

for continuous distributions.*

The *expected value* of a mathematical function $g(\mathbf{X}_1,\mathbf{X}_2,\ldots,\mathbf{X}_n)$ of $\mathbf{X}_1$, $\mathbf{X}_2, \ldots, \mathbf{X}_n$ is defined as

$$(89)\qquad E[g(\mathbf{X}_1,\mathbf{X}_2,\ldots,\mathbf{X}_n)] = \sum_i g(x_1^{(i)},x_2^{(i)},\ldots,x_n^{(i)})\Pr\left[\bigcap_{j=1}^{k}(\mathbf{X}_j = x_j^{(i)})\right]$$

*In (87) and (88), subscripts should follow F and p. We have omitted them for convenience, but it should be kept in mind that F and p are not fixed functions, but rather tools for calculating quantities defined by the joint distribution of the random variables.

for discrete distributions and as

$$(90) \qquad E[g(X_1,X_2,\ldots,X_n)]$$
$$= \int_{-\infty}^{\infty}\int_{-\infty}^{\infty}\cdots\int_{-\infty}^{\infty} g(x_1,\ldots,x_n)p(x_1,\ldots,x_n)\,dx_1\ldots dx_n$$

for continuous distributions. In particular (with $n = 1$)

$$E[g(X)] = \sum_j g(x_j)\Pr[X = x_j] \quad \text{or} \quad \int_{-\infty}^{\infty} g(x)p(x)\,dx.$$

If K is a constant, then

$$E[K] = K,$$
$$E[Kg(X)] = KE[g(X)],$$

and

$$E[g_1(X_1) + g_2(X_2)] = E[g_1(X_1)] + E[g_2(X_2)].$$

More generally

$$(91) \qquad E\left[\sum_{j=1}^{M} K_j g_j(X_1,X_2,\ldots,X_n)\right] = \sum_{j=1}^{M} K_j E[g_j(X_1,X_2,\ldots,X_n)].$$

These results apply to both discrete and continuous random variables.

Conditional expected values are defined similarly, and formulas like (89) and (90) are valid for them.

Random variables X_1, X_2 are *mutually independent* if, for all real x_1, x_2 the events $(X_1 \le x_1)$, $(X_2 \le x_2)$ are mutually independent. The set $\{X_1,X_2,\ldots X_n\}$ is a mutually independent set if for all real $x_1, \ldots x_n$, the set of events $\{(X_1 \le x_1), \ldots, (X_n \le x_n)\}$ is a mutually independent set. In this last case

$$(92) \qquad E\left[\prod_{j=1}^{k} g_j(X_j)\right] = \prod_{j=1}^{k} E[g_j(X_j)].$$

5. Moments, Cumulants and Generating Functions

The expected value of X^r, for r any real number is termed the *r-th moment* (*about zero*) of X. The words "about zero" are commonly omitted, unless they are needed to ensure clarity. The rth moment is often denoted by the symbols $\mu_r'(X)$. We shall omit the symbol (X) when there is no confusion.

The *r-th moment about K* (*a constant*) is the expected value of $(X - K)^r$. The special case, $K = E[X] = \mu_1'$ is by far the most frequently encountered. The corresponding moments are called *central moments*, and we usually denote the rth central moment by μ_r Thus

$$(93) \qquad \mu_r' = E[X^r]$$
$$(94) \qquad \mu_r = E[(X - E[X])^r].$$

The first central moment μ_1 must be zero. The second central moment, μ_2, is called the *variance* of X and sometimes written var (X). The square root of this quantity (taken positively) also has a special name — the *standard deviation.* It is often denoted by the symbols $\sigma(X)$.

As well as these natural moments there are *absolute moments,* defined as expected values of absolute values (moduli) of various functions of X. Thus the *r-th absolute moment* (about zero) of X is:

$$(95) \qquad \nu_r'(X) = E[|X|^r]$$

while the *r-th absolute central moment* is

$$(96) \qquad \nu_r(X) = E[|X - E[X]|^r].$$

If r is even, $\nu_r' = \mu_r'$ and $\nu_r = \mu_r$, but not if r is odd. In fact, we have already noted that $\mu_1 = 0$, but in general $\nu_1 > 0$. ν_1 is called the *mean deviation* of X.

Commonly used indices of the shape of a distribution are the *shape-factors* or *moment ratios.* The most important of these are:

$$(97) \qquad \alpha_3(X) = \sqrt{\beta_1(X)} = \mu_3\{\mu_2\}^{-\frac{3}{2}} \qquad \text{(an index of } \textit{skewness}\text{)}$$

$$(98) \qquad \alpha_4(X) = \beta_2(X) = \mu_4\{\mu_2\}^{-2} \qquad \text{(an index of } \textit{kurtosis}\text{).}$$

The α and β notations are both in use.

More generally

$$\alpha_r(X) = \mu_r\{\mu_2\}^{-\frac{1}{2}r}$$

Note that these moment-ratios have the same value for any linear function $(A + BX)$ with $B > 0$. If $B < 0$, the absolute values are not altered but ratios of odd order have their signs reversed.

It is often convenient to calculate the central moments (μ_r) from moments about zero (μ_r') and, though less often, *vice-versa.* Formulas by which this can be done are

$$(99) \qquad \mu_r = E[(X - E[X])^r] = \sum_{j=0}^{r} (-1)^j \binom{r}{j} \mu_1'^j \mu_{r-j}'.$$

The last two terms of the expansion combine to give $(-1)^{r-1}(r - 1)\mu_1'^r$. In particular

$$(100) \qquad \begin{cases} \mu_2 = \mu_2' - \mu_1'^2 \\ \mu_3 = \mu_3' - 3\mu_2'\mu_1' + 2\mu_1'^3 \\ \mu_4 = \mu_4' - 4\mu_3'\mu_1' + 6\mu_2'\mu_1'^2 - 3\mu_1'^4. \end{cases}$$

For the inverse calculation

$$(101) \qquad \mu_r' = E[\{(X - E[X]) + E[X]\}^r] = \sum_{j=0}^{r} \binom{r}{j} \mu_1'^j \mu_{r-j}.$$

In particular (noting that $\mu_1 = 0$)

$$(102)\qquad \begin{cases} \mu_2' = \mu_2 + \mu_1'^2 \\ \mu_3' = \mu_3 + 3\mu_1'\mu_2 + \mu_1'^3 \\ \mu_4' = \mu_4 + 4\mu_1'\mu_3 + 6\mu_1'^2\mu_2 + \mu_1'^4. \end{cases}$$

It is often convenient (particularly in studying discrete distributions) to first calculate *factorial moments.* The more commonly used are descending factorial moments. The *r-th descending factorial moment* of **X** is the expected value of $\mathbf{X}^{(r)}$. It is usually written $\mu_{(r)}$. Similarly, the *r-th ascending factorial moment* of **X** is

$$(103)\qquad \mu_{[r]} = E[\mathbf{X}^{[r]}].$$

Expressions for moments about zero in terms of descending factorial moments follow from equation (18), whence

$$(104)\qquad \mu_r' = \sum_{j=0}^{r} (\Delta^j 0^r/j!)\mu_{(r)}.$$

In particular

$$(105)\qquad \begin{cases} \mu_1' = \mu_{(1)} \\ \mu_2' = \mu_{(2)} + \mu_{(1)} \\ \mu_3' = \mu_{(3)} + 3\mu_{(2)} + \mu_{(1)} \\ \mu_4' = \mu_{(4)} + 6\mu_{(3)} + 7\mu_{(2)} + \mu_{(1)} \end{cases}$$

It is rare that factorial moments need to be calculated from moments about zero. When they do, the following formulas are used. The absolute values of the coefficients are Stirling numbers of the first kind.

$$(106)\qquad \begin{cases} \mu_{(1)} = \mu_1' \\ \mu_{(2)} = \mu_2' - \mu_1' \\ \mu_{(3)} = \mu_3' - 3\mu_2' + 2\mu_1' \\ \mu_{(4)} = \mu_4' - 6\mu_3' + 11\mu_2' - 6\mu_1'. \end{cases}$$

The formulas corresponding to (105) for ascending factorial moments are

$$(107)\qquad \begin{cases} \mu_1' = \mu_{[1]} \\ \mu_2' = \mu_{[2]} - \mu_{[1]} \\ \mu_3' = \mu_{[3]} - 3\mu_{[2]} + \mu_{[1]} \\ \mu_4' = \mu_{[4]} - 6\mu_{[3]} + 7\mu_{[2]} - \mu_{[1]}. \end{cases}$$

Many results in probability theory are expeditiously obtained by the use of *generating functions.* Seal [16] has given an interesting historical commentary on generating functions, as developed in the work of De Moivre in 1730 and earlier. We shall introduce some of the important kinds here, but the *probability generating function*, which is connected with (discrete) lattice distributions, will be introduced in Chapter 2.

The expected value of $e^{t\mathbf{X}}$, if it exists (i.e. is finite) is called the *moment-generating function* of **X**. If it does exist for some interval $|t| < T$, where $T > 0$, then the coefficient of $t^r/r!$ in its expansion as a Taylor series is μ'_r. Since

$$\exp [t(\mathbf{X} - E(\mathbf{X}))] = e^{-t\mu'_1}E[e^{t\mathbf{X}}]$$

it follows that the coefficient of $t^r/r!$ in the Taylor series expansion of

$$e^{-t\mu'_1}E[e^{t\mathbf{X}}] \tag{108}$$

is the rth central moment of **X**. The function (108) is therefore called the *central moment generating function* of **X**.

The logarithm of the moment generating function of **X** is the *cumulant generating function* of **X**. If the moment generating function exists, so does the cumulant generating function. The coefficient of $t^r/r!$ in the Taylor series expansion of the latter function is the *r-th cumulant* of **X**, and is denoted by the symbols $\kappa_r(\mathbf{X})$ or, when no confusion is likely to arise, simply κ_r. Evidently κ_r is a function of the moments of **X**.

These generating functions are mathematical functions of t, not of **X**. They do, of course, reflect certain properties of the distribution of **X**. When it is desirable that this be indicated in the notation it is common to write

$$E[e^{t\mathbf{X}}] = \phi_{\mathbf{X}}(t) \qquad \text{(moment generating function)}$$

and

$$\log \phi_{\mathbf{X}}(t) = \psi_{\mathbf{X}}(t) \qquad \text{(cumulant generating function).}$$

Since, for any constant A,

$$\phi_{\mathbf{X}+A}(t) = E[e^{t(\mathbf{X}+A)}] = e^{tA}\phi_{\mathbf{X}}(t)$$

it follows that

$$\psi_{\mathbf{X}+A}(t) = tA + \psi_{\mathbf{X}}(t). \tag{109}$$

Hence for $r \geq 2$, the coefficients of $t^r/r!$ in $\psi_{\mathbf{X}+A}(t)$ and $\psi_{\mathbf{X}}(t)$ are the same; that is

$$\kappa_r(\mathbf{X} + A) = \kappa_r(\mathbf{X}) \qquad \text{for } r \geq 2 \tag{110}$$

while

$$\kappa_1(\mathbf{X} + A) = \kappa_1(\mathbf{X}) + A. \tag{111}$$

Thus the cumulants (for $r \geq 2$) are not affected by adding a constant to **X**, but κ_1 is changed by the addition of the same constant. For this reason the cumulants are sometimes called *seminvariants*. Putting $A = -\mu'_1$, it follows that, for $r \geq 2$, the cumulants κ_r are functions of the central moments μ_r. In fact

$$\begin{aligned}\kappa_1 &= \mu'_1 \\ \kappa_2 &= \mu_2 \\ \kappa_3 &= \mu_3 \\ \kappa_4 &= \mu_4 - 3\mu_2^2.\end{aligned} \tag{112}$$

There is a set of moment ratios, analogous to those of (97) and (98), but defined in terms of the cumulants. The general formula is

$$\gamma_j = \kappa_{j+2}/\kappa_2^{\frac{1}{2}j+1} \qquad (j = 1,2,\ldots). \tag{113}$$

Note that $\gamma_1 = \alpha_3$; $\gamma_2 = \alpha_4 - 3$.

If $\mathbf{X}_1, \mathbf{X}_2, \ldots, \mathbf{X}_n$ are independent random variables and if the relevant functions exist, then

$$\phi_{\sum_1^n \mathbf{X}_j}(t) = \prod_{j=1}^{n} \phi_{\mathbf{X}_j}(t) \tag{114}$$

and

$$\psi_{\sum_1^n \mathbf{X}_j}(t) = \sum_{j=1}^{n} \psi_{\mathbf{X}_j}(t). \tag{115}$$

It follows from (115) that

$$\kappa_r\left(\sum_{j=1}^{n} \mathbf{X}_j\right) = \sum_{j=1}^{n} \kappa_r(\mathbf{X}_j) \qquad \text{for all } r, \tag{116}$$

that is, the cumulant of the sum equals the sum of the cumulants, which makes the name "cumulant" appropriate.

The descending factorial moment $\mu_{(r)}$ is the coefficient of $t^r/r!$ in the expansion of $E[(1 + t)^{\mathbf{X}}]$, if it can be so expanded. This expected value is therefore called the *(descending) factorial moment generating function*. Its logarithm is called the *(descending) factorial cumulant generating function*. The coefficient of $t^r/r!$ in the expansion of this function is the *r-th (descending) factorial cumulant* of **X**, and is written $\kappa_{(r)}(\mathbf{X})$ or $\kappa_{(r)}$. Formulas connecting $\{\kappa_{(r)}\}$ and $\{\mu_{(r)}\}$ exactly parallel those (112) connecting $\{\kappa_r\}$ and $\{\mu'_1, \mu_r\}$. Similarly, *ascending factorial cumulants*, $\kappa_{[r]}$, are the coefficients of $t^r/r!$ in the expansion of $\log\{E[(1 - t)^{-\mathbf{X}}]\}$.

Golomb [6] has defined the expected value of the *u*th power of the likelihood (defined in (123) below) as the *information generating function*, with argument *u*.*

Under fairly general conditions, the derivative of the information generating function, at $u = 1$, equals the expected value of the (natural) logarithm of the likelihood. The *negative* of this quantity is called the *entropy* of the distribution.

*The same quantity has been used in estimation procedures by Sichel [17], who called it the $(u - 1)$th *frequency moment*.

For some distributions the moment generating function $E[e^{t\mathbf{X}}]$ may be infinite. However, the function (of t) $E[e^{it\mathbf{X}}]$, where $i = \sqrt{-1}$, always exists and is finite. This is called the *characteristic function* of the distribution. It has properties similar to the moment generating function, and in addition (a) the cumulative distribution function $F(x)$ is uniquely determined by the characteristic function; in fact, if $F(x)$ is continuous for $a - k \leq x \leq a + k$, then

$$F(a + k) - F(a - k) = \lim_{T\to\infty} \pi^{-1} \int_{-T}^{T} \{t^{-1} \sin (kt)\} e^{-ita} \phi_{\mathbf{X}}(it)\, dt,$$

(b) if $\lim_{j\to\infty} \phi_{\mathbf{X}_j}(it) = \phi_{\mathbf{X}}(it)$, where $\phi_{\mathbf{X}}(it)$ is the characteristic function of a random variable with cumulative distribution function $F_{\mathbf{X}}(x)$, then

$$\lim_{j\to\infty} F_{\mathbf{X}_j}(x) = F_{\mathbf{X}}(x).$$

An alternative approach, proposed by Sarmanov [15], uses *characteristic coefficients* $\{\lambda(\kappa)\}$, defined for a random variable $\mathbf{X}$ with cumulative distribution $F_{\mathbf{X}}(x)$, as:

$$\lambda(\kappa) = E\{e^{2i\kappa \arctan \mathbf{X}}\} = \omega(\kappa) + i\tilde{\omega}(\kappa), \qquad \kappa = 1, 2, \ldots$$

where

$$\omega(\kappa) = E\left[(1 + \mathbf{X}^2)^{-\kappa} \sum_{j=0}^{\kappa} (-1)^j \binom{2\kappa}{2j} \mathbf{X}^{2j}\right]$$

and

$$\tilde{\omega}(\kappa) = E\left[(1 + \mathbf{X}^2)^{-\kappa} \sum_{j=1}^{\kappa} (-1)^{j-1} \binom{2\kappa}{2j-1} \mathbf{X}^{2j-1}\right]$$

They are the values of the characteristic function $E(e^{it\mathbf{Z}})$ of the random variable $\mathbf{Z} = 2 \arctan \mathbf{X}$ for $t = 1, 2, \ldots$. These characteristic coefficients exist for *all* distributions. Like the ordinary moments, but under more general conditions, they completely determine the given distribution of the random variable.

Moments of joint distributions, quantities like $E\left[\prod_{j=1}^{n} \mathbf{X}_j^{a_j}\right]$ are called *product moments* (*about zero*) and denoted $\mu'_{a_1a_2\ldots a_n}$. Quantities like

$$E\left[\prod_{j=1}^{n} \{\mathbf{X}_j - E[\mathbf{X}_j]\}^{a_j}\right] = \mu_{a_1a_2\ldots a_n} \tag{117}$$

are called *central product moments.**

*Sometimes they are called *central mixed moments.*

The central product moment:

$$E[\{\mathbf{X}_j - E[\mathbf{X}_j]\}\{\mathbf{X}_{j'} - E[\mathbf{X}_{j'}]\}]$$

is called the *covariance* of $\mathbf{X}_j$ and $\mathbf{X}_{j'}$, and denoted $\text{cov}(\mathbf{X}_j,\mathbf{X}_{j'})$. The *correlation* between $\mathbf{X}_j$ and $\mathbf{X}_{j'}$ is defined as

$$\text{corr}(\mathbf{X}_j,\mathbf{X}_{j'}) = \frac{\text{cov}(\mathbf{X}_j,\mathbf{X}_{j'})}{[\text{var}(\mathbf{X}_j)\text{var}(\mathbf{X}_{j'})]^{1/2}}. \tag{118}$$

(Such a correlation is often written $\rho(\mathbf{X}_j,\mathbf{X}_{j'})$ or even $\rho_{jj'}$.) It can be shown that $-1 \leq \text{corr}(\mathbf{X}_j,\mathbf{X}_{j'}) \leq 1$. If $\mathbf{X}_j$ and $\mathbf{X}_{j'}$ are mutually independent then $\text{cov}(\mathbf{X}_j,\mathbf{X}_{j'}) = 0 = \text{corr}(\mathbf{X}_j,\mathbf{X}_{j'})$, but the converse is not necessarily true.

The *joint moment generating function* of $\mathbf{X}_1, \mathbf{X}_2, \ldots, \mathbf{X}_n$ is defined as a function of n variables $t_1, t_2, \ldots, t_n$:

$$\phi_{\{\mathbf{X}_j\}}(t_1,t_2\ldots t_n) = E\left[\exp \sum_{j=1}^{n} t_j\mathbf{X}_j\right]. \tag{119}$$

The *joint central moment generating function* is

$$E\left[\exp\left(\sum_{j=1}^{n} t_j\{\mathbf{X}_j - E[\mathbf{X}_j]\}\right)\right] = \exp\left[-\sum_{j=1}^{n} t_jE[\mathbf{X}_j]\right]\phi_{\{\mathbf{X}_j\}}(t_1,t_2,\ldots,t_n). \tag{120}$$

The *joint cumulant generating function* is $\log \phi_{\{\mathbf{X}_j\}}(t_1,t_2,\ldots,t_n)$.

Use of these generating functions is similar to that of the single variable function.

Finally, we define the *regression function* of a random variable $\mathbf{X}$ on s other random variables $\mathbf{X}_1, \mathbf{X}_2, \ldots, \mathbf{X}_s$ as

$$E[\mathbf{X} \mid \mathbf{X}_1,\mathbf{X}_2,\ldots,\mathbf{X}_s].$$

If this is a linear function of $\mathbf{X}_1, \mathbf{X}_2, \ldots, \mathbf{X}_s$ the regression is called *linear* (or *multiple linear*). The variance of the conditional distribution of $\mathbf{X}$, given $\mathbf{X}_1, \mathbf{X}_2, \ldots, \mathbf{X}_s$ is called the *scedasticity*. If $\text{var}(\mathbf{X} \mid \mathbf{X}_1,\mathbf{X}_2,\ldots\mathbf{X}_s)$ does not depend on $\mathbf{X}_1, \mathbf{X}_2, \ldots, \mathbf{X}_s$ the conditional distributions are called *homoscedastic*.

6. Statistical Techniques (Estimation)

While this is not a text-book on statistical method, we shall often refer to well-established statistical procedures and concepts. To aid understanding of the terms we use, we shall now briefly describe some of the more important methods.

If a cumulative distribution function depends on the values of a finite number of quantities $\theta_1, \theta_2, \ldots, \theta_s$, these are called *parameters*. $F(x_1,x_2,\ldots,x_n)$ is then also a function of the θ's and may be written $F(x_1,\ldots,x_n \mid \theta_1,\ldots,\theta_s)$. Often we want to estimate the values of these parameters. This is done by functions of the random variables $\mathbf{T}_j \equiv \mathbf{T}_j(\mathbf{X}_1,\mathbf{X}_2,\ldots,\mathbf{X}_n)$ called *statistics*.

When a statistic is used to estimate a parameter, θ_j, it is called an *estimator* of θ_j. The inaccuracy of an estimator is measured by its *bias*

$$(121) \qquad b_j(\mathbf{T}_j \mid \theta_1, \ldots, \theta_s) = E[\mathbf{T}_j - \theta_j \mid \theta_1, \ldots, \theta_s]$$

and by its variance, or by its *mean square error*, which is a combination of these.

$$(122) \quad \text{Mean square error} = [b_j(\mathbf{T}_j \mid \theta_1, \ldots, \theta_s)]^2 + \operatorname{var}(\mathbf{T}_j \mid \theta_1, \ldots, \theta_s).$$

If $b_j(\mathbf{T}_j \mid \theta_1, \ldots, \theta_s) = 0$, $\mathbf{T}_j$ is an *unbiased* estimator of θ_j.

If an measure of overall inaccuracy is required when several parameters $\theta_1, \ldots, \theta_q$ are being estimated by the unbiased estimators $\mathbf{T}_1, \ldots, \mathbf{T}_q$ respectively the *generalized variance* may be used. This is a determinant in which the element in the jth row and j'th column is $\operatorname{cov}(\mathbf{T}_j, \mathbf{T}_{j'} \mid \theta_1, \ldots, \theta_s)$.

For many distributions the *method of maximum likelihood* gives about as accurate estimators as are possible, from a given set of data. If observed values corresponding to $\mathbf{X}_1, \mathbf{X}_2, \ldots, \mathbf{X}_n$ are $x_1, x_2, \ldots, x_n$ the *likelihood* is

$$(123) \qquad L(x_1 \ldots x_n \mid \theta_1 \ldots \theta_s) = \begin{cases} \Pr\left[\bigcap_{j=1}^{n} (\mathbf{X}_j = x_j) \middle| \theta_1, \ldots, \theta_s\right] & \text{for discrete distributions} \\ p(x_1, x_2, \ldots x_n \mid \theta_1, \ldots, \theta_s) & \text{for continuous distributions.} \end{cases}$$

In either case, the values $\hat{\boldsymbol{\theta}}_1, \hat{\boldsymbol{\theta}}_2, \ldots \hat{\boldsymbol{\theta}}_s$ maximizing the likelihood are called *maximum likelihood estimators.* (Note that $\hat{\boldsymbol{\theta}}_j$'s are random variables.) If $\mathbf{X}_1, \mathbf{X}_2, \ldots, \mathbf{X}_n$ are mutually independent and have identical distributions, then under rather general conditions

$$(124) \qquad \lim_{n \to \infty} E(\hat{\boldsymbol{\theta}}_j \mid \theta_1, \theta_2, \ldots, \theta_s) = \theta_j \qquad (j = 1, \ldots, s)$$

$$(125) \quad \lim_{n \to \infty} n[\operatorname{var}(\hat{\boldsymbol{\theta}}_j \mid \theta_1, \theta_2, \ldots, \theta_s)] = j\text{th diagonal element of the matrix}$$

inverse to $\left\| -E\left[\frac{\partial^2 L(x_1 \mid \theta_1, \ldots, \theta_s)}{\partial \theta_j \, \partial \theta_{j'}}\right] \right\|$

$$(126) \quad \lim_{n \to \infty} n[\operatorname{cov}(\hat{\boldsymbol{\theta}}_j, \hat{\boldsymbol{\theta}}_{j'} \mid \theta_1, \theta_2, \ldots, \theta_s)] = \text{element in } j\text{th row, } j'\text{th column of}$$

this matrix.

In many cases $n\,[\operatorname{var}(\hat{\boldsymbol{\theta}}_j \mid \theta_1, \theta_2, \ldots, \theta_s)]$ cannot be less than the value given by (125) for any unbiased estimator of θ. This is known as the *Cramer-Rao lower bound.* The generalized variance is also minimized by (125) and (126) in these cases.

Despite the apparent advantages of maximum likelihood estimators, their determination is so often troublesome that other methods may be considered, requiring less onerous calculation. A common method, for example, is based on equating the *sample r-th moment* (about zero), $m'_r = n^{-1} \sum_{j=1}^{r} x_j^r$, to the corresponding value of μ'_r for a number of different values of r. Alternatively *sample central moments* $m_r = n^{-1} \sum_{j=1}^{r} (x_j - \bar{x})^r$, where $\bar{x} = n^{-1} \sum_{j=1}^{r} x_j$ is the *sample arithmetic mean*, may be equated to μ_r. There are many variants on this kind of approach.

In all cases, it is customary to compare the accuracy of such estimators with each other and with that given by (125) and (126). These are called respectively calculations of *relative efficiency* and *efficiency*. Usually efficiency is measured (often as a percentage) as an inverse ratio of variances, that is,

$$\text{(127)} \qquad \text{efficiency of } \mathbf{T} \text{ relative to } \mathbf{T}' = \frac{\text{var}(\mathbf{T}')}{\text{var}(\mathbf{T})} \times 100\%$$

If two *sets* of estimators (of different parameters $\theta_1, \theta_2, \ldots, \theta_s$) are being compared the 'efficiency' is sometimes measured by the inverse ratio of generalized variances. (All these comparisons are only appropriate when bias is zero, or can be neglected.)

In recent years an increasing amount of research is being devoted to statistical inference and theoretical as well as practical aspects of estimation procedures.

To mention a few: Bayesian methods of estimation continue to receive special attention and are being widely used in various applications (in particular in business and engineering models). In addition, 'structural' methods of inference and estimation have been developed by Fraser [5]. Moreover, Wolfowitz and his school in several papers (Kaufman [9], Weiss and Wolfowitz [19], and Wolfowitz [20]) have introduced and investigated the so-called 'generalized maximum likelihood estimators.' Somewhat similar work is being done by Soviet statisticians (e.g. Ibramhalilov [8]).

Methods of estimation using 'prior distributions,' or using 'fiducial and structural distributions,' of unknown parameters will not be discussed in this book, except incidentally. It is always possible to introduce a 'prior distribution' of unknown parameters, and to work out the consequences. The bewildering variety of possible distributions so obtained, however, would not be of importance commensurate with the space needed to describe them. While we will not describe fiducial or structural estimation directly, we will include descriptions of one or two of the more notable fiducial and structural distributions.

Although the above mentioned investigations are not reflected in the succeeding chapters, they might quite possibly have an important impact on the theory and practice of estimation in various distributions discussed in these volumes.

The subject of tests of significance is not an essential part of this book. The reader will encounter occasional references to the subject but relevant definitions will not be included.

7. Miscellaneous

7.1 *Geometrical Concepts*

Occasionally, we shall use multidimensional geometrical concepts. These are derived by extension of familiar ideas in two and three dimensional Cartesian coordinate geometry. They are introduced to aid comprehension. They are not essential to the formal development of the theory, and can be replaced by analytical treatment. Thus when we say

"The point with coordinates $(X_1,X_2,\ldots,X_n)$ is in the region R"

we mean that $(X_1,X_2,\ldots,X_n)$ satisfy a certain set of inequalities. The two statements are equivalent

We shall use the concepts *hypersphere*, a region bounded by the *surface* $\sum_{j=1}^{n}(X_j-\xi_j)^2 = R^2$ ($\xi_1,\ldots,\xi_n$, R are constants), and *inside* which, $\sum_{j=1}^{n}(X_j-\xi_j)^2 < R^2$, *straight line* $X_j - a_j = \omega\ell_j(\sum\ell_j^2 = 1$; ℓ_j is the *direction cosine* of the line relative to the axis of X_j) and *(hyper) ellipsoid* which is bounded by the surface $\sum_{j=1}^{n}\lambda_j(X_j-\xi_j)^2 = R^2$ with $\lambda_1,\ldots,\lambda_n > 0$; *inside* the ellipsoid $\sum_{j=1}^{n}\lambda_j(X_j-\xi_j)^2 < R^2$. Other concepts such as *hyperplane*, *simplex*, *cylinder*, etc, will seldom be used in this book.

7.2 *Truncation and Censoring — Order Statistics*

If values of the random variables $\mathbf{X}_1, \mathbf{X}_2,\ldots\mathbf{X}_n$ in a given region R are excluded, then the conditional joint cumulative distribution of the variables is

$$(128)\qquad F(x_1,x_2,\ldots,x_n \mid \bar{R}) = \Pr\left[\bigcap_{j=1}^{n}(\mathbf{X}_j \le x_j) \mid (\mathbf{X}_1\ldots,\mathbf{X}_n) \text{ not in } R\right]$$

$$= \frac{\Pr\left[\bigcap_{j=1}^{n}(\mathbf{X}_j \le x_j)\cap(\mathbf{X}_1,\ldots,\mathbf{X}_n) \text{ not in } R\right]}{\Pr[(\mathbf{X}_1,\ldots,\mathbf{X}_n) \text{ not in } R]}.$$

$\bar{R}$ is the complement of R, that is all the points not in R. All the quantities on the right hand side can be calculated from the (unconditional) joint cumulative distribution function $F(x_1,x_2,\ldots,x_n)$. The distribution (128) is called a *truncated* distribution.

We will usually be concerned with truncated distributions of single variables, for which $\bar{R}$ is a finite or infinite interval. If $\bar{R}$ is a finite interval with end points A and B, the distribution is *doubly truncated* (or *right and left truncated*, or *truncated above and below*). A and B are *truncation points.*

If $\overline{R}$ consists of all values greater than A the distribution is *truncated from below* or *left truncated.* If $\overline{R}$ consists of all values less than B the distribution is *truncated from above* or *right truncated.* (The same terms are also used when $\overline{R}$ includes values equal to A or B, as the case may be.)

If $\mathbf{X}'$ is a variable having a distribution formed by doubly truncating the distribution of a continuous random variable $\mathbf{X}$, then the probability density function of $\mathbf{X}'$, in terms of the probability density and cumulative distribution functions of $\mathbf{X}$, is

$$(129) \qquad p_{\mathbf{X}'}(x') = p_{\mathbf{X}}(x')/[F_{\mathbf{X}}(B) - F_{\mathbf{X}}(A)]^{-1} \qquad (A \leq x' \leq B).$$

If $\mathbf{X}'_j$ is defined to be equal to the jth smallest of $\mathbf{X}_1, \ldots \mathbf{X}_n$, then $\mathbf{X}'_1, \mathbf{X}'_2, \ldots \mathbf{X}'_n$ are called the *order-statistics* corresponding to $\mathbf{X}_1, \mathbf{X}_2, \ldots \mathbf{X}_n$. Evidently $\mathbf{X}'_1 \leq \mathbf{X}'_2 \leq \cdots \leq \mathbf{X}'_n$. If the X's are continuous then the probability of any equalities occurring is zero. However, even if there are any equalities the definition is unambiguous, provided we interpret 'jth smallest' to mean "not more than $(j - 1)$ smaller, *and* not more than $(n - j)$ larger."

$\mathbf{X}'_1$ corresponds to the smallest, and $\mathbf{X}'_n$ to the largest, among the n values of $\mathbf{X}_1, \mathbf{X}_2 \ldots \mathbf{X}_n$. The difference $w = \mathbf{X}'_n - \mathbf{X}'_1$ is called the *range* of $\mathbf{X}_1, \mathbf{X}_2, \ldots \mathbf{X}_n$. If n is odd, the 'middle' value, $\mathbf{X}'_{(n+1)/2}$, is called the *median.*

Sometimes a predetermined number of order-statistics are omitted. This is called *censoring.* If the smallest s_1 values $\mathbf{X}'_1, \ldots \mathbf{X}_{s_1}$ are omitted it is *censoring from below,* or *left censoring*; if the largest s_2 values $\mathbf{X}'_{n-s_2+1}, \ldots, \mathbf{X}'_n$ are omitted it is *censoring from above* or *right censoring.* If both sets, $\{\mathbf{X}'_1, \ldots, \mathbf{X}'_{s_1}\}$ and $\{\mathbf{X}'_{n-s_2+1}, \ldots, \mathbf{X}'_n\}$ are omitted we have *double censoring.* There is clearly a close analogy between censoring and truncation, but the differences are evident. Censoring modifies the selection of the random variables; truncation directly modifies the distribution. In other words, censoring is an agreement to ignore observed values because they are larger (or smaller) than a certain number of other observed values, while truncation is omission of values outside predetermined, *fixed,* limits.

7.3 *Mixture Distributions*

If $\{F_j(x_1, x_2, \ldots x_n)\} (j = \ldots, -1, 0, 1, 2, \ldots)$ represent different (proper) cumulative distribution functions and if $a_j \geq 0$; $\sum_{j=-\infty}^{\infty} a_j = 1$, then

$$(130) \qquad F(x_1, \ldots x_n) = \sum_{j=-\infty}^{\infty} a_j F_j(x_1, \ldots x_n)$$

also is a proper cumulative distribution function. This is called a *mixture* of the distributions $\{F_j\}$.

In many of the cases discussed in this book, the functions $F_j(x_1, \ldots x_n)$ are of the form $F(x_1, x_2, \ldots x_n; \theta_{1j}, \ldots, \theta_{sj})$. The a_j's may then be regarded as probabilities in a (discrete) joint distribution of the θ's. The idea may then be extended to suppose that the θ's have a joint continuous distribution with probability density function $p(\theta_1,\ldots\theta_s)$. Then the cumulative distribution function of the mixture is

$$(131)\quad F(x_1,x_2,\ldots,x_n) = \int_{-\infty}^{\infty}\int_{-\infty}^{\infty}\cdots\int_{-\infty}^{\infty} p(\theta_1,\ldots,\theta_s)F(x_1,\ldots,x_n;\theta_1,\ldots,\theta_s)\,d\theta_1\ldots d\theta_s.$$

In either case (discrete or continuous distribution of θ's) we can write

$$(132)\quad F(x_1,x_2,\ldots,x_n) = E_{\boldsymbol{\theta}_1,\ldots\boldsymbol{\theta}_s}F(x_1,x_2,\ldots,x_n;\boldsymbol{\theta}_1,\ldots,\boldsymbol{\theta}_s).$$

(The subscripts to E indicate that expectation is taken with respect to the variables $\boldsymbol{\theta}_1\ldots\boldsymbol{\theta}_s$.)

Note that the parameters $\theta_1,\ldots,\theta_s$ do not appear in the mixture distribution — they have been integrated, or summed out. Of course there may be other parameters in the F_j's which are not eliminated in this way.

7.4 *Sufficiency*

If the conditional distribution of $\mathbf{X}_1, \mathbf{X}_2 \ldots \mathbf{X}_n$ given a statistic $\mathbf{T}$ (usually $\mathbf{T}$ is a function of $\mathbf{X}_1,\ldots,\mathbf{X}_n$) does not depend on a parameter θ then $\mathbf{T}$ is said to be *sufficient* for the parameter θ. An extended form of the definition applies when $\mathbf{T}$ is replaced by a set of statistics, θ by a set of parameters, or both.

If $\mathbf{T}$ is sufficient for θ then the conditional distribution of the $\mathbf{X}$'s, given $\mathbf{T}$, cannot be used to provide information about the value of θ. With this background it is possible to appreciate the meaning of Blackwell's [2] (see also Rao [13]) theorem which states: — "If $\mathbf{T}$ is a sufficient statistic for θ and $\mathbf{t}$ is an unbiased estimator of θ then $\mathbf{t}' = E[\mathbf{t} \mid \mathbf{T}]$ is also an unbiased estimator of θ, and $\text{var}(\mathbf{t}' \mid \theta) \leq \text{var}(\mathbf{t} \mid \theta)$." Since $\mathbf{t}'$ is a function of $\mathbf{T}$, it follows that an unbiased estimator of θ with minimum variance can be a function of $\mathbf{T}$ alone. If, further, $\mathbf{T}$ is a *complete sufficient statistic*, i.e. if it is impossible to find a function $f(\mathbf{T})$ of $\mathbf{T}$, not identically zero, such that $E[f(\mathbf{T}) \mid \theta] = 0$ (for all θ) then $\mathbf{t}' = E[\mathbf{t} \mid \mathbf{T}]$ is a minimum variance unbiased estimator of θ.

7.5 *Statistical Differentials*

Given the moments of a random variable $\mathbf{X}$, it is sometimes desired to obtain the moments of a mathematical function of $\mathbf{X}$,

$$\mathbf{Y} = f(\mathbf{X}).$$

If exact values can be obtained and are convenient to use, this should, of course, be done. However, in some cases it is necessary to use approximate methods. One approximate method is to expand $f(\mathbf{X})$ in a Taylor series about $E(\mathbf{X})$

$$(133)\quad \mathbf{Y} \doteqdot f(E[\mathbf{X}]) + (\mathbf{X} - E[\mathbf{X}])f'(E[\mathbf{X}]) + \frac{1}{2!}(\mathbf{X} - E[\mathbf{X}])^2 f''(E[\mathbf{X}]) + \cdots$$

Then, taking expected values of both sides of (133)

$$E[\mathbf{Y}] \doteqdot f(E[\mathbf{X}]) + \tfrac{1}{2}\operatorname{var}(\mathbf{X}) \cdot f''(E[\mathbf{X}]). \tag{134}$$

This method has been widely used, under a number of different names. We call it the *method of statistical differentials*; quantities like $(\mathbf{X} - E(\mathbf{X}))$ are called *statistical differentials.*

This method can give inaccurate results. It assumes that the expected value of the remainder term in (133) is small. It is natural to expect that the method will be more accurate for smaller values of var(**X**), but accuracy also depends on the size of higher order central moments.

Following the same method, it can be shown that

$$\operatorname{var}(\mathbf{Y}) \doteqdot [f'(E[\mathbf{X}])]^2 \cdot \operatorname{var}(\mathbf{X}). \tag{135}$$

This equation is a basis for suggesting *variance-equalizing transformations.* If var(**X**) is a function, $g(E[\mathbf{X}])$ say, of $E[\mathbf{X}]$, then var(**Y**) might be expected to be more nearly constant if

$$[f'(E[\mathbf{X}])]^2 g(E[\mathbf{X}]) = \text{constant}.$$

This will be so if

$$f(\mathbf{X}) \propto \int^{\mathbf{X}} \frac{dt}{[g(t)]^{\frac{1}{2}}}. \tag{136}$$

This suggests $\mathbf{Y} = f(\mathbf{X})$ as a variance-equalizing transformation. Such transformations are often effective as *normalizing* transformations in that the distribution of **Y** is nearer to normality (Chapter 12) than is that of **X**.

Application of the method of statistical differentials to the ratio $\mathbf{X}_1/\mathbf{X}_2$ of two random variables yields the results

$$E[\mathbf{X}_1/\mathbf{X}_2] \doteqdot (\xi_1/\xi_2)\{1 + (\sigma_2/\xi_2)^2\} - \rho(\sigma_1/\xi_1)(\sigma_2/\xi_2) \tag{137}$$

$$\operatorname{var}(\mathbf{X}_1/\mathbf{X}_2) \doteqdot (\xi_1/\xi_2)^2\{(\sigma_1/\xi_1)^2 + (\sigma_2/\xi_2)^2 - 2\rho(\sigma_1/\xi_1)(\sigma_2/\xi_2)\} \tag{138}$$

where

$$\xi_j = E[\mathbf{X}_j];\ \sigma_j^2 = \operatorname{var}(\mathbf{X}_j) \qquad (j = 1,2)$$

and

$$\rho = \operatorname{corr}(\mathbf{X}_1,\mathbf{X}_2).$$

These formulas are conveniently expressed in terms of the *coefficients of variation* $V_j = \sigma_j/\xi_j$;

$$E[\mathbf{X}_1/\mathbf{X}_2] \doteqdot (\xi_1/\xi_2)(1 + V_2^2) - \rho V_1 V_2 \tag{137$'$}$$

$$\operatorname{var}(\mathbf{X}_1/\mathbf{X}_2) \doteqdot (\xi_1/\xi_2)^2(V_1^2 + V_2^2 - 2\rho V_1 V_2). \tag{138$'$}$$

REFERENCES

[1] Abramowitz, M. and Stegun, I. A. (Ed.) (1964). *Handbook of Mathematical Functions*, National Bureau of Standards, *Applied Mathematics Series No. 55*, Washington, D.C.: U.S. Government Printing Office.

[2] Blackwell, D. (1947). Conditional expectation and unbiased sequential estimation, *Annals of Mathematical Statistics*, **18,** 105–110.

[3] David, F. N., Kendall, M. G. and Barton, D. E. (1966). *Symmetric Function and Allied Tables*, London: Cambridge University Press.

[4] Davis, H. T. (1933, 1935). *Tables of the Higher Mathematical Functions*, 2 vols., Bloomington, Indiana: Principia Press.

[5] Fraser, D. A. S. (1968). *The Structure of Inference*, New York: John Wiley & Sons, Inc.

[6] Golomb, S. W. (1966). The information generating function of a probability distribution, *Transactions of IEEE — Information Theory*, **11,** 75–77.

[7] Harter, H. L. (1964). *New Tables of Incomplete Gamma Function Ratio*, Aerospace Research Laboratories, United States Air Force.

[8] Ibramhalilov, I. S. (1964). An estimate for the parameters of a distribution, *Izvestiya Akademii Nauk Azerbaidjan SSR, Seria Fisiko-Tekhnicheskih Matematicheskih Nauk*, **2,** 31–41. (in Russian)

[9] Kaufman, S. (1966). Asymptotic efficiency of the maximum likelihood estimator, *Annals of the Institute of Statistical Mathematics, Tokyo*, **18,** 155–178.

[10] Milne-Thomson, L. M. (1935). *The Calculus of Finite Differences*, (Chapter VI), London: The Macmillan Company.

[11] Pearson, K. (Ed.) (1922). *Tables of the Incomplete Γ-Function*, London: H. M. Stationery Office.

[12] Pearson, K. (Ed.) (1934). *Tables of the Incomplete Beta-Function*, London: Cambridge University Press.

[13] Rao, C. R. (1949). Sufficient statistics and minimum variance estimates, *Proceedings of the Cambridge Philosophical Society*, **45,** 213–238.

[14] Rushton, S. and Lang, E. D. (1954). Tables of the confluent hypergeometric function, *Sankhyā*, **13,** 369–411.

[15] Sarmanov, O. V. (1965). Characteristic coefficients of random distributions, *Doklady Akademii Nauk SSSR*, **162,** 281–284. (in Russian)

[16] Seal, H. L. (1949). The historic development of the use of generating functions in probability theory, *Bulletin de l'Association des Actuaires Suisses*, **49,** 209–228.

[17] Sichel, H. S. (1949). The method of frequency-moments and its application to Type IV populations, *Biometrika*, **36,** 404–425.

[18] Vogler, L. E. (1964). *Percentage Points of the Beta Distribution*, National Bureau of Standards, *Technical Note No. 215*.

[19] Weiss, L. and Wolfowitz, J. (1966). Generalized maximum likelihood estimators, *Teoriya Veroyatnostei i ee Primeneniya*, **11,** 68–93.

[20] Wolfowitz, J. (1965). Asymptotic efficiency of the maximum likelihood estimator, *Teoriya Veroyatnostei i ee Primeneniya*, **10,** 267–281.

2

Discrete Distributions—General

1. Lattice Distributions

According to the definition given in Chapter 1, the class of discrete distributions has considerable variety. For example, the distribution defined by

$$\Pr[\mathbf{x} = r/s] = (e - 1)^2(e^{r+s} - 1)^{-2}, \tag{1}$$

with r and s, relatively prime positive integers, is a discrete distribution. Its expected value is $e[1 - \log(e - 1)]$ and all positive moments are finite. However, it is not possible to write down the values r/s of $\mathbf{x}$ in ascending order of magnitude, though it is possible to enumerate them according to the values of r and s.

Most of the discrete distributions used in statistics, belong to a much narrower class, the *lattice distributions.* In these distributions the intervals between values of any one random variable for which there are non-zero probabilities, are all integral multiples of one quantity (which depends on the random variable). Points with these coordinates thus form a lattice. By an appropriate linear transformation it can be arranged that all variables take values which are integers.

For most of the discrete distributions which we will discuss, the values that are taken by the random variables cannot be negative. For such non-negative lattice variables there are available certain special methods of analysis which we will now describe.

2. Probability Generating Functions

Suppose we have a single non-negative lattice random variable $\mathbf{x}_1$ which can (without loss of generality) be supposed to have non-zero probabilities only at non-negative integral values. Let

$$(2) \qquad \Pr[\mathbf{x}_1 = k] = P_k \qquad (k = 0,1,2,\ldots).$$

If the distribution is proper, then $\sum_{j=0}^{\infty} P_j = 1$, and hence

$$(3) \qquad g(t) = \sum_{j=0}^{\infty} P_j t^j$$

converges for $|t| \leq 1$.* The function $g(t)$ is defined by the P_j's and, in turn, defines the P_j's since a polynomial expansion is unique. It is therefore called the *probability generating function* of the distribution (2), or of the random variable $\mathbf{x}_1$. In terms of expected values

$$(4) \qquad g(t) = E[t^{\mathbf{x}_1}].$$

The probability generating function is closely related to the moment generating function. In fact,

$$(5) \qquad \phi_{\mathbf{x}_1}(t) = E[e^{t\mathbf{x}_1}] = g(e^t).$$

(Although it would be logical to use the notation $g_{\mathbf{x}_1}(t)$ for the probability generating function of $\mathbf{x}_1$, this will not usually be done in this book).

If $\mathbf{x}_1, \mathbf{x}_2, \ldots, \mathbf{x}_n$ are mutually independent random variables with probability generating functions $g_1(t), g_2(t), \ldots, g_n(t)$ respectively, then since

$$(6) \qquad \prod_{j=1}^{n} g_j(t) = \prod_{j=1}^{n} E[t^{\mathbf{x}_j}] = E[t^{\sum_1^n \mathbf{x}_j}]$$

it follows that the probability generating function of $\sum_{i=1}^{n} \mathbf{x}_j$ is $\prod_{j=1}^{n} g_j(t)$. This is analogous to the formula ((114) of chapter 1).

$$\phi_{\sum_1^n \mathbf{x}_j}(t) = \prod_{j=1}^{n} \phi_{\mathbf{x}_j}(t)$$

for moment generating functions.

The joint probability generating function of n lattice variables $\mathbf{x}_1, \mathbf{x}_2, \ldots \mathbf{x}_n$ is

$$(7) \qquad g(t_1, t_2, \ldots t_n) = E\left[\prod_{j=1}^{n} t_j^{\mathbf{x}_j}\right],$$

where $\quad \Pr\left[\bigcap_{j=1}^{n} (\mathbf{x}_j = a_j)\right] = P_{a_1, a_2, \ldots a_n} \qquad (a_j = 0,1,2,\ldots).$

*This is also true when the distribution is not proper, since then $0 < \sum_{j=0}^{\infty} P_j < 1$. However, we will be concerned only with proper distributions.

By repeated differentiation of (3) we obtain

$$\frac{d^r g(t)}{dt^r} = g^{(r)}(t) = \sum_{j=r}^{\infty} j^{(r)} P_j t^{j-r}$$

(provided the right hand side exists). Putting $t = 1$,

$$g^{(r)}(1) = \mu_{(r)} \tag{8}$$

(provided $\mu_{(r)}$ is finite).

Similarly, from (7), under similar conditions

$$\left.\frac{\partial^{\Sigma r_j} g(t_1, \ldots, t_n)}{\partial t_1^{r_1} \partial t_2^{r_2} \ldots \partial t_n^{r_n}}\right|_{t_1 = \ldots = t_n = 1} = \mu_{(r_1, r_2, \ldots, r_n)}. \tag{9}$$

Gurland [5], in connection with certain techniques of estimation (see Section 5 of this Chapter) has introduced a set of quantities $\{k_j\}$ related to the quantities $\{j!P_j/P_0\}$ in exactly the same way as cumulants $\{\kappa_r\}$ are related to moments $\{\mu'_r\}$. From (3)

$$\log g(t) = \log P_0 + \log \left\{1 + \sum_{j=1}^{\infty} (j!P_j/P_0) \cdot (t^j/j!)\right\}.$$

Hence if

$$\log g(t) = \log P_0 + \sum_{j=1}^{\infty} k_j \cdot (t^j/j!) \tag{10}$$

then k_j will be the same function of $P_1/P_0, 2!P_2/P_0, \ldots, j!P_j/P_0$ as κ_j is of $\mu'_1, \mu'_2, \ldots, \mu'_j$. The k_j's might be called *probability-ratio cumulants* and $\log g(t)$ might be called the *probability-ratio cumulant generating function* (though these terms are not in general use).

3. Generalized Power Series Distributions

There are a number of ways of classifying non-negative lattice distributions. The first one we study is based directly on the form of P_j, as a function of j. If P_j can be written in the form

$$P_j = a_j \theta^j / f(\theta) \qquad (j = 0,1,\ldots; \theta > 0) \tag{11}$$

where $a_j \geq 0$, and $f(\theta) = \sum_{j=0}^{\infty} a_j \theta^j$, then the distribution is a *power series distribution*. Noack [10] drew attention to this class of distributions in 1950. Khatri [9] extended the definition to multivariate distributions in 1959. For our purposes we can use the definition

$$P_{j_1 j_2 \ldots j_n} = \Pr\left[\bigcap_{i=1}^{n} (\mathbf{x}_i = j_i)\right] = a_{j_1 j_2 \ldots j_n} \theta_1^{j_1} \theta_2^{j_2} \ldots \theta_n^{j_n} / f(\theta_1, \ldots, \theta_n) \tag{12}$$

$$(j_i = 0,1,2,\ldots; a_{j_1 \ldots j_n} \geq 0)$$

where $$f(\theta_1,\ldots,\theta_n) = \sum_{j_1=0}^{\infty}\cdots\sum_{j_n=0}^{\infty} a_{j_1j_2\ldots j_n}\theta_1^{j_1}\theta_2^{j_2}\ldots\theta_n^{j_n}.$$

In both (11) and (12) the set of values of j, or $\{j_i\}$, for which a_j, or $a_{j_1\ldots j_n}$, is greater than zero, is called the *range* of the distribution, and $f(\cdot)$ is called its *series function.*

Patil [15], [16] extended the set of values which j, or $j_1, j_2, \ldots j_n$ can take to be any non-empty enumerable set of real numbers without a limit point and called this extended class *generalized power series distributions.*

Among distributions of major importance belonging to this class are the binomial (Chapter 3), Poisson (Chapter 4), negative binomial (Chapter 5) and logarithmic series (Chapter 7) distributions, and their related multivariate distributions. (It is clear that the marginal distributions implied by (12) are generalized power series distributions.) Furthermore, if a generalized power series distribution be truncated, the truncated distribution is also a generalized power series distribution. Also the sum of n mutually independent random variables each having the same generalized power series distribution, has a distribution of the same class, with series function $[f(\theta)]^n$.

The probability generating function of the distribution (11) is

(13) $$g(t) = f(t\theta)/f(\theta).$$

The moment generating function is

(14) $$\phi_x(t) = g(e^t) = f(\theta e^t)/f(\theta)$$

and the factorial moment generating function is

(15) $$g(1 + t) = f(\theta(1 + t))/f(\theta).$$

From (15) the rth factorial moment is

(16) $$\mu_{(r)} = \theta^r f^{(r)}(\theta)/f(\theta)$$

and the rth factorial cumulant is

(17) $$\kappa_{(r)} = \theta^r(d/d\theta)^r \log f(\theta).$$

The factorial cumulants satisfy the recurrence relation

(18) $$\kappa_{(r+1)} = \theta\frac{d\kappa_{(r)}}{d\theta} - r\kappa_{(r)}$$

and the cumulants satisfy the recurrence relation

(19) $$\kappa_{r+1} = \theta\frac{d\kappa_r}{d\theta}.$$

For the multivariate distribution (12)

$$\kappa_{(r_1\ldots,r_{i-1},r_i+1,r_{i+1}\ldots r_n)} = \theta_i[\partial\kappa_{(r_1,\ldots,r_n)}/\partial\theta_i] - r_i\kappa_{(r_1,\ldots,r_n)} \tag{20}$$

and

$$\kappa_{r_1\ldots,r_{i-1},r_i+1,r_{i+1}\ldots,r_n} = \theta_i[\partial\kappa_{r_1\ldots,r_n}/\partial\theta_i]. \tag{21}$$

From (18) or (19) it can be seen that if $\kappa_1 = \kappa_{(1)} = \mu_1'$ is known *as a function of* θ, then all the cumulants (and so all the moments) of the distribution are determined from this one function. Alternatively the variance, or a higher cumulant might be given as a function of θ (Tweedie and Veevers [21]). An even more remarkable result was obtained by Khatri [9]. According to this, knowledge of the first two moments (or equivalently, the first two cumulants or factorial cumulants), as functions of a parameter ω is sufficient to determine the whole distribution, given that it *is* a generalized power series distribution.

For suppose that

$$\kappa_1 = y_1(\omega); \; \kappa_2 = y_2(\omega).$$

Then, from (19)

$$y_2(\omega) = \theta(dy_1/d\theta) = \theta \cdot (dy_1/d\omega) \cdot (d\omega/d\theta) \tag{22}$$

and from (17)

$$y_1(\omega) = \theta \cdot d\{\log f(\theta)\}/d\theta. \tag{23}$$

These equations are equivalent to the equations

$$d \log \theta/d\omega = (dy_1/d\omega)/y_2(\omega) \tag{24.1}$$

$$d\{\log f(\theta)\}/d\omega = y_1(\omega) \cdot (dy_1/d\omega)/y_2(\omega). \tag{24.2}$$

Apart from multiplicative constants, θ is determined, as a function of ω, from (24.1) and $f(\theta)$ from (24.2).

Khatri points out that no other pair of consecutive cumulants κ_j, κ_{j+1} possesses this property.

Now suppose that $\mathbf{x}_1, \mathbf{x}_2, \ldots, \mathbf{x}_n$ are mutually independent random variables; each having the same distribution, defined by (11). The likelihood function is then

$$\prod_{j=1}^{n} \left\{\frac{a_{\mathbf{x}_j}\theta^{\mathbf{x}_j}}{f(\theta)}\right\} = \theta^{\sum_{j=1}^{n}\mathbf{x}_j}[f(\theta)]^{-n} \prod_{j=1}^{n} a_{\mathbf{x}_j}.$$

The maximum likelihood estimator, $\hat{\boldsymbol{\theta}}$, of θ satisfies the equation

$$\hat{\boldsymbol{\theta}}^{-1} \sum_{j=1}^{n} \mathbf{x}_j - nf'(\hat{\boldsymbol{\theta}})/f(\hat{\boldsymbol{\theta}}) = 0 \tag{25}$$

and is thus a function of $\sum_{j=1}^{n} \mathbf{x}_j$, and does not depend on the $\mathbf{x}$'s in any other way.

The conditional distribution of $\mathbf{x}_1, \ldots \mathbf{x}_n$, given $\sum_{j=1}^{n} \mathbf{x}_j$, does not depend on θ, i.e., $\mathbf{T} = \sum_{j=1}^{n} \mathbf{x}_j$ is sufficient for θ. It is also complete, since it has a generalized power series distribution and the equation

$$\sum_{j=0}^{\infty} A_j[f(T)\theta]^j = 0 \qquad \text{for all } \theta \tag{26}$$

implies $f(T) = 0$. A minimum variance unbiased estimator of θ is

$$\begin{cases} b\left(\sum_{j=1}^{n} \mathbf{x}_j - 1\right) \Big/ b\left(\sum_{j=1}^{n} \mathbf{x}_j\right) & \text{if } \sum_{j=1}^{n} \mathbf{x}_j > 0 \\ 0 & \text{if } \sum_{j=1}^{n} \mathbf{x}_j = 0 \end{cases} \tag{27}$$

where $b(k)$ is the coefficient of θ^k in the expansion of $(f(\theta))^n$ (Roy and Mitra [19]).

The *general Dirichlet series distribution* (Siromoney [20]) is defined by the formula

$$\Pr[\mathbf{x} = k] = a_k e^{-\lambda_k \theta}/f(\theta) \qquad (k = 0,1,2,\ldots) \tag{28}$$

where $f(\theta) = \sum_{j=0}^{\infty} a_j e^{-\lambda_j \theta}$ (supposing the series to converge). Putting $\lambda_j = j$ (and replacing $e^{-\theta}$ by θ) a power series distribution is obtained.

4. Systems Defined by Difference Equations

Pearson [18] noted that for the hypergeometric distribution (see Chapter 6), the ratio $(P_{j+1} - P_j)/(P_{j+1} + P_j)$ is of the form

$$\frac{\text{linear function of } j}{\text{quadratic function of } j}. \tag{29}$$

He used this as a starting point for obtaining (by a limiting process) the differential equation defining the Pearson system of *continuous* distribution functions. (see Chapter 12). It may also be used as a basis for defining a system of *discrete* distributions. Guldberg [4] pointed out that if

$$\frac{P_{j+1}}{P_j} = K \cdot \frac{(j - a_1) \cdots (j - a_r)}{(j - b_1) \cdots (j - b_s)} \tag{30}$$

then recurrence relations among the moments can be established.

Binet [2] has considered the case $r = s = 2$, while Ord [11, 12] has studied the case $r = 1$, $s = 2$, which is that introduced by Pearson. Ord shows that the factorial moments of a distribution for which

$$\Delta P_{j-1} = \frac{(a-j)P_{j-1}}{b_0 + b_1 j + b_2 j^{(2)}}$$

$$\left(\text{or equivalently} \qquad \Delta P_{j-1} = \frac{(a-j)P_j}{(a+b_0) + (b_1 - 1)j + b_2 j^{(2)}}\right)$$

satisfy the equations

$$\begin{aligned}\{(j+2)b_2 - 1\} - \mu'_{(j+1)} + \{(j+1)(b_1 + 2jb_2) + a - 2j - 1\}\mu'_{(j)}\\ + \{b_0 + b_1 j + b_2 j^{(2)} + a - j\}\mu'_{(j-1)} = E_{j+1}\end{aligned}$$

(for $j = 0,1,2,\ldots$) with $\mu'_{(-1)} = 0$, $\mu'_{(0)} = 1$ and

$$\begin{aligned}E_{j+1} &= \{a + b_0 + (v+1)(b_1 - 1 + vb_2)\}(v+1)^{(j)}P_v\\ &\quad - \{b_0 + b_1 u + b_2 u^{(2)}\}u^{(j)}P_u\end{aligned}$$

where (u,v) are, respectively, the lower and upper limits of variation of j for which $P_j \neq 0$. These were taken to be either $(-\infty,+\infty)$ or $(0,N)$ where N may be a positive integer or infinite.

The types of distribution devised by Ord are summarized in Table 1; the corresponding regions of the (β_1,β_2) plane are shown in Figure 1 (taken from [11]).

Katz [8] referred to his own earlier work on systems of distributions for which

$$P_{j+1}/P_j = (\text{linear function of } j)/(\text{quadratic function of } j)$$

and discussed the simpler systems for which

$$\frac{P_{j+1}}{P_j} = \frac{\alpha + \beta j}{1 + j} \qquad (j = 0,1,2,\ldots; \alpha > 0; \beta < 1). \tag{31}$$

It is understood that if $\alpha + \beta j < 0$ then $P_{j+r} = 0$ for all $r > 0$. From (31)

$$(j+1)P_{j+1} = (\alpha + \beta j)P_j \tag{32}$$

whence

$$(j+1)^{r+1}P_{j+1} = (\alpha + \beta j)(j+1)^r P_j.$$

Summing both sides with respect to r (and remembering that $\sum_{j=0}^{\infty} P_j = 1$)

$$\mu'_{r+1} = \alpha + \sum_{j=1}^{r}\left\{\alpha\binom{r}{j} + \beta\binom{r}{j-1}\right\}\mu'_j + \beta\mu'_{r+1}. \tag{33}$$

Putting $r = 0$, and remembering that $\mu'_0 = 1$, we obtain

$$\mu'_1 = \alpha/(1-\beta). \tag{34}$$

TABLE 1 *The Types of Distributions Derived by Ord [11]*

Type	Name	P_j	Criteria	Range	Comments
I (*a*)	Hypergeometric	$\binom{Np}{j}\cdot\binom{Nq}{n-j}\Big/\binom{N}{n}$	$I < 1, \kappa > 1$	$[0,m]$ $m = \min(n,Np)$	J- or Bell-shaped
I (*b*)	Negative hypergeometric or beta-binomial	$\binom{k+j-1}{j}\binom{N-k-j}{Np-j}\Big/\binom{N}{Np}$	$\kappa < 0$	$[0,Np]$	J- or Bell-shaped
I (*e*)	—	$\binom{A}{j}\binom{C}{B-j}\Big/\binom{A+C}{B}$	$\kappa > 1$	$[0,\infty]$	*A*, *C* non-integer, but have the same integral part
I (*u*)	—	$\alpha\left\{\binom{A}{C+j}\binom{B}{D-j}\right\}^{-1}$	$\kappa > 1$	$[0,n]$	U-shaped
VI	Beta-Pascal	$\frac{A}{(k+A)}\binom{k+j-1}{j}\binom{A+B-1}{A}\Big/\binom{k+A+B+j-1}{k+A}$	$I > 1, \kappa > 1$	$[0,\infty)$	J- or Bell-shaped
IV	—	$\alpha Q(j,a,d)/Q(j,k+a,b), j > 0$; similar expression for $j < 0$	$0 < \kappa < 1$	$(-\infty,\infty)$	*k* a positive integer. Bell-shaped
II (*a*), II (*b*), II (*u*)	As for type I(·)		$I < 1, \kappa = 1$; $\kappa = 0$; $\kappa = 1$	As type I(·)	Symmetric forms of type I(·)
V	—	As type IV, but $b = 0$	$\kappa = 0$	$[0,\infty)$ or $(-\infty,\infty)$	Limiting form of IV
III (*B*)	Binomial	$\binom{n}{j}p^j(1-p)^{n-j}\quad(0 < p,q;\ p+q = 1)$	$I < 1, \kappa \to \infty$	$[0,n]$	Limiting form of I (*a*), I (*b*)
III (*N*)	Negative binomial or Pascal	$\binom{k+j-1}{j}p^k(1-p)^j$	$I > 1, \kappa \to \infty$	$[0,\infty)$	Limiting form of I (*b*), VI
III (*P*)	Poisson	$e^{-m}m^j/j!$	$I = 1, \kappa \to \infty$	$[0,\infty)$	Limiting form of III (*B*), III (*N*)
VII	Discrete Student's *t*	$\alpha\left[\prod_{j=1}^{k}\{(j+r+a)^2+b^2\}\right]^{-1}$	$0 < \kappa < 1$	$(-\infty,\infty)$	'Nearly' symmetric form of IV

Notes
(1) $Q(r,a,d) = (a^2+d^2)\{(a+1)^2+d^2\}\cdots\{(a+r)^2+d^2\}$.
(2) α is a constant such that the total probability is 1.
(3) $\kappa = (b_1 - b_2 - 1)^2[4b_2(b_0 + a)]^{-1}$
(4) I = (variance)/(mean)

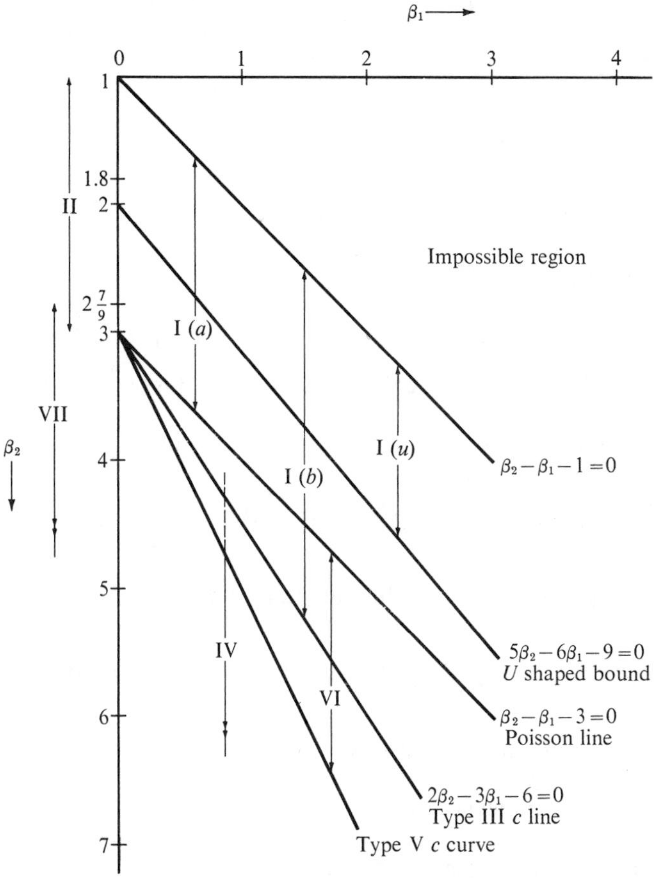

FIGURE 1

(β_1, β_2) Diagram for the Discrete System

Notes: (*i*) The binomial is described by the family of straight lines $\beta_2 - \beta_1 - 3 + 2/n = 0$ and the negative binomial by $\beta_2 - \beta_1 - 3 - 2/k = 0$.

(*ii*) The type III*d* distributions may be shown to satisfy $2\beta_2 - 3\beta_1 - 6 + 1/\mu_2 = 0$.

(*iii*) III*c*, V*c* correspond to *continuous* Pearson curves (see Chapter 12).

(*iv*) Type II*d* covers the β_2 axis over the interval [0,3], while type VII*d* covers this axis over $[2\frac{7}{9}, \infty)$.

(*v*) The boundaries of the type IV*d* and type V*d* regions are uncertain, designated by lines without arrowheads, dotted above the type V*c* curve.

(*vi*) *c* and *d* (without parentheses) are used to denote continuous, discrete respectively.

Similarly, $\mu'_2 = \alpha + (\alpha + \beta)\mu'_1 + \beta\mu'_2$, whence

(35) $$\mu_2 = \alpha/(1 - \beta)^2.$$

Further

(36) $$\mu_3 = \mu_2(2c - 1)$$

and

$$\mu_4 = 3\mu_2^2 + \mu_2(6c^2 - 6c + 1)$$

where $$c = \mu_2/\mu'_1 = (1 - \beta)^{-1}.$$

The moment ratios are

(37) $$\alpha_3 = \sqrt{\beta_1} = (2 - c^{-1})\alpha^{-\frac{1}{2}} = (1 + \beta)\alpha^{-\frac{1}{2}}$$

(38) $$\alpha_4 = \beta_2 = 3 + (6 - 6c^{-1} + c^{-2})\alpha^{-1} = 3 + (1 + 4\beta + \beta^2)\alpha^{-1}.$$

If (32) be summed over all values of j not less than $[\mu'_1]$, we obtain

$$\sum_{j=[\mu'_1]}^{\infty} (j + 1)P_{j+1} = \alpha \sum_{j=[\mu'_1]}^{\infty} P_j + \beta\left[P_{[\mu'_1]} + \sum_{j=[\mu'_1]}^{\infty} (j + 1)P_{j+1}\right].$$

Noting that $\mu'_1 = \alpha(1 - \beta)^{-1}$, this can be written

$$(1 - \beta) \sum_{j=[\mu'_1]}^{\infty} (j + 1 - \mu'_1)P_{j+1} = (\alpha + \beta[\mu'_1])P_{[\mu'_1]} \quad .$$

Hence, for this class of distributions,

(39) $$\text{mean deviation} = 2P_{[\mu'_1]}(\alpha + \beta[\mu'_1])(1 - \beta)^{-1}.$$

Noting that $\alpha + \beta\mu'_1 = \mu_2(1 - \beta)$ we see that

(40) $$\text{mean deviation} \doteqdot 2P_{[\mu'_1]}\mu_2$$

with an error of $2P_{[\mu'_1]}\beta(1 - \beta)^{-1}\{\mu'_1 - [\mu'_1]\}$. The formula is exact if μ'_1 is an integer, or if $\beta = 0$. (Equation (40) has been discussed by Bardwell [1] and Kamat [7], *inter alia.*)

The probability generating function $g(t)$ satisfies the equation (derived from (32))

$$g'(t) = \alpha g(t) + \beta t g'(t)$$

that is $$\frac{d \log g(t)}{dt} = \frac{\alpha}{1 - \beta t}.$$

Hence, if $\beta \neq 0$

(41) $$g(t) = [(1 - \beta t)/(1 - \beta)]^{-\alpha/\beta}$$

(remembering that $g(1) = 1$). The factorial moment generating function is therefore

$$[(1 - \beta - \beta t)/(1 - \beta)]^{-\alpha/\beta}$$

and the rth factorial moment is

$$\mu_{(r)} = (-\beta/(1 - \beta))^r(-\alpha/\beta)^{(r)}. \tag{42}$$

We note that

$$\mu_{(r+1)}/\mu_{(r)} = (-\beta/(1 - \beta))(-\alpha/\beta - r). \tag{43}$$

If $\beta = 0$ then (31) becomes

$$P_{j+1} = \alpha P_j/(j + 1)$$

and so

$$P_j = (\alpha^j/j!)P_0.$$

The corresponding distribution is the Poisson distribution (Chapter 4) with

$$P_j = (\alpha^j/j!)e^{-\alpha}. \tag{44}$$

(The value of P_0 is determined by the condition $\sum_{j=0}^{\infty} P_j = 1$.)

A similar analysis shows that if $0 < \beta < 1$ then

$$P_j = \binom{\alpha/\beta + j - 1}{j}\beta^j(1 - \beta)^{\alpha/\beta} \qquad (j = 0,1,2,\ldots) \tag{45}$$

which is a negative binomial distribution (Chapter 5) with parameters α/β, $\beta/(1 - \beta)$.

If $\beta < 0$ and $(-\alpha/\beta)$ is an integer, then $0 < -\beta/(j - \beta) < 1$ and

$$P_j = \binom{-\alpha/\beta}{j}\left(\frac{-\beta}{1 - \beta}\right)^j\left(1 - \frac{-\beta}{1 - \beta}\right)^{-\frac{\alpha}{\beta}-j}. \tag{46}$$

This is a binomial distribution (Chapter 3) with parameters $-\alpha/\beta$ and $-\beta/(1 - \beta)$.

If α/β is not an integer, then the range of j is $0, 1, 2, \ldots, [-\alpha/\beta] + 1$, and

$$P_j = P'_j/\sum P'_j \tag{47}$$

with $$P'_j = \frac{\Gamma(-\alpha/\beta + 1)}{j!\Gamma(-\alpha/\beta + 1 - j)}\left(-\frac{\beta}{1 - \beta}\right)^j\left(1 - \frac{-\beta}{1 - \beta}\right)^{-\frac{\alpha}{\beta}-j};$$

the summation in the denominator of (47) is over $i = 0, 1, 2, \ldots [-\alpha/\beta] + 1$.

Figure 2 (taken from Katz [8]) shows those parts of the (α,β) plane occupied by the Poisson, negative binomial and binomial distributions. Note that the latter correspond to integer values only of the ratio $-\alpha/\beta$, shown as broken lines in the diagram. The intermediate spaces correspond to distributions like (47).

It is clear that one way of discriminating among the three kinds of distribution is to consider the sequence of values $T_j = (j + 1)P_{j+1}/P_j$.

According to (31) this ratio is equal to $(\alpha + \beta j)$. Hence the three types of ratio should produce plots of T_j against j of the forms shown in Figures 3a–c.

Ottestad [13] has pointed out that the ratios $\mu_{(j+1)}/\mu_{(j)}$ (see (43)) plotted against j have a similar appearance and can also be used to discriminate among the three distributions. Katz [8] proposes an alternative criterion, based on the ratio $(\mu_2 - \mu'_1)/\mu'_1 = c - 1 = \beta/(1 - \beta)$. The ratio is zero for the Poisson distribution, positive for the negative binomial, and negative for the binomial or (47) distribution. Given n independent random variables with any common distribution (having μ'_1 and μ_2 finite), unbiased estimators of μ'_1 and μ_2 are the sample mean $\bar{\mathbf{x}} = n^{-1} \sum_{j=1}^{n} \mathbf{x}_j$ and variance

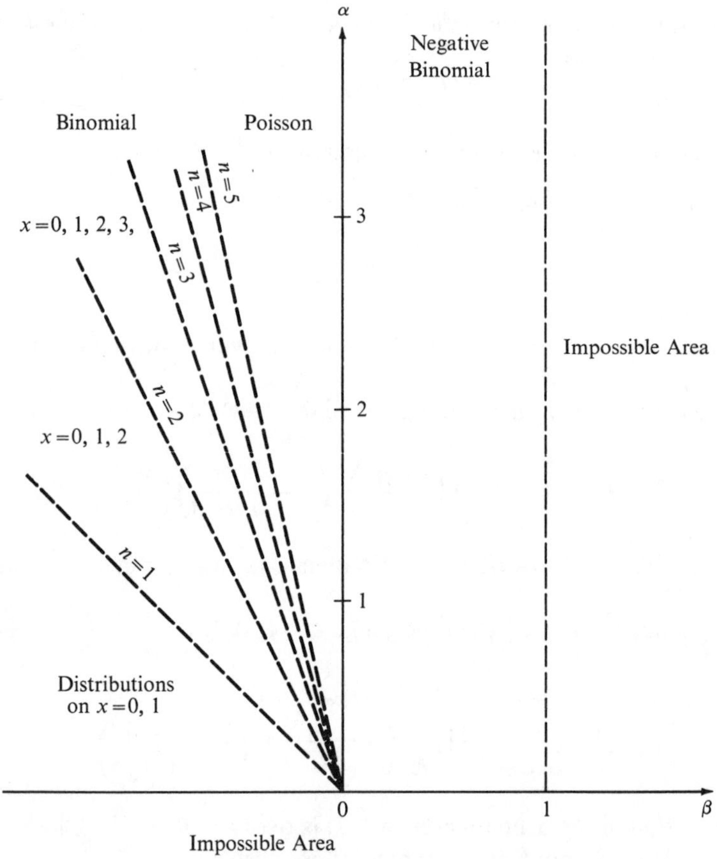

FIGURE 2

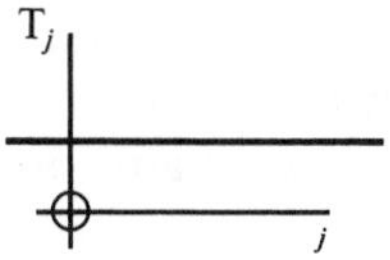

FIGURE 3a

Poisson

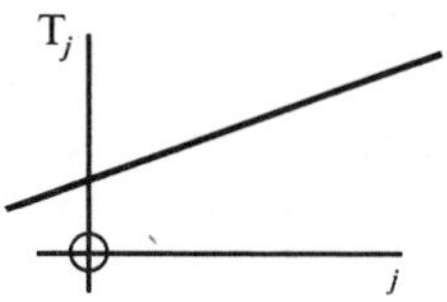

FIGURE 3b

Negative Binomial

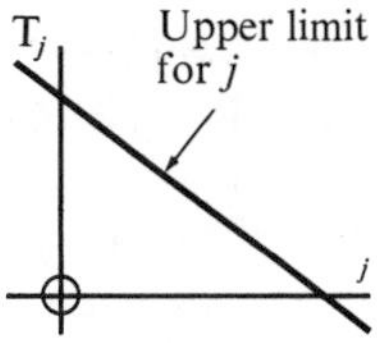

FIGURE 3c

Binomial

$\mathbf{s}^2 = (n-1)^{-1} \sum_{j=1}^{n} (\mathbf{x}_j - \bar{\mathbf{x}})^2$. Katz showed that the statistic $(\mathbf{s}^2 - \bar{\mathbf{x}})/\bar{\mathbf{x}}$ has variance approximately equal to $2/n$ if c is not far from 1. Its expected value is approximately $(c - 1)$.

The above criteria help in deciding which of the three kinds of distribution to use, *if one of them is appropriate*. In order to test whether this is so, Guldberg [4] made use of the identity

$$\frac{P_{j+1}}{P_j}(j+1) + \frac{\mu_1' - \mu_2}{\mu_2} j = \alpha + \beta j + \frac{(1-\beta)^{-1} - (1-\beta)^{-2}}{(1-\beta)^{-2}} j \tag{48}$$
$$= \alpha$$
$$= \mu_1'^2/\mu_2.$$

Hence the quantities

$$\left[T_j + \frac{\mu_1' - \mu_2}{\mu_2} j\right] \frac{\mu_2}{\mu_1'^2} \tag{49}$$

are constant and equal to one. The sample values

$$\left[\mathbf{t}_j + \frac{\bar{\mathbf{x}} - \mathbf{s}^2}{\mathbf{s}^2} j\right] \frac{\mathbf{s}^2}{\bar{\mathbf{x}}^2} = \mathbf{t}_j \frac{\mathbf{s}^2}{\bar{\mathbf{x}}^2} + \frac{\bar{\mathbf{x}} - \mathbf{s}^2}{\bar{\mathbf{x}}^2} j$$

should be approximately equal to 1 if the distribution is such that (31) is satisfied.

Another special form of difference equation is

$$\frac{P_{j+1}}{P_j} = \frac{\theta}{\lambda + j} \qquad (j = 0,1,2,\ldots). \tag{50}$$

This defines the family of *hyper-Poisson distributions* (see Chapter 11, Section 4), Equation (48) is not satisfied by distributions of this family (except when $\lambda = 1$. corresponding to a Poisson distribution). The equation is not satisfied, either, if $(P_{j+1}/P_j)(j+1)$ is replaced by $(P_{j+1}/P_j)(j+\lambda)$.

We finally note that all the distributions studied in this Section are also generalized power series distributions (as defined in Section 3).

5. Some Methods of Estimation

Apart from the methods of maximum likelihood and moments, which have already been outlined, estimation of parameters of discrete distribution is often based on observations of relative frequencies (proportions) of certain values of the variable. By equating observed proportions $\{f_j\}$ with the corresponding formulas for $\{P_j\}$, possibly with some corrective factors, we obtain equations for estimation.

Since the number of parameters to be estimated rarely exceeds three, there is no need for a large number of estimating equations. In particular only one or two observed proportions are usually included in the equations from which the estimators are derived. These are usually the proportions of the lowest possible values (e.g. 0 and 1) for the variable. Use of these proportions is likely to give better results, generally speaking, the greater the values of the corresponding P_j's.

Hinz and Gurland [6] have carried out a systematic comparative study of various methods of estimation, using moments or proportions or combinations of the two, when applied to the negative binomial (Chapter 5) Poisson binomial and Poisson negative binomial (Poisson Pascal) (Chapter 8) and Neyman Type A (Chapter 9) distributions. Their results are presented in the form of asymptotic efficiencies or joint efficiencies (as defined in Chapter 1, Section 7), relative to maximum likelihood estimation. They consider methods in which the number of estimating equations, obtained by equating sample values with corresponding formulas for the appropriate distribution, exceeds the number of parameters to be estimated. Estimation is carried out by a generalized least square procedure (called minimum chi-square), using large sample approximations for the variances and covariances of the sample statistics employed.

The properties of the distributions used by Gurland and Hinz are

(i) ratios of factorial cumulants $\eta_{(j)} = \kappa_{(j+1)}/\kappa_{(j)}$ (note that Ottestad [10] suggested the use of ratios of factorial *moments* and that $\eta_{(0)} = \kappa_1$)

(ii) the probability of a zero value, P_0

(iii) ratios of probabilities P_{j+1}/P_j, in the form of the ratios of probability cumulants $\tau_j = k_{j+1}/k_j$ (k_j as defined in Section 2 of this chapter).

Values of $\eta_{(j)}$ and τ_j for the four distributions are summarized below:

Distribution	Negative Binomial	Neyman Type A	Poisson Pascal	Poisson Binomial
Parameters	N, P	λ, ϕ	λ, N, P	λ, n, p (n known)
$\eta_{(j)}$	jP	ϕ	$(N+j)P$	$(n-j)p$
τ_j	$jP(1+P)^{-1}$	ϕ	$(N+j)P(1+P)^{-1}$	$(n-j)p(1-p)^{-1}$

If $\eta_{(j)}$ and τ_j are plotted against j, the four distributions give different pictures as shown in Figures 4a–d. Hinz and Gurland [6] suggest that plots of the corresponding sample values might help in discriminating among the four

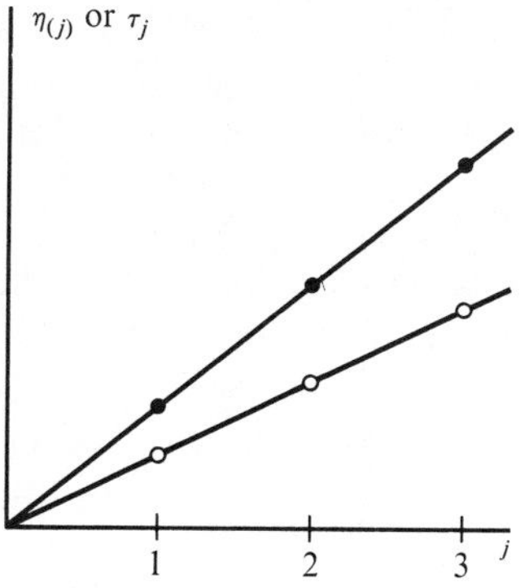

FIGURE 4a

Negative Binomial Distribution

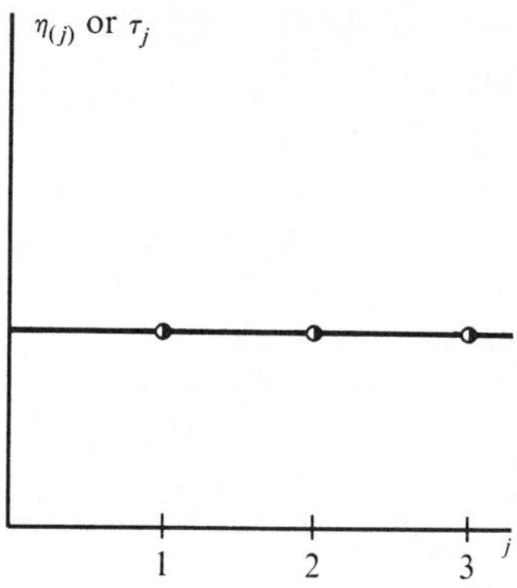

FIGURE 4b

Neyman Type A Distribution

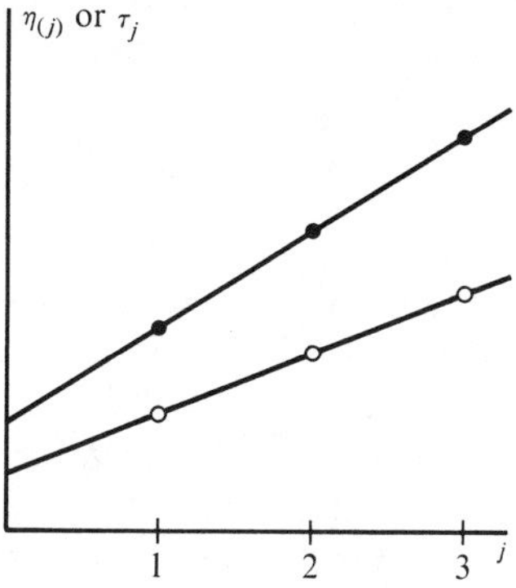

FIGURE 4c

Poisson Pascal Distribution

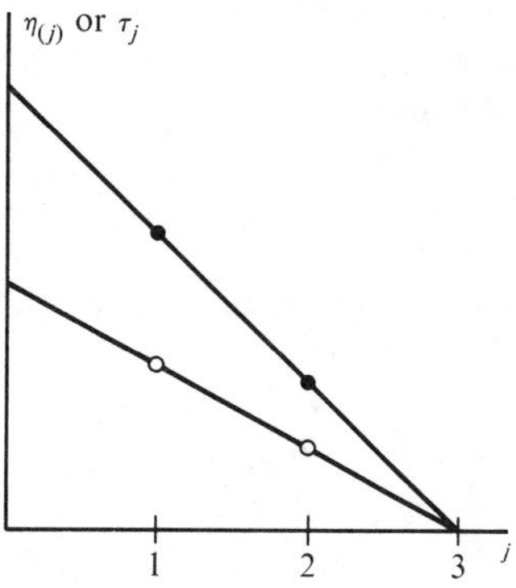

FIGURE 4d

Poisson Binomial Distribution

distributions, in the same way as the methods based on Figures 3a–c might help in making a choice among Poisson, binomial and negative binomial distributions. To decide whether one of the four former or one of the three latter distributions is more appropriate one might use a criterion such as Guldberg's (49).

Hinz and Gurland [6] found that, for a given number (s) of estimating equations the best combination to use is generally that corresponding to $\eta_{(0)}, \eta_{(1)}, \ldots, \eta_{(s-2)}$ and P_0. This gives at least 96.5% efficiency with $s = 2$ for the Neyman Type A distribution with $\lambda \leq 2$, $\theta \leq 2$; for the negative binomial with $N \leq 10$, $P \leq 0.5$; and for the Poisson binomial with $n \leq 5$; $p \leq 0.1$ and $\lambda \leq 2$ (and for $p = 0.3$, $n = 3$ or 5, $\lambda \leq 0.5$). With an increase of 1 in the number of equations ($s = 3$) these ranges of high efficiency are increased, and joint efficiency of at least 96.5% attained for the Poisson Pascal distribution with $N \leq 10$, $P \leq 0.5$ and $\lambda \leq 5$ (and for $\lambda = 5$, $P \leq 2$). No higher efficiency can be attained for the Poisson-Pascal with only three equations, than would be attained using three moments, i.e. $\eta_{(0)}$, $\eta_{(1)}$ and $\eta_{(2)}$.

These calculations were based on *asymptotic* values. Hinz and Gurland [6] suggest caution be used in transferring the comparisons to finite sized samples.

For all methods of estimation investigated only low joint efficiencies were indicated for the Poisson binomial distribution with p in excess of 0.5 and $\lambda \geq 0.3$, if $s = 1$, or $p \geq 0.8$, $\lambda \geq 1$ if $s = 2$.

When the value of P_0 is small, use of $\eta_{(j)}$'s only tends to be better than combination of $\eta_{(j)}$'s with P_0, or use of τ_j's alone. This is the case, for example, for the negative binomial with $N \geq 2.5$, $P \geq 2$.

The estimators based on equations for the τ_j's are generally not as accurate as those based on the $\eta_{(j)}$'s or on $\eta_{(j)}$'s and P_0. The τ_j's appear to give better results for the Poisson binomial distribution with $N = 2$, $\lambda = 0.1$, but this is an exceptional case.

In the following chapters we will not usually be considering 'minimum chi-square' estimators of the kind just described. However the relative merits of the different kinds of estimators — in particular the indicated advantages of using the equations

$$\mathbf{f}_0 = P_0, \quad \bar{\mathbf{x}} = \kappa_1, \quad \mathbf{s}^2 = \kappa_2,$$

etc, as estimating equations — are more probably relevant in assessing the merits of the estimators we do describe. In all discussion of the merits of estimators, it should be remembered that there is an implied assumption that the form of distribution used *really does* apply to the physical situation under consideration. For this reason comparison of goodness of fit to empirical data (real distribution unknown) of a distribution of specified form with different methods of estimation of parameters is an inadequate guide to the merits of the methods of estimation. Comparisons of this kind between different *forms* of fitted distribution may be relevant to the problem of choosing an approximate form of distribution.

6. Approximations

In the following chapters there will be described various approximations specific to particular distributions. Here we wish to describe a general technique for approximating the values $\{P_j\}$, given the probability generating function $g(t)$, developed by Douglas [3].

The problem is: given

$$g(t) = \sum_{j=0}^{\infty} P_j t^j,$$

obtain approximations to the coefficients $\{P_j\}$. The method of steepest descent gives an expansion with the leading term

$$P_j \doteqdot (2\pi)^{-\frac{1}{2}} g(t_j) \cdot t_j^{-(j+1)} [\phi''(t_j)]^{-\frac{1}{2}} \tag{51}$$

where

$$\phi(t) = \log [g(t)/t^j]$$

and t_j satisfies

$$\phi'(t_j) = 0 \qquad \text{(provided } \phi''(t_j) > 0\text{)}.$$

Douglas gives a number of examples which show that the accuracy of approximation can vary widely. However, used with some caution, formula (51) can give useful information about the magnitude of P_j. Such formulas are particularly useful when, as in the Neyman Type A distribution, no simple general explicit formula for P_j is available.

Occasionally, useful approximations to discrete distributions can be obtained using *Gram-Charlier Type B expansions.** The basic idea is to express P_j as a linear series in the backward differences (with respect to j) of the Poisson probabilities

$$\omega_j = \begin{cases} e^{-\theta}\theta^j/j! & (j \geq 0) \\ 0 & (j < 0). \end{cases} \tag{52}$$

Thus

$$P_j = \sum_{i=0}^{\infty} a_i \nabla^i \omega_j \qquad (\text{where } \nabla\omega_j = \omega_j - \omega_{j-1}). \tag{53}$$

Note that if (53) is summed from $j = c$ to $j = \infty$ then

$$\sum_{j=c}^{\infty} P_j = \sum_{i=0}^{\infty} a_i \nabla^i \left(\sum_{j=c}^{\infty} \omega_j\right) \quad \text{where } \nabla \text{ now operates on } c \tag{54}$$

(i.e. the cumulative sum of P_j's is expressed in terms of corresponding sums of Poisson probabilities). With the definition (52), $\sum_{j=c}^{\infty} \omega_j = 1$ for all $c \leq 0$.

Since the factorial moments are determined by the probability generating function, and (provided the moments exist) conversely, we might expect the values of P_j to be approximately determined by a sufficient number of values $\mu_{(1)}, \mu_{(2)}, \ldots$.

In fact if $P_j = 0$ for $j > N$, then

$$P_k = (k!)^{-1} \sum_{j=0}^{N-k} (-1)^j (j!)^{-1} \mu_{(k+j)} \qquad (k = 0,1,2,\ldots N) \tag{55}$$

and P_k is included between any two successive values obtained by terminating the sum at $j = s$, $j = s + 1$ respectively.

*The *Type A* expansions, applicable to continuous distributions, are of considerably wider use — see Chapter 12.

REFERENCES

[1] Bardwell, G. E. (1960). On certain characteristics of some discrete distributions, *Biometrika*, **47**, 473–475.

[2] Binet, F. E. (1969). On the analogue of the Pearson-Elderton family in enumeration distributions, *Proceedings of the Hungarian Academy of Science*.

[3] Douglas, J. B. (1963). Asymptotic expansions for some contagious distributions, **Proceedings of the International Symposium on Discrete Distributions, Montreal*, 291–302.

[4] Guldberg, A. (1931). On discontinuous frequency-functions and statistical series, *Skandinavisk Aktuarietidskrift*, **14**, 167–187.

[5] Gurland, J. (1963). A method of estimation for some generalized Poisson distributions, *Proceedings of the International Symposium on Discrete Distributions, Montreal*, 141–158.

[6] Hinz, P. and Gurland, J. (1967). Simplified techniques for estimating parameters of some generalized Poisson distributions, *Biometrika*, **54**, 555–566.

[7] Kamat, A. R. (1963). Incomplete and absolute moments of some discrete distributions, *Proceedings of the International Symposium on Discrete Distributions, Montreal*, 45–64.

[8] Katz, L. (1963). Unified treatment of a broad class of discrete probability distributions, *Proceedings of the International Symposium on Discrete Distributions, Montreal*, 175–182.

[9] Khatri, C. G. (1959). On certain properties of power-series distributions, *Biometrika*, **46**, 486–490.

[10] Noack, A. (1950). A class of random variables with discrete distributions, *Annals of Mathematical Statistics*, **21**, 127–132.

[11] Ord, J. K. (1967). Graphical methods for a class of discrete distributions, *Journal of the Royal Statistical Society, Series A*, **130**, 232–238.

[12] Ord, J. K. (1967). On a system of discrete distributions, *Biometrika*, **54**, 649–656.

[13] Ottestad, P. (1939). On the use of the factorial moments in the study of discontinuous frequency distributions, *Skandinavisk Aktuarietidskrift*, **22**, 22–31.

[14] Patil, G. P. (1961). Asymptotic bias and variance of ratio estimates in generalized power series distributions and certain applications, *Sankhyā, Series A*, **23**, 269–280.

[15] Patil, G. P. (1962). Certain properties of the generalized power series distribution, *Annals of the Institute of Statistical Mathematics, Tokyo*, **14**, 179–182.

[16] Patil, G. P. (1963). On the multivariate generalized power series distribution and its application to the multinomial and negative multinomial, *Proceedings of the International Symposium on Discrete Distributions, Montreal*, 183–194.

*These Proceedings have been published as *Classical and Contagious Discrete Distributions* (Edited by G. P. Patil) by Statistical Publishing Society, Calcutta and Pergamon Press, London and New York 1965. We will however use the date (1963) of presentation, and refer to it in the present style throughout this book.

[17] Patil, G. P. (1964). Estimation for the generalized power series distribution with two parameters and its application to the binomial distribution, *Contributions to Statistics on the* 70*th Birthday of P. C. Mahalanobis*, Oxford: Pergamon Press, Calcutta: Statistical Publishing Society, 335–344.

[18] Pearson, K. (1895). Contributions to the mathematical theory of evolution II. Skew variation in homogeneous material, *Philosophical Transactions of the Royal Society of London, Series A*, **186,** 343–414.

[19] Roy, J. and Mitra, S. K. (1957). Unbiased minimum variance estimation in a class of discrete distributions, *Sankhyā*, **18,** 371–378.

[20] Siromoney, G. (1964). The general Dirichlet's series distribution, *Journal of the Indian Statistical Association*, **2,** 69–74.

[21] Tweedie, M. C. K. and Veevers, A. (1968). The inversion of cumulant operators for power-series distributions, and the approximate stabilization of variance by transformations, *Journal of the American Statistical Association*, **63,** 321–328.

3

Binomial Distribution

1. Definition

The binomial distribution can be defined in terms of the expansion of the binomial $(q + p)^N$, where $q + p = 1; p > 0, q > 0$ and N is a positive integer. The $(k + 1)$th term in the expansion of $(q + p)^N$ is

$$\binom{N}{k} p^k q^{N-k} = \frac{N!}{k!(N-k)!} p^k q^{N-k}.$$

The *binomial distribution with parameters N, p* is defined as the distribution of a random variable **x** for which

$$(1) \qquad \Pr[\mathbf{x} = k] = \binom{N}{k} p^k q^{N-k} \qquad (k = 0,1,2, \ldots ,N).$$

Occasionally, a more general form is used in which on the right hand side (only) of (1), k is replaced by $a + bk$, a and b being real numbers, with b not equal to zero.

2. Genesis

If N independent trials are made, and in each there is probability p that the outcome E will occur, then the number of trials in which E occurs may be represented by a random variable **x** with the binomial distribution with parameters N, p.

Approximations of such situations often occur in applied statistics. The assumptions of independence and constant probability may not be strictly correct, but they often give a sufficiently accurate representation. Even when they do not, the binomial model provides a reference mark from which departures can be measured. This use of the binomial distribution is discussed in section 13.

3. Historical Remarks

The binomial distribution is one of the oldest to have been the subject of study. The distribution was derived by James Bernoulli in his treatise *Ars Conjectandi* published in 1713. At a considerably earlier date, binomial coefficients are to be found in the works of Pascal. Earlier references are given in an article by Boyer [11].

4. Moments and Generating Functions

The characteristic function of the distribution defined by (1) is $(q + pe^{it})^N$; the moment generating function is $(q + pe^t)^N$; the probability generating function is $(q + pt)^N$. There are a number of convenient formulas for the moments of the distribution.

The *r-th factorial moment is*

$$\text{(2)} \qquad \mu_{(r)}(\mathbf{x}) = E[\mathbf{x}^{(r)}] = E[\mathbf{x}(\mathbf{x}-1)(\mathbf{x}-2)\ldots(\mathbf{x}-r+1)] = N^{(r)}p^r.$$

Since $\mathbf{x}^r = \sum_{j=0}^{r} (\mathbf{x}^{(j)}/j!)\, \Delta^j 0^r$, it follows that the rth moment about zero is

$$\text{(3)} \qquad \mu'_r(\mathbf{x}) = E[\mathbf{x}^r] = \sum_{j=1}^{r} (\mu_{(j)}(\mathbf{x})/j!)\, \Delta^j 0^r$$

$$= \sum_{j=1}^{r} \binom{N}{j} p^j\, \Delta^j 0^r \equiv (1 + p\Delta)^N 0^r.$$

(Note that $\Delta^j 0^r = 0$ if $j > r$, or $j = 0$.)

In particular

$$\text{(4.1)} \qquad \mu'_1(\mathbf{x}) = Np$$

$$\text{(4.2)} \qquad \mu'_2(\mathbf{x}) = Np + N(N-1)p^2$$

$$\text{(4.3)} \qquad \mu'_3(\mathbf{x}) = Np + 3N(N-1)p^2 + N(N-1)(N-2)p^3$$

$$\text{(4.4)} \qquad \mu'_4(\mathbf{x}) = Np + 7N(N-1)p^2 + 6N(N-1)(N-2)p^3 + N(N-1)(N-2)(N-3)p^4.$$

The lower order central moments are

$$\text{(5.1)} \qquad \text{var}(\mathbf{x}) = \mu_2(\mathbf{x}) = Npq$$

$$\text{(5.2)} \qquad \mu_3(\mathbf{x}) = Npq(q-p)$$

$$\text{(5.3)} \qquad \mu_4(\mathbf{x}) = 3(Npq)^2 + Npq(1 - 6pq).$$

The moment ratios $\sqrt{\beta_1}$ and β_2 are

$$\sqrt{\beta_1} = (q - p)(Npq)^{-\frac{1}{2}} \tag{6.1}$$

$$\beta_2 = 3 + (1 - 6pq)(Npq)^{-1}. \tag{6.2}$$

For a fixed value of p (and so of q) the (β_1,β_2) points fall on the straight line

$$(\beta_2 - 3)/\beta_1 = (1 - 6pq)/(q - p)^2 = 1 - 2pq/(q - p)^2. \tag{7}$$

As $N \to \infty$ the point approaches the point (0, 3) as a limit.

Note that the same straight line is obtained if p is replaced by $(1 - p)$. In fact the two distributions are mirror images of each other, so they have identical values of β_2, and the same absolute value of $\sqrt{\beta_1}$. The slope $[(\beta_2 - 3)/\beta_1]$ is always less than 1. The limit of the ratio as p approaches 0 or 1 is 1. For $p = q = \frac{1}{2}$ the distribution is symmetrical (about $x = \frac{1}{2}$) and $\beta_1 = 0$. For $N = 2$, the point (β_1,β_2) lies on the line $\beta_2 - \beta_1 - 1 = 0$. (Note that, for *any* distribution, $\beta_2 - \beta_1 - 1 \geq 0$.)

Romanovsky [67] derived a recursion formula for the central moments $\{\mu_j\}$, namely

$$\mu_{r+1} = pq[Nr\mu_{r-1} + d\mu_r/dp]. \tag{8}$$

A similar relation holds for moments about zero, namely

$$\mu'_{r+1} = pq[(N/q)\mu'_r + d\mu'_r/dp]. \tag{9}$$

Formula (9) also holds for the incomplete moments

$$\mu'_{j,K} = \sum_{k=K}^{N} k^j \binom{N}{k} p^k q^{N-k}.$$

The mean deviation is

$$\nu_1 = E[|\mathbf{x} - Np|] = N\binom{N-1}{[Np]} p^{[Np]+1} q^{N-[Np]} \tag{10}$$

(Frisch [23], Frame [21]). Using Stirling's approximation for $N!$

$$\nu_1 \doteqdot \sqrt{\frac{2}{\pi}Npq}\left[1 + \frac{(Np - [Np])(Nq - [Nq])}{2Npq} - \frac{1 - pq}{12Npq}\right]. \tag{11}$$

From (11) it can be seen that the ratio of the mean deviation to the standard deviation, $\sqrt{Npq}$, approaches the limiting value $\sqrt{2/\pi} \doteqdot 0.798$ as N tends to infinity.

Any *inverse moment* of the binomial distribution (i.e. $E(\mathbf{x}^{-r})$ with $r > 0$) is infinite, because $\Pr[\mathbf{x} = 0] > 0$. Inverse moments of the *positive* binomial distribution (formed by truncation, omitting $\mathbf{x} = 0$) are discussed in Section 10.

The *inverse factorial* moment $E[\{(\mathbf{x} + r)^{(r)}\}^{-1}]$ is equal to

(12) $$\{(N + r)^{(r)}\}^{-1} p^{-r}\left[1 - \sum_{y=0}^{r-1} \binom{N+r}{y} p^y q^{N+r-y}\right].$$

(Note that $(N + r)^{(r)} = (N + 1)^{[r]}$.)

5. Properties

The distribution defined by (1) consists of $(N + 1)$ nonzero probabilities associated with the values 0, 1, 2, ..., N of the random variable $\mathbf{x}$. The ratio

(13) $$\frac{\Pr[\mathbf{x} = k + 1]}{\Pr[\mathbf{x} = k]} = \frac{N - k}{k + 1} \cdot \frac{p}{q} \qquad (k = 0,1,\ldots,(N - 1))$$

shows that $\Pr[\mathbf{x} = k]$ increases with k so long as $k < Np - q = (N + 1)p - 1$, and decreases with k if $k > Np - q$. Hence as k increases the values of the probability $\Pr[\mathbf{x} = k]$ increase to a maximum value at the integer k satisfying

$$(N + 1)p - 1 < k \leq (N + 1)p$$

and thereafter they decrease. If $(N + 1)p$ is an integer then

$$\Pr[\mathbf{x} = (N + 1)p - 1] = \Pr[\mathbf{x} = (N + 1)p].$$

The common value is the maximum among the values of $\Pr[\mathbf{x} = k]$.

Uhlmann [85] has shown that (for $N \geq 2$), denoting $\Pr[\mathbf{x} \leq c]$ by $L_{N,c}(p)$,

$$L_{N,c}\left(\frac{c}{N - 1}\right) > \frac{1}{2} > L_{N,c}\left(\frac{c + 1}{N + 1}\right) \quad \text{for } 0 \leq c < \tfrac{1}{2}(N - 1)$$

$$L_{N,c}\left(\frac{c}{N - 1}\right) = \frac{1}{2} = L_{N,c}\left(\frac{c + 1}{N + 1}\right) \quad \text{for } c = \tfrac{1}{2}(N - 1)$$

and $$L_{N,c}\left(\frac{c + 1}{N + 1}\right) > \frac{1}{2} > L_{N,c}\left(\frac{c}{N - 1}\right) \quad \text{for } \tfrac{1}{2}(N - 1) < c \leq N.$$

The *skewness* of the distribution is positive if $p < \frac{1}{2}$ and is negative if $p > \frac{1}{2}$ (see Section 4). The distribution is symmetrical if and only if $p = \frac{1}{2}$. Figures 1a–c represent binomial distributions with a common value of p (see also Figures 1a–c of Chapter 4, which also represent binomial distributions).

The distribution of the *standardized binomial variable*

$$\mathbf{x}' = (\mathbf{x} - Np)/\sqrt{Npq}$$

tends to the unit normal distribution as N tends to infinity. That is, for any real numbers α, β (with $\alpha < \beta$)

(14) $$\lim_{N\to\infty} \Pr[\alpha < \mathbf{x}' < \beta] = \frac{1}{\sqrt{2\pi}} \int_\alpha^\beta e^{-\frac{1}{2}u^2}\, du.$$

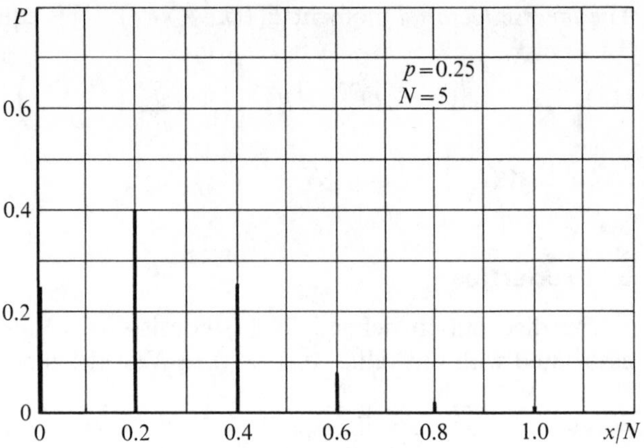

FIGURE 1a

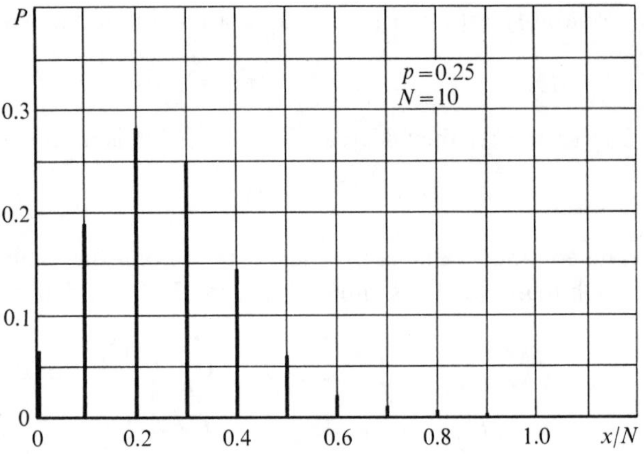

FIGURE 1b

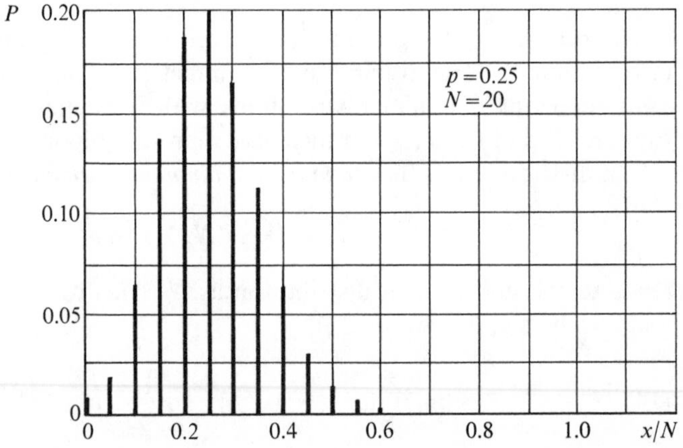

FIGURE 1c

The result is known as *Laplace's theorem*. It forms a starting point for a number of approximations in the calculation of binomial probabilities, which will be discussed in Section 8.

In Section 4 we noted that the probability generating function of $\mathbf{x}$ is $(q + pt)^N$. If $\mathbf{x}_1$, $\mathbf{x}_2$ are independent random variables having binomial distributions with parameters N_1, p and N_2, p respectively, then the probability generating function of $(\mathbf{x}_1 + \mathbf{x}_2)$ is $(q + pt)^{N_1}(q + pt)^{N_2} = (q + pt)^{N_1+N_2}$. It follows that $(\mathbf{x}_1 + \mathbf{x}_2)$ has a binomial distribution with parameters $(N_1 + N_2)$, p. This property of the binomial distribution is also apparent on interpreting $(\mathbf{x}_1 + \mathbf{x}_2)$ as the number of occurrences of an outcome E having constant probability, p, in each of $(N_1 + N_2)$ independent trials.

The distribution of $\mathbf{x}_1$, conditional on $\mathbf{x}_1 + \mathbf{x}_2 = X$, is

$$(15) \qquad \Pr[\mathbf{x}_1 = k \mid X] = \frac{\binom{N_1}{k} p^k q^{N_1-k} \binom{N_2}{X-k} p^{X-k} q^{N_2-(X-k)}}{\binom{N_1+N_2}{X} p^X q^{N_1+N_2-X}}$$

$$= \frac{\binom{N_1}{k}\binom{N_2}{X-k}}{\binom{N_1+N_2}{X}}$$

$$(\max(0, X - N_2) \le k \le \min(N_1, X)).$$

This is a *hypergeometric distribution* (see Chapter 6 and section 12 of this chapter).

The distribution of the difference $(\mathbf{x}_1 - \mathbf{x}_2)$ is

$$(16) \qquad \Pr[\mathbf{x}_1 - \mathbf{x}_2 = k] = \sum_{x_1} \binom{N_1}{x_1}\binom{N_2}{x_1 - k} p^{2x_1-k} q^{N_1+N_2-(2x_1-k)}$$

where the summation is between the limits $\max(0,k) \le x_1 \le \min(N_1, N_2 + k)$.

When $p = q = \frac{1}{2}$,

$$\Pr[\mathbf{x}_1 - \mathbf{x}_2 = k] = \binom{N_1 + N_2}{N_2 + k}\Big/ 2^{N_1+N_2} \quad (-N_2 \le k \le N_1)$$

so that $(\mathbf{x}_1 - \mathbf{x}_2)$ has a binomial distribution of the more general form mentioned in Section 1.

There is no simpler way of expressing (16). From Laplace's theorem, and the independence of $\mathbf{x}_1$ and $\mathbf{x}_2$, it follows that the distribution of the standardized difference

$$\{\mathbf{x}_1 - \mathbf{x}_2 - p(N_1 - N_2)\}\{pq(N_1 + N_2)\}^{-\frac{1}{2}}$$

tends to the unit normal distribution as $N_1, N_2 \to \infty$ (in whatever ratio). A similar result also holds when $\mathbf{x}_1, \mathbf{x}_2$ have binomial distributions with parameters N_1, p_1 and N_2, p_2 with $p_1 \neq p_2$, but the conditional distribution of

$\mathbf{x}_1$, given $\mathbf{x}_1 + \mathbf{x}_2 = X$, is no longer hypergeometric. Hannan and Harkness [35] have developed approximations for this distribution (see Section 13).

6. Order Statistics

As is the case for most discrete distributions, order statistics based on observed values of random variables with a common binomial distribution are not often used. Mention may be made, however, of discussions of binomial order statistics by Gupta [31], Khatri [42] and Siotani [74]. Tables of the cumulative distribution of the smallest and largest order statistic and of the range (in random samples of sizes $n = 1(1)20$) are in Gupta [30] and in Siotani and Ozawa [75]. Gupta [30] also has tables of the expected value and variance of these statistics for $p = 0.05(0.05)0.50$, $N = 1(1)20$ and sample sizes $n = 1(1)10$. In [32], Gupta and Panchapakesan have given tables which include the cumulative distribution functions of these statistics for the same values of p, N and n.

These tables can be applied in selecting the largest binomial probability among a set of k, based on k independent series of trials. This problem has been considered by Somerville [78] and by Sobel and Huyett [77].

7. Estimation of Parameters

7.1 *Point Estimation*

If $\mathbf{x}_1, \mathbf{x}_2, \ldots, \mathbf{x}_n$ are independent random variables, and $\mathbf{x}_j$ has a binomial distribution with parameters N_j, p $(j = 1,2,\ldots,n)$ then the maximum likelihood estimator of p is the overall relative frequency

$$\hat{\mathbf{p}} = \left(\sum_{j=1}^{n} \mathbf{x}_j\right) \Big/ \left(\sum_{j=1}^{n} N_j\right).$$

In fact $\sum_{j=1}^{n} \mathbf{x}_j$ is a sufficient statistic for $\hat{\mathbf{p}}$, and since this sum has a binomial distribution with parameters $\sum_{j=1}^{n} N_j$, p one need consider only the analysis for a single binomial variable in this connection.

The estimator $\hat{\mathbf{p}}$ is unbiased. Its variance is $pq\left(\sum_{j=1}^{n} N_j\right)^{-1}$ which is equal to the Cramer-Rao lower bound for unbiased estimators of p; $\hat{\mathbf{p}}$ is, in fact, the minimum variance unbiased estimator of p.

The above discussion assumes that $N_1, N_2, \ldots, N_n$ (or at least $\sum_{j=1}^{n} N_j$) are known. It is rarely necessary to estimate the value of the N_j's, but the problem has been studied by Student [82], Fisher [20], Hoel [38], and Binet [4]. Given a single observation of a random variable, $\mathbf{x}$, having a binomial distribution with parameters N, p, then if p is known, a natural estimator for N is x/p. This is unbiased, and has variance Nq/p. If neither N nor p is known, it is not possible to estimate either N or p from a single observation. However, if $\mathbf{x}_1, \mathbf{x}_2, \ldots, \mathbf{x}_n$ all have the same binomial distribution with parameters N, p (i.e. as above, with $N_1 = N_2 = \cdots = N_n = N$) then, equating observed and expected first and second moments, gives estimators $\tilde{\mathbf{N}}$, $\tilde{\mathbf{p}}$ of N, p respectively by solving the equations

$$(17.1) \qquad \bar{\mathbf{x}} = n^{-1} \sum_{j=1}^{n} \mathbf{x}_j = \tilde{\mathbf{N}}\tilde{\mathbf{p}}$$

$$(17.2) \qquad \mathbf{s}^2 = (n-1)^{-1} \sum_{j=1}^{n} (\mathbf{x}_j - \bar{\mathbf{x}})^2 = \tilde{\mathbf{N}}\tilde{\mathbf{p}}\tilde{\mathbf{q}}$$

whence

$$(18.1) \qquad \tilde{\mathbf{p}} = 1 - \mathbf{s}^2/\bar{\mathbf{x}}$$

$$(18.2) \qquad \tilde{\mathbf{N}} = \bar{\mathbf{x}}/\tilde{\mathbf{p}} = \bar{\mathbf{x}}^2/(\bar{\mathbf{x}} - \mathbf{s}^2).$$

Note that if $\bar{\mathbf{x}}$ is less than $\mathbf{s}^2$ then $\tilde{\mathbf{N}}$ is negative. This will indicate that the binomial distribution is not an appropriate model to use. Ignoring the limitation that N must be an integer, the maximum likelihood estimators $\hat{\mathbf{N}}$, $\hat{\mathbf{p}}$ satisfy the equations

$$(19.1) \qquad \hat{\mathbf{N}}\hat{\mathbf{p}} = \bar{\mathbf{x}}$$

$$(19.2) \qquad \sum_{j=0}^{\mathbf{R}-1} (\hat{\mathbf{N}} - j)^{-1}\mathbf{f}_j = -n \log_e (1 - \bar{\mathbf{x}}/\hat{\mathbf{N}})$$

where $\mathbf{f}_j$ = number of $\mathbf{x}$'s which exceed j and $\mathbf{R} = \max(\mathbf{x}_1, \ldots, \mathbf{x}_n)$. (Note similarity of (18.2) and (19.1).) If n is large the variance of $\hat{\mathbf{N}}$ is approximately

$$(20) \qquad N\left[\sum_{j=2}^{N} (p/q)^j \frac{(j-1)!}{j(n-1)^{(j-1)}}\right]^{-1} \Big/ n.$$

The asymptotic efficiency of $\tilde{\mathbf{N}}$, relative to $\hat{\mathbf{N}}$, is

$$(21) \qquad \left\{1 + 2\sum_{j=1}^{N-1} (p/q)^j \frac{j!}{(j+1)(n-2)^{(j-1)}}\right\}^{-1} \qquad \text{(Fisher [20]).}$$

Binet [4] states that $\tilde{\mathbf{N}}$ is negligibly less accurate than $\hat{\mathbf{N}}$ if either

(i) $N > 21$,
(ii) $N > 2$ and $p < \frac{1}{21}$

or

(iii) $N > 2$, $(p^{-1} - 1)(n - 2) > 20$.

Binet further points out that both $\hat{\mathbf{N}}$ and $\tilde{\mathbf{N}}$ may be improved by taking into account the facts that

(a) N must be an integer

and

(b) $N \geq \mathbf{R} = \max(\mathbf{x}_1, \mathbf{x}_2, \ldots, \mathbf{x}_n)$.

He suggests taking as the improved estimate the nearest integer to $\hat{\mathbf{N}}$ (or $\tilde{\mathbf{N}}$) if this integer is greater than **R**, and otherwise taking **R**.

If p is known, the equation for the maximum likelihood estimator, $\hat{\mathbf{N}}$ of N is obtained by replacing $\bar{\mathbf{x}}/\hat{\mathbf{N}}$, on the right-hand side of (19.2) by p. For n large

$$\sqrt{n}\,\mathrm{Var}(\hat{\mathbf{N}}) \doteqdot \left\{\sum_{j=1}^{N} \Pr[\mathbf{x} = j] \sum_{i=0}^{j-1} (N - i)^{-2}\right\}^{-1}.$$

From now on we shall assume that the first parameter (N) is known, and that only estimation of the parameter p (or some function of p) is under consideration.

Sometimes, it is required to estimate $\Pr[\alpha < \mathbf{x} < \beta]$. The minimum variance unbiased estimator of this polynomial function of p is

$$\text{(22)} \qquad \left[\sum_{\alpha<\xi<\beta} \binom{N}{\xi}\binom{N(n-1)}{\mathbf{X}-\xi}\right] \Big/ \binom{Nn}{\mathbf{X}} \qquad \text{with } \mathbf{X} = \sum_{j=1}^{n} \mathbf{x}_j$$

(see Barton [3]). From (22) it can be seen that the minimum variance unbiased estimator of a probability $\Pr[\mathbf{x} \in \omega]$, where ω is *any* sub-set of the integers $(0,1,\ldots,N)$, is of form (22) with "$\alpha < \xi < \beta$" replaced by "$\xi \in \omega$."

Rutemiller [68] has studied the estimator of $\Pr[\mathbf{x} = 0]$ in some detail, giving tables of its bias and variance.

Another function of p for which estimators have been constructed is

$$\min\,(p, 1 - p).$$

The natural estimator to use, given an observed value of a random variable $\mathbf{x}$ with distribution (1) is

$$\text{(23)} \qquad \min\,(\mathbf{x}/N, 1 - \mathbf{x}/N).$$

The moments of this statistic have been studied by Greenwood and Glasgow [29], and the cumulative distribution by Sandelius [69].

Median-unbiased estimators of p have been discussed by Birnbaum [5]. They differ little from the relative frequency, except for small values of N.

7.2 *Confidence Intervals*

The binomial distribution is a discrete distribution. Generally it is not possible to construct a confidence interval for p with an exactly specified confidence coefficient, when all that is used are the values of independent random variables $\mathbf{x}_1, \mathbf{x}_2, \ldots, \mathbf{x}_n$ having binomial distributions with first parameters $N_1, N_2, \ldots, N_n$ and common second parameter p. Approximate $100(1 - \alpha)\%$ limits may be obtained by solving the following equations for p_L and p_U.

$$\text{(24.1)} \qquad \sum_{j=\mathbf{X}}^{N} \binom{N}{j} (p_L)^j (1 - p_L)^{N-j} = \tfrac{1}{2}\alpha$$

(24.2)
$$\sum_{j=0}^{\mathbf{X}} \binom{N}{j} (p_U)^j (1 - p_U)^{N-j} = \tfrac{1}{2}\alpha$$

where

$$N = \sum_{j=1}^{n} N_j, \qquad \text{and} \qquad \mathbf{X} = \sum_{j=1}^{n} \mathbf{x}_j.$$

The values of p_L and p_U depend on X, and the interval (p_L,p_U) is an *approximate* $100(1 - \alpha)\%$ confidence interval for p. Values of p_L and p_U are found in the following publications:

Mainland [49] gives values to two or three significant figures; for $\alpha = 0.01, 0.05, 0.20$; $X = 0(1)20$ and various values of N up to 1000 and (for the same values of α); $100(X/N) = 0.1(0.2)0.7, 1(1)3, 5, 7.5, 10(5)50$ and various values of N up to 100,000.

Clark [14] gives values of p_L only to four significant figures. $\alpha = 0.01, 0.02, 0.05, 0.10$; $N = 10(1)50$ (and all possible values of X).

Pachares [59] gives values to four significant figures for $\alpha = 0.01, 0.02, 0.05, 0.10$; $N = 55(5)100$; and all possible values of X.

Lachenbruch [44] gives values, to four decimal places, for $\alpha = 0.1k^{-1}, 0.05k^{-1}, 0.01k^{-1}$ with $k = 1(1)10$, and $N = 10(2)20(5)40(10)100$.

Tables of the incomplete beta function ratio (Pearson [61]) can be used to solve (24.1) provided max $(\mathbf{X},N - \mathbf{X} + 1) \leq 50$, and to solve (24.2) provided max $(\mathbf{X} + 1,N - \mathbf{X}) \leq 50$. The identity

(25)
$$I_p(k,N - k + 1) = \sum_{j=k}^{N} \binom{N}{j} p^j q^{N-j}$$

is used (see Section 8).

Since the lower $100\beta\%$ point $F_{\nu_1,\nu_2,\beta}$ of the F distribution with ν_1, ν_2 degrees of freedom (Chapter 26, equation (9)) satisfies the equation

(26)
$$I_{\nu_1 F/(\nu_2+\nu_1 F)}(\tfrac{1}{2}\nu_1,\tfrac{1}{2}\nu_2) = \beta$$

while (24.1) can be written in the form

(27)
$$I_{p_L}(\mathbf{X},N - \mathbf{X} + 1) = \tfrac{1}{2}\alpha$$

it follows that

(28)
$$p_L = \nu_1 F_{\nu_1,\nu_2,\alpha/2}/(\nu_2 + \nu_1 F_{\nu_1,\nu_2,\alpha/2})$$

with $\nu_1 = 2\mathbf{X}$, $\nu_2 = 2(N - \mathbf{X} + 1)$.

There is a similar formula for p_U with $\alpha/2$ replaced by $1 - \alpha/2$ and $\nu_1 = 2(\mathbf{X} + 1)$, $\nu_2 = 2(N - \mathbf{X})$. Tables of percentage points of the F distribution can therefore be used to obtain values of p_L and p_U.

Charts from which confidence limits for p can be read off are given in the Biometrika tables [60]. One such chart is shown in Figure 2.

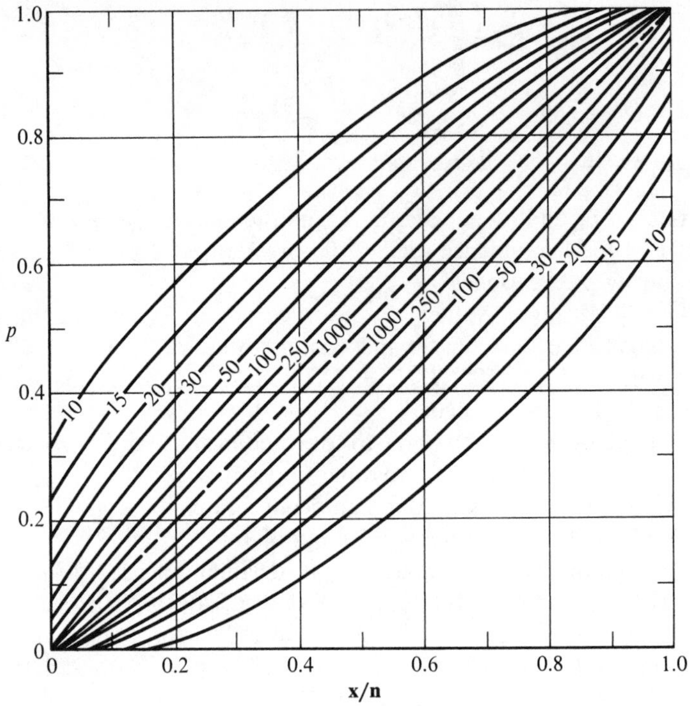

FIGURE 2

Confidence Limits for Binomial p.
Confidence Coefficient = 0.95

Approximate values of p_L and p_U can be obtained using the normal approximation (see Section 8) to the binomial distribution. The required values are the roots of

$$[X - Np]^2 = \lambda_{\alpha/2}^2 \cdot Np(1 - p) \tag{29}$$

where

$$(2\pi)^{-1/2} \int_{\lambda_{\alpha/2}}^{\infty} e^{-\frac{1}{2}u^2}\, du = \alpha/2.$$

More accurate values can be obtained using continuity corrections. However, since even the exact solutions of (24.1) and (24.2) do not provide exact confidence intervals, the matter will not be pursued here.

Nayatani and Kurahara [55] consider that the simple approximate confidence limits

$$\frac{X}{N} \pm \lambda_{\alpha/2} \sqrt{\frac{X}{N^2}\left(1 - \frac{X}{N}\right)} \tag{30}$$

can be used provided $Np > 5$; $p < 0.5$. They also discuss the use of a Poisson approximation, giving approximate confidence limits

$$(2N)^{-1}\chi^2_{2X,\alpha/2}, \quad (2N)^{-1}\chi^2_{2(X+1),1-\alpha/2}$$

where $\chi^2_{\nu,\epsilon}$ is defined by the equation $\Pr[\chi^2_\nu < \chi^2_{\nu,\epsilon}] = \epsilon$. That is $\chi^2_{\nu,\alpha/2}$ is the lower, and $\chi^2_{\nu,1-\alpha/2}$ the upper $100(\alpha/2)\%$ point of the distribution of χ^2 with ν degrees of freedom.

They suggest that this approximation might be used if p is less than 0.10. A better rule would be based on N being large — say at least 50, although if p is near to 0.5, it would probably be more convenient, and as accurate to use the normal approximation limits (30).

If p is known, we may obtain approximate confidence limits (N_L, N_U) for N by solving the equations

$$\sum_{j=0}^{X} \binom{N_U}{j} p^j(1-p)^{N_U-j} = \tfrac{1}{2}\alpha$$

$$\sum_{j=0}^{X} \binom{N_L}{j} p^j(1-p)^{N_L-j} = 1 - \tfrac{1}{2}\alpha.$$

Hald and Kousgaard [34] give values of N satisfying

$$\sum_{j=0}^{c} \binom{N}{j} p^j(1-p)^{N-j} = P$$

to four significant figures for $c = 0(1)50$; $P = 0.001, 0.005, .01, 0.025, 0.05, 0.1, 0.2, 0.5, 0.8, 0.9, 0.95, 0.975, 0.99, 0.995$ and 0.999; and $p = 0.001, 0.01, 0.02, 0.05, 0.07, 0.10(0.05)0.50$. For $c = 1, 2, 3$ values are also given for $p = 0.005(0.010)0.035, 0.06, 0.085, 0.125(0.05)0.475$.

8. Approximations and Bounds

8.1 *Approximations*

The binomial distribution is of such importance in applied probability and statistics that it is frequently necessary to calculate probabilities based on this distribution. Although the calculation of sums of the form

$$\sum_x \binom{N}{x} p^x q^{N-x}$$

is straightforward, it can be tedious, especially when N and x are large, and when there is a large number of terms in the summation. It is not surprising that attention, and ingenuity have been applied to constructing useful approximations for sums of this kind.

The approximation based on Laplace's theorem

$$\Pr[\alpha < (\mathbf{x} - Np)(Npq)^{-\frac{1}{2}} < \beta] \doteqdot (\sqrt{2\pi})^{-1} \int_\alpha^\beta e^{-\frac{1}{2}u^2}\, du \tag{31}$$

has been indicated in section 5. This is a relatively crude approximation, but useful if N is large.* A definite improvement is obtained by replacing α, β on the right-hand side of (31) by

$$\begin{aligned} &\{[\alpha\sqrt{Npq} + Np] - \tfrac{1}{2} - Np\}(Npq)^{-\frac{1}{2}};\\ &\{[\beta\sqrt{Npq} + Np] + \tfrac{1}{2} - Np\}(Npq)^{-\frac{1}{2}} \end{aligned} \tag{32}$$

respectively, (where [ ] denotes "largest integer not greater than"). The corrections to α, β represented by (32) are called *continuity corrections.*

Another (equivalent) form, which shows the nature of these corrections is

$$\Pr[r_1 \le \mathbf{x} \le r_2] \doteqdot (2\pi)^{-\frac{1}{2}} \int_{(r_1-\frac{1}{2}-Np)/\sqrt{Npq}}^{(r_2+\frac{1}{2}-Np)/\sqrt{Npq}} e^{-\frac{1}{2}u^2}\, du. \tag{33}$$

An approximately equivalent form is one given by Laplace (*Théorie Analytique de Probabilités*, livre 2, Chapter 3, 3rd edition, Paris, Courcier, 1820)

$$\Pr[\mathbf{x} \le r] = \Phi\left(\frac{r - Np}{(Npq)}\right) + \frac{1}{2\sqrt{Npq}} Z\left(\frac{r - Np}{(Npq)}\right)$$

where

$$Z(u) = (2\pi)^{-\frac{1}{2}} e^{-\frac{1}{2}u^2}$$

$$\Phi(u) = \int_{-\infty}^{u} Z(t)\, dt.$$

Uspensky [87], has shown that the error for this formula is numerically less than

$$(\tfrac{1}{5} + \tfrac{1}{4}|q - p|)(Npq)^{-1} + \exp[-3/2(Npq)^{\frac{1}{2}}]$$

provided $Npq \ge 25$.

Kalinin [40] has obtained the asymptotic expansion

$$\Pr[r_1 \le \mathbf{x} \le r_2] = (2\pi)^{-\frac{1}{2}} \int_{y_1}^{y_2} \exp(-\tfrac{1}{2}u^2)\, du + \sum_{j=1}^{\infty} (Npq)^{-\frac{1}{2}j} Q_j$$

where

$$Q_j = -\frac{B_j(\theta)}{j!} H_{j-1}(y)\left[\left(\frac{1}{\sqrt{2\pi}} e^{-\frac{1}{2}y^2}\right)\right]^{y-y_2} + (2\pi)^{-\frac{1}{2}} \int_{y_1}^{y_2} e^{-\frac{1}{2}y^2} T_j(y)\, dy$$

*Numerical comparisons have been published in a number of text books (e.g. Hald [30], pp. 682–683).

$$+ \sum_{i=1}^{j-1} \frac{(-1)^i B(\theta)}{i!} \left[\frac{d^{i-1}}{dy^{i-1}} \left\{ \left(\frac{1}{\sqrt{2\pi}} e^{-\frac{1}{2}y^2} \right) T_{j-i}(y) \right\} \right]_{y=y_1}^{y=y_2}$$

with $y_1 = (r_1 - (1 - \theta) - Np)/\sqrt{Npq}$; $y_2 = (r_2 + \theta - Np)/\sqrt{Npq}$. $T_j(y)$ is a rather complicated polynomial. It is the sum, over all sets

$$\left\{ \nu_1, \nu_2, \ldots \nu_j \middle| \sum_{i=1}^{j} i\nu_i = j, \nu_i \text{ an integer and } \nu_i \geq 0 \right\}$$

of $\prod_{i=1}^{j} \{[G_i(y)]^{\nu_i}/\nu_i!\}$, where

$$G_i(y) = \sum_{h=0}^{[(i+1)/2]} \frac{B_h \cdot \binom{j-h+2}{h} y^{i-2h+2}}{(i-h+1)(i-h+2)} [p^{i-h+1} + (-1)^i q^{i-h+1}]$$

if $i \neq 4\mu - 2$, μ an integer,

$$G_i(y) = \sum_{h=0}^{2\mu-1} \left\{ \frac{B_h \cdot \binom{4\mu - h}{h} y^{4\mu-2h}(p^{4\mu-h-1} + q^{4\mu-h-1})}{(4\mu - h - 1)(4\mu - h)} \right\}$$

$$= -\frac{B_{2\mu}}{2\mu(2\mu - 1)} (p^{2\mu-1} + q^{2\mu-1} - [pq]^{2\mu-1}) \qquad \text{if } i = 4\mu - 2$$

where $B_j(\cdot)$, B_j are Bernoulli polynomials and numbers respectively, and $H_j(\cdot)$ are Hermite polynomials as defined in Chapter 1.

This formula is valid uniformly (as $N \to \infty$) over any finite interval containing y_1 and y_2.

The formula is complicated but of interest because explicit expressions for the coefficients Q_j are available. The continuity correction, θ, can be assigned arbitrarily. The usual correction corresponds to $\theta = \frac{1}{2}$.

A number of approximations to binomial probabilities are based on the equation

$$\text{(34)} \qquad \Pr[\mathbf{x} \geq k] = \sum_{x=k}^{N} \binom{N}{x} p^x q^{N-x}$$

$$= B(k,N - k + 1)^{-1} \int_0^p t^{k-1}(1 - t)^{N-k}\, dt$$

$$= I_p(k,N - k + 1).$$

(Formula (34) can be established by integration by parts.) Methods of approximation are then applied either to the integral, or to the incomplete beta function ratio $I_p(k,N - k + 1)$.

Bizley [6] and Jowett [39] have pointed out that since there is an exact correspondence between sums of binomial probabilities and probability integrals of certain central F distributions (see Section 7), it follows that where the values of the parameters are appropriate, approximations developed for one distribution are applicable to the other.

The *Camp-Paulson* approximation (Chapter 26) was developed with reference to the F distribution. When applied to the binomial distribution it gives

$$\text{(35)} \qquad \Pr[\mathbf{x} \le k] \doteqdot (2\pi)^{-\frac{1}{2}} \int_{-\infty}^{Y(3\sqrt{Z})^{-1}} e^{-\frac{1}{2}u^2}\,du,$$

where

$$Y = [(N-k)p(k+1)^{-1}q^{-1}]^{\frac{1}{3}}[9-(N-k)^{-1}] - 9 + (k+1)^{-1}.$$
$$Z = [(N-k)p(k+1)^{-1}q^{-1}]^{\frac{2}{3}}(N-k)^{-1} + (k+1)^{-1}.$$

The maximum absolute error in this approximation cannot exceed $0.007(Npq)^{-\frac{1}{2}}$.

By comparison the maximum absolute error using the normal approximation with continuity correction

$$\text{(36)} \qquad \Pr[\mathbf{x} \le k] \doteqdot (2\pi)^{-\frac{1}{2}} \int_{-\infty}^{[k+\frac{1}{2}-Np]/\sqrt{Npq}} e^{-\frac{1}{2}u^2}\,du$$

is less than $0.140(Npq)^{-\frac{1}{2}}$, and can approach quite close to this limit. For fixed N, this maximum value is minimized when $p = q = \frac{1}{2}$. In general, the normal approximation (with or without continuity correction) is overall more accurate the nearer p is to $\frac{1}{2}$. It is reasonable to expect this, since the normal distribution is symmetrical, while the binomial distribution is symmetrical only if $p = \frac{1}{2}$.

A natural modification of the normal approximation, taking account of asymmetry, is to use a Gram-Charlier expansion, taking one term in addition to the leading (normal) term. This leads to the approximation (including continuity correction)

$$\text{(37)} \qquad \Pr[\mathbf{x} \le k] \doteqdot (\sqrt{2\pi})^{-1} \int_{-\infty}^{(k+\frac{1}{2}-Np)/\sqrt{Npq}} e^{-\frac{1}{2}u^2}\,du - \frac{1}{6}\frac{q-p}{(Npq)^{\frac{1}{2}}}\left[\frac{(k+\frac{1}{2}-Np)^2}{Npq} - 1\right] \cdot \frac{1}{\sqrt{2\pi}} \exp\left[-\frac{(k+\frac{1}{2}-Np)^2}{2Npq}\right].$$

The maximum absolute error is $0.056(Npq)^{-\frac{1}{2}}$. It varies with N and p in much the same way as for the normal approximations, but is usually substantially (about 50%) smaller. The relative advantage, however, depends on k, as well as N and p. (For details see Raff [64].)

Sawkins [71] first approximated

$$\Delta \log [\Pr[\mathbf{x} = k]] = \log [\{\Pr[\mathbf{x} = k+1]\}/\{\Pr[\mathbf{x} = k]\}].$$

He then used these results with the difference formula for the derivative (written symbolically $D \equiv \Delta - \frac{1}{2}\Delta^2 + \frac{1}{3}\Delta^3 - \cdots$), and solved the resulting (approximate) differential equation for $\Pr[\mathbf{x} = k]$. By this, he obtained the first two terms of the Gram-Charlier expansion (equivalent to (37)). Further approximations were obtained by using the symbolic equation for the derivative $D = (1 - \frac{1}{2}D + \frac{1}{12}D^2 - \cdots)\Delta$, which gives the formula

$$\log \Pr[\mathbf{x} = k] = \sum_{j=1}^{\infty} [j(j+1)]^{-1} N^j \{-q^{-j} B_j(y) + (-p)^j B_j(y+1)\} + C$$

where $y = (k - Np)/\sqrt{Npq}$ and C is a constant.

A remarkably accurate normalizing transformation has been proposed by Borges (see [25]). The transformed variable (approximately unit normal) is

$$(pq)^{-\frac{1}{6}}(n + \tfrac{1}{3})^{\frac{1}{2}} \int_p^{\mathbf{y}} [t(1-t)]^{-\frac{1}{3}}\, dt$$

where $\mathbf{y} = (\mathbf{x} + \frac{1}{6})/(n + \frac{1}{3})$. Comparisons by Gebhardt [25] show that this approximation is particularly good for $0.1 \leq p \leq 0.9$, though it is effective for p as small as 0.001 provided sample size is large (≥ 100, say). Tables of $\int_{0.5}^{Y} [t(1-t)]^{-\frac{1}{3}}\, dt$ $(= B_Y(\frac{2}{3}, \frac{2}{3}) - B_{0.5}(\frac{2}{3}, \frac{2}{3}))$ to 4 decimal places for $Y =$ 0.500(0.001)1.000 are given in an addendum to [25].

We now give a number of other approximations some of which are commonly used in calculations, but which are not, generally, of accuracy comparable with (35) or even (37).

The "arcsine" transformation

$$\mathbf{y} = \sin^{-1}\sqrt{(\mathbf{x} + 3/8)/(N + 3/4)}$$

produces a random variable $\mathbf{y}$ which is approximately normally distributed with expected value $\sin^{-1}\sqrt{p}$ and variance $1/4n$. This leads to the approximation

$$\Pr[\mathbf{x} \leq k] \doteqdot (2\pi)^{-1} \int_{-\infty}^{Y} e^{-\frac{1}{2}u^2}\, du \tag{38}$$

with $Y = 2N^{\frac{1}{2}}[\sin^{-1}\sqrt{(k + 3/8)/(N + 3/4)} - \sin^{-1}\sqrt{p}]$. (If a continuity correction is used then $\frac{1}{2}$ must be added to k.) We also mention the approximation suggested by Freeman and Tukey [22], according to which

$$\tfrac{1}{2}[\sin^{-1}\sqrt{\mathbf{x}/(N+1)} + \sin^{-1}\sqrt{(\mathbf{x}+1)/(N+1)}]$$

which has the same approximate distribution as $\mathbf{y}$. Tables for applying this transformation were provided by Mosteller and Youtz [53].

If N tends to infinity and p tends to zero in such a way that $Np = \theta$ remains constant then

$$\lim_{\substack{N\to\infty \\ p\to 0 \\ Np=\theta}} \Pr[\mathbf{x} = k] = e^{-\theta}\theta^k/k!. \tag{39}$$

(This is discussed in Chapter 4, on the Poisson distribution.) Equation (39) is the basis for the *Poisson approximation* to the binomial distribution

$$\Pr[\mathbf{x} \leq k] \doteqdot e^{-Np} \sum_{j=0}^{k} \frac{(Np)^j}{j!} = \sum_{j=0}^{k} P(j,Np). \tag{40}$$

The maximum absolute error is practically independent of N, and approaches zero as p approaches zero.

In more detail, Anderson and Samuels [1] have shown that the right-hand side of (40) is less than the left-hand side if $k \geq Np$, but exceeds the left-hand side if $k \leq Np/(1 + N^{-1})$. Thus the Poisson approximation tends to overestimate tail probabilities, at each end of the distribution. The absolute error of approximation increases with k for $0 \leq k \leq (Np + \frac{1}{2}) - \sqrt{(Np + \frac{1}{4})}$, and decreases with k for $(Np + \frac{1}{2}) + \sqrt{(Np + \frac{1}{4})} \leq k \leq N$.

The Poisson approximation (40) may be modified in a similar way to that in which the normal approximation (36) is modified to (37). We mention two formulas of this kind. The *Poisson Gram-Charlier approximation* is

$$\Pr[\mathbf{x} \leq k] \doteqdot \sum_{j=0}^{k} P(j,Np) + \tfrac{1}{2} \sum_{j=0}^{k} (j - Np)\, \Delta P(j,Np).^* \tag{41}$$

Kolmogorov's approximation is

$$\Pr[\mathbf{x} \leq k] = \sum_{j=0}^{k} P(j,Np) - \frac{Np^2}{2} \nabla^2 P(j,Np) \tag{42}$$

$$(\text{with } P(j,Np) = 0 \text{ if } j < 0).$$

It is a form of Gram-Charlier Type B expansion. The next term is

$$\frac{Np^3}{3} \nabla^3 P(j,Np).$$

A detailed comparison of the Poisson and Kolmogorov approximations is given by Dunin-Barkovsky and Smirnov [16]. Table 1, below, is reproduced from their work.

Bol'shev [10] developed two approximations which are the same as the Poisson approximation (40), except that Np is replaced by

$$M(2 - p)^{-1} \qquad \text{with } M = 2N - k \tag{43}$$

or

(44)

$$M(2 - p)^{-1}[1 + \tfrac{1}{6}M^{-2}\{k(k + 2) + kM(2 - p)^{-1} - 2M^2(2 - p)^{-2}\}]^{-1}.$$

*The forward difference operator Δ operates on j.

TABLE 1

*Comparison Between Poisson and Kolmogorov's Approximations for Binomial Distribution**

m	P_m	P'_m	$P_m - P'_m$	$P_m - P'_m$ as % of P_m	P''_m	$P_m - P''_m$	$P_m - P''_m$ as % of P_m
0	0.16807	0.22313	−0.05506	−32.8	0.17293	−0.00486	−2.9
1	0.36015	0.33470	+0.02545	+7.1	0.35980	+0.00035	+0.1
2	0.30870	0.25102	+0.05768	+18.7	0.29495	+0.01375	+4.5
3	0.13230	0.12551	+0.00679	+5.1	0.13492	−0.00267	−2.0
4	0.02835	0.04707	−0.01872	−66.0	0.03648	−0.00813	−28.7
5	0.00243	0.01412	−0.01169	−481.1	0.00388	−0.00145	−59.7

*P_m — binomial probabilities with $N = 5, p = 0.3$
P'_m — Poisson approximation with $Np = 1.5$
P''_m — Kolmogorov approximation with $Np = 1.5$ and $Np^2 = 0.45$

The error is $0(N^{-2})$ for (43), and $0(N^{-4})$ for (44) as $N \to \infty$; these bounds are *uniform* with respect to p $(0 < p < 1)$.

Wishart [88] suggests approximating the distribution of

$$4(q - p)^{-2}[(N + 1)p + \mathbf{x}(q - p)]$$

by that of χ^2 with $2(q - p)^{-2}(4Npq + 1)$ degrees of freedom. This gives good results when p is small, and the asymmetry of the binomial distribution makes the normal approximation inaccurate.

For *individual* binomial probabilities, the normal approximation with continuity correction (37) gives

$$\text{(45)} \qquad \Pr[\mathbf{x} = k] \doteqdot (2\pi)^{-\frac{1}{2}} \int_{(k-\frac{1}{2}-Np)/\sqrt{Npq}}^{(k+\frac{1}{2}-Np)/\sqrt{Npq}} e^{-\frac{1}{2}u^2}\, du.$$

A nearly equivalent approximation is

$$\text{(46)} \qquad \Pr[\mathbf{x} = k] \doteqdot \frac{1}{\sqrt{Npq}} \cdot \frac{1}{\sqrt{2\pi}} \exp\left[-\frac{1}{2}\frac{(k - Np)^2}{Npq}\right].$$

Prohorov [63] gave a result which is helpful in gauging the accuracy of (46). The result is

$$\text{(47)} \qquad \sum_{k=0}^{N} \left| \binom{N}{k} p^k q^{N-k} - \frac{1}{\sqrt{Npq}} \cdot \frac{1}{\sqrt{2\pi}} \exp\left[-\frac{1}{2}\frac{(k - Np)^2}{Npq}\right] \right|$$
$$= \frac{1 + 4e^{-3/2}}{3} \cdot \frac{1}{\sqrt{Npq}} \cdot \frac{1}{\sqrt{2\pi}} + 0\left(\frac{1}{Npq}\right).$$

8.2 *Bounds*

In part 8.1 of this section, we gave approximations to various binomial probabilities; we will now give inequalities for them. Generally the former are closer to the true values than are the bounds given by the inequalities. However, the inequalities are definitely correct, and often give useful limits for errors of approximations.

Feller [18] showed that if $k \geq (N+1)p$

$$\text{(48)} \qquad \Pr[\mathbf{x} = k] \leq \Pr[\mathbf{x} = m] \exp\left[-\frac{1}{2}p\frac{\{k-(N+1)p+\frac{1}{2}\}^2}{(N+1)pq} + \left\{m-(N+1)p+\frac{1}{2}\right\}^2\right]$$

where m is the integer defined by $(N+1)p - 1 < m \leq (N+1)p$ and

$$\text{(49)} \qquad \binom{N}{m}\left(\frac{m+1}{N+1}\right)^m\left(1-\frac{m+1}{N+1}\right)^{N-m} \leq \binom{N}{m}p^m q^{N-m} \leq \binom{N}{m}\left(\frac{m}{N}\right)^m\left(1-\frac{m}{N}\right)^{N-m}.$$

A more sophisticated result, also given by Feller [18], is

$$\text{(50)} \qquad \Pr[\mathbf{x} = k] = \phi[2\pi(N+1)pq]^{-\frac{1}{2}} \exp\left[-\sum_{j=2}^{\infty}\frac{p^{j-1}-(-q)^{j-1}}{j(j-1)} \cdot \frac{\{k-(N+1)p+\frac{1}{2}\}^j}{(N+1)pq^{j-1}} + \frac{1}{24(N+1)pq}\sum_{j=3}^{\infty}[p^{j-1}-(-q)^{j-1}] \cdot \frac{\{k-(N+1)p+\frac{1}{2}\}^{j-2}}{\{(N+1)pq\}^{j-2}} + \frac{1+2pq}{24(N+1)pq}\right]$$

where $1 > \phi > \exp[-\frac{1}{300}\{(N+1)pq\}^{-2}]$. The limits for ϕ are very close together (unless p is very near to 0 or 1) so (50) gives very narrow limits for $\Pr[\mathbf{x} = k]$. However, the complexity of the formula makes it of little practical use.

There are a number of formulas giving bounds on the probability $\Pr[|\mathbf{x}/N - p| \geq c]$ where c is a constant, i.e. on the probability that the difference between the relative frequency, $\mathbf{x}/N$ and its expected value, p, will have absolute value greater than c. A discussion is given by Krafft [43] and Kambo and Kotz [41]. Uspensky [87] gave

$$\text{(51)} \qquad \Pr[|\mathbf{x}/N - p| \geq c] < 2\exp(-\tfrac{1}{2}Nc^2),$$

Levy [46] gave

$$\text{(52.1)} \qquad \Pr[|\mathbf{x}/N - p| \geq c] < 2c^{-1}\exp(-2Nc^2)$$

(52.2) $$\Pr[|\mathbf{x}/N - p| \geq c] < 2c^{-1}N^{-\frac{1}{2}} \exp(-2Nc^2)$$
(provided $p, q \geq \max(4N^{-1}, 2c)$.

Okamoto [58] showed that (with $c \geq 0$)

(53.1) $$\Pr[\mathbf{x}/N - p \geq c] < \exp(-2Nc^2)$$

and

(53.2) $$\Pr[\mathbf{x}/N - p \leq -c] < \exp(-2Nc^2).$$

(See also Hoeffding [37].) He also showed that

(54.1) $$\Pr[\mathbf{x}/N - p \geq c] < \exp\left(-\frac{Nc^2}{2pq}\right) \qquad \text{for } p \geq \tfrac{1}{2}$$

and

(54.2) $$\Pr[\mathbf{x}/N - p \leq -c] < \exp\left(-\frac{Nc^2}{2pq}\right) \qquad \text{for } p \leq \tfrac{1}{2}$$

and, further

(55.1) $$\Pr[\sqrt{\mathbf{x}/N} - \sqrt{p} \geq c] < \exp(-2Nc^2)$$

and

(55.2) $$\Pr[\sqrt{\mathbf{x}/N} - \sqrt{p} \leq -c] < \exp(-2Nc^2).$$

Kambo and Kotz [41] and Krafft [43] improved Okamoto's bounds, obtaining the following formulas.

(56.1) $$\Pr[\mathbf{x}/N - p \geq c] < \exp[-2Nc^2 - \tfrac{4}{3}Nc^4]$$

(56.2) $$\Pr[\mathbf{x}/N - p \leq -c] < \exp[-2Nc^2 - \tfrac{4}{3}Nc^4] \qquad \text{if } 0 \leq c < 1 - p$$
(cf (53))

(57.1) $$\Pr[\mathbf{x}/N - p \geq c] < \exp\left[-\frac{Nc^2}{2pq} - \frac{4}{9}Nc^4\right] \qquad \text{if } p \geq \tfrac{1}{2}$$

(57.2) $$\Pr[\mathbf{x}/N - p \leq -c] < \exp\left[-\frac{Nc^2}{2pq} - \frac{4}{9}Nc^4\right] \qquad \text{if } p \leq \tfrac{1}{2} \text{ (cf (54))}$$

(58) $$\begin{aligned}\Pr[\sqrt{\mathbf{x}/N} - \sqrt{p} \geq c]& \\ &< \exp[-2Nc^2/q - \tfrac{2}{3}Nc^3(1 + \sqrt{p}/q + 2\sqrt{p}/q^2)] \\ &< \exp[-2Nc^2/q - \tfrac{2}{3}Nc^3].\end{aligned}$$

For $N > 2$ and c not too large, Kambo and Kotz [41] improved (56.1) and (56.2) to

(59.1) $$\Pr[\mathbf{x}/N - p \geq c] < q(c\sqrt{N})^{-1} \exp[-2Nc^2 - \tfrac{4}{9}Nc^4]$$
if $p + c \leq \tfrac{1}{2}$.

(59.2) $\Pr[\mathbf{x}/N - p \leq -c] < p(c\sqrt{N})^{-1} \exp[-2Nc^2 - \tfrac{4}{9}Nc^4]$
if $q + c \leq \tfrac{1}{2}$.

They also obtained the following improvement on Levy's bound (52.2)

(60) $\Pr[|\mathbf{x}/N - p| \geq c] < \sqrt{2}\,(c\sqrt{N})^{-1} \exp[-2Nc^2 - \tfrac{4}{3}Nc^4]$
if $p, q \geq \max(4N^{-1}, 2c)$ and $N > 2$.

Both upper and lower bounds for $\Pr[\mathbf{x} \geq k]$ have been obtained by Bahadur [2]. Starting from the identity

$$\text{(61)} \qquad \Pr[\mathbf{x} \geq k] = \sum_{j=k}^{N} \binom{N}{j} p^j q^{N-j} = \binom{N}{k} p^k q^{N-k} \cdot qF(N+1,1;k;p)$$

where F is the hypergeometric function

$$F(N+1,1;k;p) = 1 + \frac{N+1}{k+1}p + \frac{(N+1)(N+2)}{(k+1)(k+2)}p^2 + \cdots$$

he obtains the inequalities

$$\text{(62)} \qquad \left(1 + \frac{Npq}{(k-Np)^2}\right)^{-1} \cdot \frac{q(k+1)}{k+1-(N+1)p} \leq \frac{\Pr[\mathbf{x} \geq k]}{\binom{N}{k} p^k q^{N-k}} \leq \frac{q(k+1)}{k+1-(N+1)p}.$$

Finally we note (i) the following inequalities for the ratio of a binomial to a Poisson probability when the two distributions have the same expected value

$$\text{(63.1)} \qquad e^{Np}(1 - kN^{-1})^k(1-p)^N \leq \frac{\binom{N}{k} p^k q^{N-k}}{e^{-Np}(Np)^k/k!} \leq e^{Np}(1-p)^{N-k}$$

and (ii) the inequality (Neumann [56])

$$\text{(63.2)} \qquad \Pr[\mathbf{x} \leq Np] > \frac{1}{2} + \frac{1+q}{3\sqrt{2\pi}}(Npq)^{-\frac{1}{2}} - \frac{3q^2+12q+5}{48}(Npq)^{-1} - \frac{1+q}{36\sqrt{2\pi}}(Npq)^{-\frac{3}{2}}.$$

9. Tables and Nomographs

Many graphical methods of calculating sums of binomial probabilities have been developed. One which made its appearance in an advertisement, is reproduced in Figure 3. (See also Larson [45].) The accuracy of Larson's nomographs based on 210 samples is illustrated by Figure 3a taken from [45].* Sometimes such nomographs have been constructed and labelled for the equivalent problem of calculating values of the incomplete beta function ratio (Hartley and Fitch [36]).

There are quite a number of tables containing values of individual probabilities $\binom{N}{k} p^k q^{N-k}$, and of sums of these probabilities. We first note that tables of the incomplete beta function ratio (Pearson [61]) contain values to 8 decimal places of $\Pr[\mathbf{x} \geq k] = I_p(k, N-k+1)$ for $p = 0.01(0.01)0.99$;

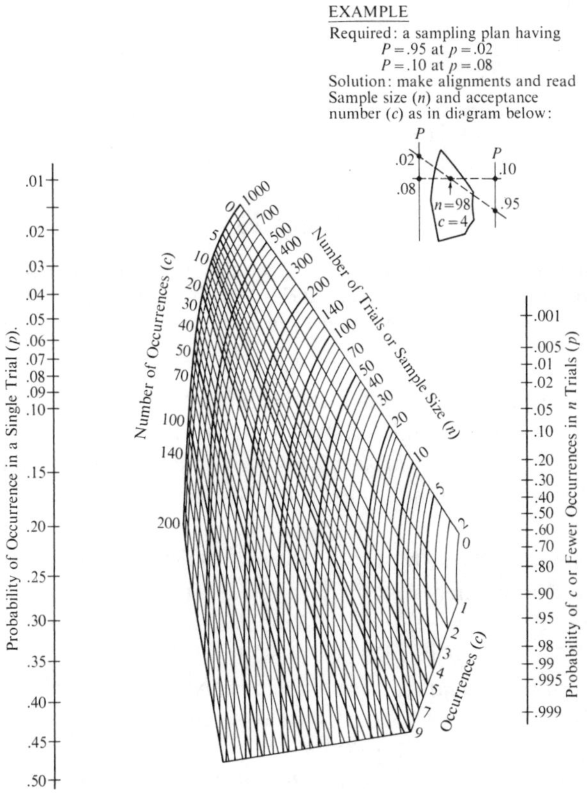
EXAMPLE
Required: a sampling plan having
P = .95 at p = .02
P = .10 at p = .08
Solution: make alignments and read
Sample size (n) and acceptance
number (c) as in diagram below:
n = 98
c = 4
Probability of Occurrence in a Single Trial (p)
Number of Occurrences (c)
Number of Trials or Sample Size (n)
Occurrences (c)
Probability of c or Fewer Occurrences in n Trials (p)

Note
If p_1 (the smallest p in a problem) is less than .01, set $k \times p$ on the p-scale and multiply the values on the n scale by k, where $k = 0.01/p_1$ (taking k to the next higher integer).

FIGURE 3

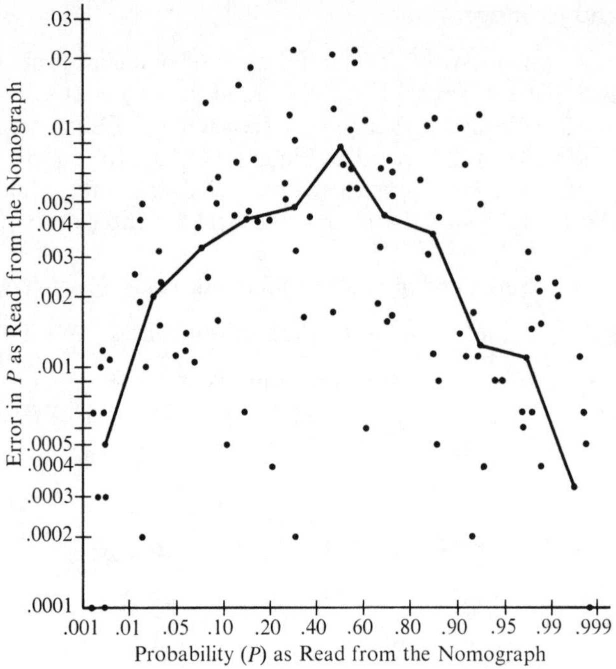

FIGURE 3a

Accuracy of the nomograph as a function of probability P. The curve passes through the median values of the errors for the indicated P-intervals. (P is the probability of c or fewer occurrences in n trials.)

max $(k, N - k + 1) \leq 50$. Other tables, specifically giving values of probabilities associated with binomial distributions are:

Tables of the Binomial Probability Distribution — (National Bureau of Standards, U.S. Government Printing Office, 1950) $p = 0.01(0.01)0.50$; $N = 2(1)49$.

50–100 *Binomial Tables* (by H. G. Romig, Wiley, 1953) $p = 0.01(0.01)0.50$; $N = 50(5)100$ (Supplements the National Bureau of Standards Tables).

Tables of the Cumulative Binomial Probabilities (U.S. Office of the Chief of Ordnance, 1952) $p = 0.01(0.01)0.50$; $N = 1(1)150$.

Tables of the Cumulative Binomial Probability Distribution (Harvard University Press, 1955) $p = 0.01(0.01)0.50$ and $\frac{1}{12}(\frac{1}{12})\frac{5}{12}$; $\frac{1}{16}(\frac{1}{16})\frac{7}{16}$; $N = 1(1)50(2)100(10)200(20)500(50)1000$.

Tables of the Cumulative Binomial Probability Distribution for Small Values of p (by Sol Weintraub, Free Press, of Glencoe, New York, (1963). $p = 0.0001(0.0001)0.001(0.001)0.1$; $N = 1(1)100$ — ten decimal places. Weintraub describes a computer routine based on the identity

$$\sum_{j=k}^{N} \binom{N}{k} p^k q^{N-k} = \sum_{j=k}^{N} (-1)^{j-k} \binom{N}{j} \binom{j-1}{k-1} p^j. \tag{64}$$

Among auxiliary tables are tables of the standard deviation $\{pq(N_1^{-1} + N_2^{-2})\}^{\frac{1}{2}}$ of the difference between two independent binomial proportions with common parameter p. Stuart [81] gives values of this function to 4 decimal places* for $p = 0.01, 0.05, 0.075, 0.1, 0.15, 0.2, 0.25, 0.3, 0.4, 0.5$ and for 20 values of each of N_1 and N_2 ranging from 25 to 5,000.

10. Truncated Binomial Distribution

A *doubly truncated binomial distribution* is formed from (1) by omitting both the values of k such that $0 \leq k < r_1$ and such that $N - r_2 < k \leq N$ (with $0 \leq r_1 \leq N - r_2 \leq N$). For the resulting distribution

$$\text{(65)} \qquad \Pr[\mathbf{x} = k] = \frac{\binom{N}{k} p^k q^{N-k}}{\sum_{j=r_1}^{N-r_2} \binom{N}{j} p^j q^{N-j}} \qquad (k = r_1, \ldots, N - r_2).$$

A *singly truncated binomial distribution* is formed if *only* the values $0, 1, \ldots, (r_1 - 1)$ $(r_1 \geq 1)$ *or* the values $N - r_2 + 1, \ldots, N$ $(r_2 \geq 1)$ are omitted. The distribution formed by omission of the value 0 only, giving

$$\text{(66)} \qquad \Pr[\mathbf{x} = k] = \frac{\binom{N}{k} p^k q^{N-k}}{1 - q^N} \qquad (k = 1, 2, \ldots, N)$$

is sometimes called the *positive binomial* distribution. However, the untruncated binomial distribution, as defined by (1), is also sometimes called the positive binomial, to distinguish it from the negative binomial distribution (Chapter 5).

The rth moment about zero of a random variable having the positive binomial distribution (66) is equal to

$$\frac{r\text{th moment of random variable with distribution (1)}}{1 - q^N}.$$

In particular, the expected value is $Np(1 - q^N)^{-1}$ and the variance is $Npq(1 - q^N)^{-1} - N^2p^2q^N(1 - q^N)^{-2}$. If r is negative ($r = -s$, with $s > 0$) it is not possible to give a short expression for $E(\mathbf{x}^{-s})$. Grab and Savage [28] note the approximation formula

$$\text{(67)} \qquad E(\mathbf{x}^{-1}) \doteqdot (Np - q)^{-1}$$

which has two significant figure accuracy for $Np > 10$. Mendenhall and Lehman [51] obtain approximate formulas by first approximating the positive binomial distribution by a (continuous) beta distribution (see Chapter 24),

*Stuart actually tables (standard deviation) × 100 that is, the standard deviation of the *percentage* — to two decimal places.

making the first and second moments agree. Among the formulas so obtained are

(68.1) $$E(\mathbf{x}^{-1}) \doteqdot (1 - 2N^{-1})(Np - q)^{-1}$$

(68.2) $$\text{var}(\mathbf{x}^{-1}) \doteqdot (1 - 2N^{-1})(1 - N^{-1})q(Np - q)^{-2}(Np - q - 1)^{-1}.$$

Formula (68.1) gives 2 significant figure accuracy for $Np > 5$.

Situations in which N is known, and it is necessary to estimate p from data represented by independent random variables with a doubly truncated binomial distribution are not of common occurrence. However methods for estimating p have been constructed, and will be briefly described later. We first mention an example of practical application given by Newell [57].

In Newell's example, four specimens of sputum are taken from each subject, and each specimen classified as "positive" or "negative" (with respect to a certain disease). For each subject, therefore, there is available the total number, 0, 1, 2, 3 or 4, "positive" specimens. It is supposed that all sputum from any one subject can be classified as (*a*) non-malignant, (*b*) grossly malignant or (*c*) partly malignant. In case (a), none, and in case (*b*), all, of the specimens give "positive" reactions. In case (*c*) reaction is uncertain. Hence cases in which there are one, two or three "positive" reactions certainly belong to class (*c*), but those with no "positive" reactions may belong to either (*a*) or (*c*), and those with four "positive" reactions may belong to either (*b*) or (*c*). Now if we make the further assumption of a constant *probability* of "positive" reaction for all sputum specimens in class (*c*) then it is reasonable to represent the values one, two or three "positive" reactions as observed values of a doubly truncated binomial distribution with $N = 4$, $r_1 = r_2 = 1$.

Finney [19] has shown how to calculate the maximum likelihood estimator of p; this is the method used by Newell for the data of his example.

Shah [73] gives a method of estimating p, using the sample moments calculated from n observed values $x_1, x_2, \ldots, x_n$ represented by random variables each having distribution (65). The first three moments about zero of (65) are

(69.1) $$\mu_1' = X - Y + Np$$

(69.2) $$\mu_2' = r_1 X - (N - r_2 + 1)Y + \mu_1'(N - 1)p + Np$$

(69.3) $$\mu_3' = r_1^2 X - (N - r_2 + 1)^2 Y + \mu_2'(N - 2)p + \mu_1'(2N - 1)p + Np$$

where

$$X = r_1 q \binom{N}{r_1} p^{r_1} q^{N-r_1} \left[\sum_{j=r_1}^{N-r_2} \binom{N}{j} p^j q^{N-j}\right]^{-1}$$

and

$$Y = (N - r_2 + 1)q \binom{N}{N - r_2 + 1} p^{N-r_2+1} q^{r_2-1} \left[\sum_{j=r_1}^{N-r_2} \binom{N}{j} p^j q^{N-j}\right]^{-1}.$$

Eliminating X and Y between (69.1)–(69.3) leads to

(70)

$$p = \frac{\mu_3' - (r_1 + N - r_2 + 1)\mu_2' + r_1(N - r_2 + 1)\mu_1'}{(N-2)\mu_2' + \{(N-1)(r_2 - r_1) - N(N-2)\}\mu_1' + N(N - r_2)(r_1 - 1)}.$$

The moment estimator of p is obtained by replacing μ_s' $(s = 1,2,3)$ on the right hand side of (70) by the sample moment $n^{-1} \sum_{j=1}^{n} x_j^s$.

Shah [73] calculates the asymptotic efficiency of this moment estimator relative to the maximum likelihood estimator for the case $r_1 = r_2 = 1$ and obtains:

p	0.1 or 0.9	0.2 or 0.8	0.3 or 0.7	0.4 or 0.6	0.5
Asymptotic relative efficiency	92.1%	90.3%	91.2%	92.9%	93.5%

These values seem remarkably high, especially since the third sample moment is used in the moment estimator of p.

The singly truncated (positive) binomial (66) is of frequent occurrence in demographic enquires wherein families are chosen for investigation on the basis of an observed "affected" individual, so that there is at least one such individual in each family included in the study. The maximum likelihood estimator, $\hat{\mathbf{p}}$, of p, based on n independent random variables $\mathbf{x}_1, \mathbf{x}_2, \ldots, \mathbf{x}_n$ each having distribution (66) satisfies the equation

$$\bar{\mathbf{x}} = n^{-1} \sum_{j=1}^{n} \mathbf{x}_j = N\hat{\mathbf{p}}(1 - \hat{\mathbf{q}}^N)^{-1}$$

where $\hat{\mathbf{q}} = 1 - \hat{\mathbf{p}}$. An alternative estimator proposed by Mantel [50] is

$$\tilde{\mathbf{p}} = (\bar{\mathbf{x}} - \mathbf{f}_1)/(N - \mathbf{f}_1)$$

where $\mathbf{f}_1 = n^{-1}$ (number of $\mathbf{x}_j$'s equal to 1).

For n large

$$\text{var}(\hat{\mathbf{p}}) \doteqdot (Nn)^{-1}pq(1 - q^N)^2(1 - q^N - Npq^{N-1})^{-1}$$

$$\text{var}(\tilde{\mathbf{p}}) \doteqdot (Nn)^{-1}pq(1 - q^N)(1 - 2q^{N-1} + Npq^{N-1} + q^N)(1 - q^{N-1})^{-2}$$

The asymptotic efficiency $[\lim_{n\to\infty} \{\text{var}(\hat{\mathbf{p}})/\text{var}(\tilde{\mathbf{p}})\}]$ of $\tilde{\mathbf{p}}$ is generally quite high. Some values are given in Table 2 (taken from [24]). Note that for p fixed there is a minimum efficiency as N increases, and for N fixed a minimum efficiency as p increases.

TABLE 2

Asymptotic efficiency (%) of $\tilde{\mathbf{p}}$

N \ p	0.1	0.2	0.3	0.4	0.5	0.6	0.7	0.8	0.9
3	99.26	98.72	98.53	98.31	98.44	99.20	99.43	99.64	99.93
4	98.61	97.79	97.59	97.72	98.26	99.53	99.74	99.88	99.99
5	98.05	97.16	97.06	97.77	98.63	99.81	99.92	99.97	100.00
6	97.58	96.77	96.85	98.12	99.08	99.93	99.98	99.99	100.00
7	97.17	96.58	96.87	98.56	99.44	99.98	99.99	100.00	100.00
8	96.84	96.55	97.04	98.96	99.68	99.99	100.00	100.00	100.00
9	96.56	96.63	97.31	99.29	99.83	100.00	100.00	100.00	100.00
10	96.35	96.80	97.62	99.53	99.91	100.00	100.00	100.00	100.00
15	95.96	98.11	99.06	99.95	100.00	100.00	100.00	100.00	100.00
20	96.32	99.15	99.72	100.00	100.00	100.00	100.00	100.00	100.00

For $N = 2$, $\hat{\mathbf{p}}$ and $\tilde{\mathbf{p}}$ are identical

When p is large the effect of truncation is small and $\tilde{\mathbf{p}}$ differs but little from $\hat{\mathbf{p}}$.

Thomas and Gart [84] have carried out a thorough study of the small sample properties of $\hat{\mathbf{p}}$ and $\tilde{\mathbf{p}}$, and conclude that $\tilde{\mathbf{p}}$ is of comparable accuracy to, and less biased than $\hat{\mathbf{p}}$. For practical, use they recommend the "corrected value"

$$\mathbf{p}^* = \tilde{\mathbf{p}} + N^{-1}(1 - \tilde{\mathbf{q}}^N)\{\bar{\mathbf{x}}(1 - \tilde{\mathbf{q}}^N) - N\tilde{\mathbf{p}}\}/\{1 - N\tilde{\mathbf{p}}\tilde{\mathbf{q}}^{N-1} - \tilde{\mathbf{q}}^N\}$$

where $\tilde{\mathbf{q}} = 1 - \tilde{\mathbf{p}}$. If reduction of bias is a primary aim, the estimator

$$\mathbf{p}' = \tilde{\mathbf{p}} + \frac{(N-1)\tilde{\mathbf{p}}^2\tilde{\mathbf{q}}^{N-1}(1 - \tilde{\mathbf{q}}^N)}{nN(1 - \tilde{\mathbf{q}}^{N-1})^2}$$

is recommended.

If $\mathbf{x}_1, \ldots, \mathbf{x}_n$ have different values $N_1, \ldots, N_n$ instead of a constant N (but a common p) an estimator analogous to $\tilde{\mathbf{p}}$ is $(\bar{\mathbf{x}} - \mathbf{f}_1)/(\bar{N} - \mathbf{f}_1)$ where $\bar{N} = n^{-1}\sum_{j=1}^{n} N_j$.

11. Mixtures of Binomial Distributions

The most common mixture of binomial distributions is one in which the component distributions have a common (usually known) value of N, but different values of p. The mixture distribution has probabilities of form

$$\Pr[\mathbf{x} = k] = \binom{N}{k} \sum_{j=1}^{\infty} \omega_j p_j^k (1 - p_j)^{N-k} \tag{71.1}$$

$$\left(\text{with } \omega_j \geq 0, \sum_{j=1}^{\infty} \omega_j = 1\right)$$

or

$$(71.2)\qquad \Pr[\mathbf{x} = k] = \binom{N}{k} \int_0^1 f(p) \cdot p^k (1-p)^{N-k}\, dp$$

$$\left(\text{with } f(p) \geq 0, \int_0^1 f(p)\, dp = 1\right)$$

according as p has a discrete or a continuous distribution.

If in (71.1) there are only a few non-zero ω_j's and particularly if the number of non-zero ω_j's is known, it may be possible to obtain useful estimates of both the ω_j's and the p_j's. Blischke ([8], [9]) shows that estimation is possible provided

$$R = \text{number of non-zero } \omega_j\text{'s} \leq \tfrac{1}{2}(N+1)$$

and there are available at least $(2R-1)$ observations represented by random variables having the distribution.

If $\mathbf{x}_1, \mathbf{x}_2, \ldots, \mathbf{x}_n$ each have distribution (71.1) with the summation from $j = 1$ to $j = R$, and $n \geq 2R - 1$ then the first $2R - 1$ reduced factorial moments

$$(72)\qquad \mathbf{F}_s = \frac{1}{n} \sum_{j=1}^{n} \frac{\mathbf{x}_j(\mathbf{x}_j - 1)\cdots(\mathbf{x}_j - s + 1)}{N(N-1)\cdots(N-s+1)} = n^{-1} \sum_{j=1}^{n} \{\mathbf{x}_j^{(s)}/N^{(s)}\}$$

$$(s = 1,2,\ldots,2R-1)$$

are calculable from $\mathbf{x}_1, \mathbf{x}_2, \ldots, \mathbf{x}_n$, and

$$(73)\qquad E[\mathbf{F}_s] = \sum_{j=1}^{R} \omega_j p_j^s = E(\mathbf{x}^s) = \mu_s'.$$

The method is then, essentially, to equate the observed values $\mathbf{F}_1, \mathbf{F}_2, \ldots, \mathbf{F}_{2R-1}$ to the expected values $\sum_{j=1}^{R} \omega_j p_j, \sum_{j=1}^{R} \omega_j p_j^2, \ldots, \sum_{j=1}^{R} \omega_j p_j^{2R-1}$ respectively and solve the resulting equations for $\omega_1, \omega_2, \ldots, \omega_R, p_1, p_2, \ldots, p_R$ (subject to the condition $\sum_{j=1}^{R} \omega_j = 1$). This is done by (i) eliminating the ω_j's from the equations by solving the first $(R-1)$ equations for the ω_j's in terms of the p_j's and inserting the solutions in the remaining R equations and (ii) reducing the resulting set of equations to a single polynomial of degree R, the roots of which are the required estimators of $p_1, p_2, \ldots, p_R$.

In obtaining approximations to the variances and covariances of the estimators of the ω's and the p's, use may be made of the results

$$(74)\qquad \operatorname{var}(\mathbf{F}_i) = \{nN^{(i)}\}^{-1}(N-i)^{(i-1)} \times \left[(N-2i+1)\sum_{j=1}^{R} \omega_j p_j^{2i} + i^2 \sum_{j=1}^{R} \omega_j p_j^{2i-1}\right] - n^{-1}\left[\sum_{j=1}^{R} \omega_j p_j^i\right]^2 + 0(N^{-2})$$

and

$$\text{(75)}\qquad \operatorname{cov}(\mathbf{F}_i,\mathbf{F}_{i'}) = \{nN^{(i)}N^{(i')}\}^{-1}N^{(i+i'-1)} \times\left[(N-i-i'+1)\sum_{j=1}^{R}\omega_j p_j^{i+i'} + ii'\sum_{j=1}^{R}\omega_j p_j^{i+i'-1}\right] - n^{-1}\left(\sum_{j=1}^{R}\omega_j p_j^{i}\right)\left(\sum_{j=1}^{R}\omega_j p_j^{i'}\right) + 0(N^{-2}).$$

Blischke [8] found that the estimators based on moments have an asymptotic efficiency, relative to the maximum likelihood estimator, of 100% if N has the lowest possible value $(2R - 1)$. The asymptotic efficiency also tends to 100% as N tends to infinity. (These asymptotic results refer to limits as n tends to infinity.)

Rider [66] and Blischke [7] have considered the special case $R = 2$ in more detail. This corresponds to a mixture of two binomials with

$$\text{(76)}\qquad \Pr[\mathbf{x} = k] = \alpha\binom{N_1}{k}p_1^k q_1^{N-k} + (1-\alpha)\binom{N_2}{k}p_2^k q_2^{N-k}$$

with $0 < \alpha < 1$, $0 < q_j = 1 - p_j < 1$ $(j = 1,2)$.

In the case $N_1 = N_2 = N$ (known), simple explicit formulas for the parameters α, p_1, p_2 in terms of the first three moments can be obtained. These give the estimators

$$\tilde{\mathbf{p}}_1, \tilde{\mathbf{p}}_2 = \tfrac{1}{2}\mathbf{A} \mp \tfrac{1}{2}(\mathbf{A}^2 - 4\mathbf{A}\mathbf{F}_1 + 4\mathbf{F}_2)^{\frac{1}{2}} \qquad \left(\text{where } \mathbf{A} = \frac{\mathbf{F}_3 - \mathbf{F}_1\mathbf{F}_2}{\mathbf{F}_2 - \mathbf{F}_1^2}\right)$$

if the values for $\tilde{\mathbf{p}}_1$, $\tilde{\mathbf{p}}_2$ satisfy $0 \le \tilde{\mathbf{p}}_j \le 1$ $(j = 1,2)$; otherwise $\tilde{\mathbf{p}}_1 = \tilde{\mathbf{p}}_2 = \mathbf{F}_1$, that is only a *single* binomial distribution should be fitted.

Mixtures of binomial distributions (with common N) were studied by Lexis [47]. These distributions are sometimes called *Lexian distributions*.* Lexis obtained the general formulas

$$\text{(77.1)}\qquad E(\mathbf{x}) = NE(\mathbf{p})$$

$$\text{(77.2)}\qquad \operatorname{var}(\mathbf{x}) = NE(\mathbf{p})[1 - E(\mathbf{p})] + N(N-1)\operatorname{var}(\mathbf{p}).$$

Note that the variance of $\mathbf{x}$ is always greater than the value for a binomial distribution with parameters N, $E(\mathbf{p})$ (cf Section 13).

We now consider continuous mixtures of binomial distributions (with a common N). The "mixing distribution" of p which has been studied most fully is the beta distribution (Chapter 24) with probability density function

$$\text{(78)}\qquad f(p) = \frac{\Gamma(a+b)}{\Gamma(a)\Gamma(b)}p^{a-1}q^{b-1} \qquad (a > 0; b > 0; 0 < p < 1).$$

*Govindarajulu [26] calls such distributions "Poisson binomial." This should not be confused with the Poisson-binomial distribution considered in Chapter 8.

This is often said to be a "natural" distribution to use for p. However, there is little substantial reason for this, and in fact mathematical convenience has strongly contributed to the popularity of (78) as a mixing distribution.

Inserting this function $f(p)$ in (71.2) the distribution

$$(79)\qquad \Pr[\mathbf{x} = k] = \frac{\Gamma(N+1)}{\Gamma(k+1)\Gamma(N-k+1)} \cdot \frac{\Gamma(k+a)\Gamma(N-k+b)}{\Gamma(N+a+b)} \qquad (k = 0,1,\ldots,N)$$

is obtained. The expected value of a random variable having this distribution is $Na/(a+b)$, the variance is

$$Nab(a+b)^{-2} + N(N-1)ab(a+b)^{-2}(a+b+1)^{-1}.$$

Further discussion of distributions of this kind will be found in Chapter 8.

12. Applications

The binomial distribution is often used as an approximation to the hypergeometric distribution. The latter distribution (see Chapter 6) is appropriate to describe the variation of a random variable, $\mathbf{x}$, representing the number of individuals with a property E in a set of N members chosen at random from a population of M ($\geq N$) individuals, of which D have the property E. Then

$$(80)\qquad \Pr[\mathbf{x} = k] = \binom{D}{k}\binom{M-D}{N-k}\bigg/\binom{M}{N} \qquad (\max(0,N-M+D) \leq k \leq \min(D,N)).$$

If $M \to \infty$ and $D \to \infty$ with $D/M = p$ remaining fixed, then as in (1).

$$(81)\qquad \lim_{\substack{M\to\infty\\ D\to\infty\\ D/M=p}} \Pr[\mathbf{x} = k] = \binom{N}{k} p^k q^{N-k}$$

The following table indicates how rapidly the limit in (81) is approached when $p = \frac{1}{2}$; $N = 10$.

Convergence is not so rapid for values of p near 1 or 0, but the approximation is good enough in many practical cases. (The binomial approximation is often called the "infinite population" approximation, since it corresponds to $M = \infty$.)

In Section 1, we indicated that the binomial distribution appears in many statistical models — in fact whenever assumptions of independent trials with stable probabilities are introduced. We will, therefore, not bother to give an only fragmentary catalogue of specific applications.

Although simple, the assumptions of independence and constant probability are not often precisely satisfied. Published critical appraisals of the extent of departure from these assumptions in actual data are rather rare. An interesting discussion, for mortality data, has been provided by Seal [72].

TABLE 3

Comparison of Hypergeometric and Binomial Probabilities

k \ M	50	100	300	∞ (Binomial)
0	0.0003	0.0006	0.0008	0.0010
1	0.0050	0.0072	0.0085	0.0098
2	0.0316	0.0380	0.0410	0.0439
3	0.1076	0.1131	0.1153	0.1172
4	0.2181	0.2114	0.2082	0.2051
5	0.2748	0.2539	0.2525	0.2461

13. Related Distributions

The central importance of the binomial distribution in statistics is shown by the fact that it is related to a wide variety of standard distributions. We have already encountered some of these relationships; we now summarize them.

The limiting form of the standardized binomial distribution (as $N \to \infty$) is Normal (Laplace's theorem, Section 5). If $N \to \infty$ and $p \to 0$ with $Np = \theta$ (fixed), the Poisson distribution is obtained (Chapter 4, Section 1). The hypergeometric distribution has already been encountered in Section 5. The binomial distribution can also be regarded as a limiting form of the hypergeometric distribution (Section 12). Relations between binomial and limiting Poisson and hypergeometric probabilities have been studied by Uhlmann [85] (see also Chapters 4 and 6).

In Section 11, mixtures of binomial distributions have been described. These can be regarded as arising when p changes from one sequence of N trials to another, but remains constant within each sequence. In contradistinction are the distributions studied by Poisson [62] in which the probability of outcome O is the same, p_i, for the ith trial in each sequence, but $p_1, p_2, \ldots, p_N$ are not all equal. In this case the mean and variance are $NE(p)$ and

$$NE(p)(1 - E(p)) - N\,\mathrm{var}(p)$$

respectively (where var(p) now denotes the 'variance' among $p_1, p_2, \ldots, p_N$). The variance is now *less* than that for a binomial distribution with parameters N, $E(p)$ (cf (5.1)). The Lexian and Poisson situation are sometimes described respectively as "supernormal" and "subnormal" variations.

Among special distributions associated with binomial distributions may be mentioned the conditional distribution of a binomial variable $\mathbf{x}_1$ having parameters N_1, p_1 given that $\mathbf{x}_1 + \mathbf{x}_2 = \mathbf{y}$, where $\mathbf{x}_2$ is another binomial variable independent of $\mathbf{x}_1$, having parameters N_2, p_2. This distribution has been studied by Hannan and Harkness [35] (see also Stevens [80]). They obtained the following asymptotic result:

$$\Pr[\alpha \leq \mathbf{x}_1 \leq \beta \mid \mathbf{x}_1 + \mathbf{x}_2 = y] \doteqdot \frac{1}{\sqrt{2\pi}} \int_{X_\alpha}^{X_\beta} e^{-\frac{1}{2}u^2}\, du \tag{82}$$

where $X_u = [\{(N_1 + 1)P_1Q_1\}^{-1} + \{(N_2 + 1)P_2Q_2\}^{-1}]^{\frac{1}{2}}[u - (N_1 + 1)P_1Q_1]$ $(u = \alpha, \beta)$ and P_1, P_2 $(P_1 = 1 - Q_1, P_2 = 1 - Q_2)$ satisfy the equations

$$\frac{P_1Q_2}{P_2Q_1} = \frac{p_1q_2}{p_2q_1};\ (N_1 + 1)P_1 + (N_2 + 1)P_2 = y + 1.$$

REFERENCES

[1] Anderson, T. W. and Samuels, S. M. (1965). Some inequalities among binomial and Poisson probabilities, *Proceedings of the 5th Berkeley Symposium on Mathematical Statistics and Probability*, **1,** 1–12.

[2] Bahadur, R. R. (1960). Some approximations to the binomial distribution function, *Annals of Mathematical Statistics*, **31,** 43–54.

[3] Barton, D. E. (1961). Unbiased estimation of a set of probabilities, *Biometrika*, **48,** 227–229.

[4] Binet, F. (1953). The fitting of the positive binomial distribution when both parameters are estimated from the sample, *Annual of Eugenics, London*, **18,** 117–119.

[5] Birnbaum, A. (1964). Median-unbiased estimators, *Bulletin of Mathematical Statistics*, **11,** 25–34.

[6] Bizley, M. T. L. (1951). Some notes on probability, *Journal of the Institute of Actuaries Students' Society*, **10,** 161–203.

[7] Blischke, W. R. (1962). Moment estimators for the parameters of a mixture of two binomial distributions, *Annals of Mathematical Statistics*, **33,** 444–454.

[8] Blischke, W. R. (1964). Estimating the parameters of mixtures of binomial distributions, *Journal of the American Statistical Association*, **59,** 510–528.

[9] Blischke, W. R. (1963). Mixtures of discrete distributions. *Proceedings of the International Symposium on Discrete Distributions, Montreal*, 351–372.

[10] Bol'shev, L. N. (1963). Asymptotically Pearson transformations, *Teoriya Veroyatnostei i ee Primeneniya*, **8,** 129–154.

[11] Boyer, C. B. (1950). Cardan and the Pascal triangle, *American Mathematical Monthly*, **57,** 387–390.

[12] Brockwell, P. J. (1964). An asymptotic expansion for the tail of a binomial distribution and its application in queueing theory, *Journal of Applied Probability*, **1,** 161–167.

[13] Chernoff, H. (1952). A measure of asymptotic efficiency for tests of a hypothesis based on the sum of observations, *Annals of Mathematical Statistics*, **23,** 493–507.

[14] Clark, R. E. (1953). Percentage points of the incomplete beta function, *Journal of the American Statistical Association*, **48,** 831–843.

[15] Crow, E. L. (1956). Confidence limits for a proportion, *Biometrika*, **43,** 423–435.

[16] Dunin-Barkovsky, I. V. and Smirnov, N. V. (1955). *Theory of Probability and Mathematical Statistics in Engineering*, Moscow: Nauka.

[17] Edwards, A. W. F. (1960). The meaning of binomial distribution, *Nature, London*, **186,** 1074.

[18] Feller, W. (1945). On the normal approximation to the binomial distribution, *Annals of Mathematical Statistics*, **16,** 319–329.

[19] Finney, D. J. (1949). The truncated binomial distribution, *Annals of Eugenics, London*, **14,** 319–328.

[20] Fisher, R. A. (1941). The negative binomial distribution, *Annals of Eugenics, London*, **11,** 182–187.

[21] Frame, J. S. (1945). Mean deviation of the binomial distribution, *American Mathematical Monthly*, **52,** 377–379.

[22] Freeman, M. F. and Tukey, J. W. (1950). Transformations related to the angular and the square root, *Annals of Mathematical Statistics*, **21,** 607–611.

[23] Frisch, R. (1924). Solution d'un problème du calcul des probabilités, *Skandinavisk Aktuarietidskrift*, **7,** 153–174.

[24] Gart, J. J. (1968). A simple nearly efficient alternative to the simple sib method in the complete ascertainment case, *Annals of Human Genetics, London*, **31,** 283–291.

[25] Gebhardt, F. (1968). Some numerical results to R. Borges' approximation of the binomial distribution, *European Meeting of the Institute of Mathematical Statistics, Amsterdam.*

[26] Govindarajulu, Z. (1963). Normal approximations to the classical discrete distributions, *Proceedings of the International Symposium on Discrete Distributions, Montreal*, 79–108.

[27] Govindarajulu, Z. (1963). Recurrence relations for the inverse moments of the positive binomial variable, *Journal of the American Statistical Association*, **58,** 468–473.

[28] Grab, E. L. and Savage, J. R. (1954). Tables of the expected value of $1/X$ for positive Bernoulli and Poisson variables, *Journal of the American Statistical Association*, **58,** 468–473.

[29] Greenwood, R. E. and Glasgow, M. O. (1950). Distribution of maximum and minimum frequencies in a sample drawn from a multinomial population, *Annals of Mathematical Statistics*, **21,** 416–424.

[30] Gupta, S. S. (1960). *Binomial order statistics*, Bell Laboratory Report, Allentown, Pennsylvania.

[31] Gupta, S. S. (1963). Selection and ranking procedures and order statistics for the binomial distribution. *Proceedings of the International Symposium on Discrete Distributions, Montreal*, 219–230

[32] Gupta, S. S. and Panchapakesan, P. (1967). *Order statistics arising from independent binomial populations*, Department of Statistics, Purdue University, Mimeo Series No. 120.

[33] Hald, A. (1952). *Statistical Theory with Engineering Applications*, New York: John Wiley & Sons, Inc.

[34] Hald, A. and Kousgaard, E. (1967). A table for solving the binomial equation $B(c,n,p) = P$, *Matematisk-fysiske Skrifter, Det Kongelige Danske Videnskabernes Selskab*, **3,** No. 4 (48 pp.)

[35] Hannan, J. and Harkness, W. (1963). Normal approximation to the distribution of two independent binomials, conditional on a fixed sum, *Annal of Mathematical Statistics*, **34,** 1593–1595.

[36] Hartley, H. O. and Fitch, E. R. (1951). A chart for the incomplete beta function and cumulative binomial distribution, *Biometrika*, **38,** 423–426.

[37] Hoeffding, W. (1963). Probability inequalities for sums of bounded random variables, *Journal of the American Statistical Association*, **58,** 13–30.

[38] Hoel, P. G. (1947). Discriminating between binomial distributions, *Annals of Mathematical Statistics*, **18,** 556–564.

[39] Jowett, G. H. (1963). The relationship between the binomial and *F* distributions, *The Statistician*, **13,** 55–57.

[40] Kalinin, V. M. (1967). Convergent and asymptotic expansions of probability distributions, *Teoriya Veroyatnostei i ee Primeneniya*, **12,** 24–38.

[41] Kambo, N. S. and Kotz, S. (1966). On exponential bounds for binomial probabilities, *Annals of the Institute of Statistical Mathematics, Tokyo*, **18,** 277–287.

[42] Khatri, C. G. (1962). Distributions of order statistics for discrete case, *Annals of the Institute of Statistical Mathematics, Tokyo*, **14,** 167–171.

[43] Krafft, O. (1969). A note on exponential bounds for binomial probabilities, *Annals of the Institute of Statistical Mathematics, Tokyo*, **21**.

[44] Lachenbruch, P. A. (1968). Simultaneous confidence limits for the binomial and Poisson distributions, University of North Carolina, Institute of Statistics, Mimeo Series No. 596.

[45] Larson, H. R. (1966). A nomograph of the cumulative binomial distribution, *Industrial Quality Control*, **23,** 270–278.

[46] Levy, P. (1954). *Théorie de l'addition des variables aléatoires*, Paris: Gauthier Villars.

[47] Lexis, W. (1875). *Einleitung in der Theorie der Bevolkering-statistik*, Strasburg: Teubner.

[48] Li, C. C. and Mantel, N. (1968). A simple method of estimating the segregation ratio under complete ascertainment, *American Journal of Human Genetics*, **20,** 61–81.

[49] Mainland, D. (1948). Statistical methods in medical research, *Canadian Journal of Research*, **26,** (section E), 1–166.

[50] Mantel, N. (1951). Evaluation of a class of diagnostic tests, *Biometrics*, **3,** 240–246.

[51] Mendenhall, W. and Lehman, E. H. (1960). An approximation to the negative moments of the positive binomial useful in life testing, *Technometrics*, **2,** 233–239.

[52] Mosteller, F. and Tukey, J. W. (1949). The uses and usefulness of binomial probability paper, *Journal of the American Statistical Association*, **44,** 174–212.

[53] Mosteller, F. and Youtz, C. (1961). Tables of the Freeman-Tukey transformation for the binomial and Poisson distributions, *Biometrika*, **48,** 433–440.

[54] Mott-Smith, J. C. (1964). Two estimates of the binomial distribution, *Annals of Mathematical Statistics*, **35,** 809–816.

[55] Nayatani, Y. and Kurahara, B. (1964). A condition for using the approximation by the normal and the Poisson distribution to compute the confidence intervals for the binomial parameter, *Reports of Statistical Application Research, JUSE*, **11,** 99–105.

[56] Neumann, P. (1966). Uber den Median der Binomial and Poissonverteilung, *Wissenschaftliche Zeitschrift der Technischen Universität Dresden*, **15,** 223–226.

[57] Newell, D. J. (1965). Unusual frequency distributions, *Biometrics*, **21,** 159–168.

[58] Okamoto, M. (1958). Some inequalities relating to the partial sum of binomial probabilities, *Annals of the Institute of Statistical Mathematics, Tokyo*, **10,** 29–35.

[59] Pachares, J. (1960). Tables of confidence limits for the binomial distribution, *Journal of the American Statistical Association*, **55,** 521–533.

[60] Pearson, E. S. and Hartley, H. O. (Ed.) (1958). *Biometrika Tables*, **1,** London: Cambridge University Press. (2nd edition)

[61] Pearson, K. (1934). *Tables of the Incomplete Beta-Function*, London: Cambridge University Press.

[62] Poisson, Simeon Denis (1837). *Recherches sur la Probabilité des Jugements en Matière Criminelle et en Matière Civile, Précédées des Regles Générales du Calcul des Probabilitiés.* Bachelier, Imprimeur-Libraire pour les Mathematiques, la Physique, etc. Paris.

[63] Prohorov, Yu. V. (1953). The asymptotic behaviour of the binomial distribution, *Uspekhi Matematicheskii Nauk* (New Series), **8,** 135–142.

[64] Raff, M. S. (1956). On approximating the point binomial, *Journal of the American Statistical Association*, **51,** 293–303.

[65] Rider, P. R. (1955). Truncated binomial and negative binomial distribution, *Journal of the American Statistical Association*, **50,** 877–883.

[66] Rider, P. R. (1961). Estimating the parameters of mixed Poisson, binomial and Weibull distributions by the method of moments, *Bulletin de l'Institut Internationale Statistique*, **38,** 1–8.

[67] Romanovsky, V. (1923). Note on the moments of the binomial $(p + q)^N$ about its mean, *Biometrika*, **15,** 410–412. (See also *Les Principes de la Statistique Mathematique* (1933), 39–40 and 320–321.)

[68] Rutemiller, H. C. (1967). Estimation of the probability of zero failures in m binomial trials, *Journal of the American Statistical Association*, **62,** 272–277.

[69] Sandelius, M. (1952). A confidence interval for the smallest proportion of a binomial population, *Journal of the Royal Statistical Society, Series B*, **14,** 115–117.

[70] Sandiford, P. J. (1960). A new binomial approximation for use in sampling from finite population, *Journal of the American Statistical Association*, **55,** 718–722.

[71] Sawkins, D. T. (1947). A new method of approximating the binomial and hypergeometric probabilities, *Proceedings of the Royal Society of New South Wales*, **81,** 38–44.

[72] Seal, H. L. (1949). Mortality data and the binomial probability law, *Skandinavisk Aktuarietidskrift*, **32,** 188–216.

[73] Shah, S. M. (1966). On estimating the parameter of a doubly truncated binomial distribution, *Journal of the American Statistical Association*, **61,** 259–263.

[74] Siotani, M. (1956). Order statistics for discrete case with a numerical application to the binomial distribution, *Annals of the Institute of Statistical Mathematics, Tokyo*, **8,** 95–104.

[75] Siotani, M. and Ozawa, M. (1958). Tables for testing the homogeneity of k independent binomial experiments on a certain event based on the range, *Annals of the Institute of Statistical Mathematics, Tokyo*, **10,** 47–63.

[76] Skellam, J. G. (1948). A probability distribution derived from the binomial distribution by regarding the probability of success as variable between the sets of trials, *Journal of the Royal Statistical Society, Series B*, **10,** 257–261.

[77] Sobel, M. and Huyett, M. J. (1957). Selecting the best one of several binomial populations, *Bell System Technical Journal*, **36,** 537–576.

[78] Somerville, P. N. (1957). Optimum sampling in binomial populations, *Journal of the American Statistical Association*, **52,** 494–502.

[79] Stephan, F. F. (1945). The expected value of the reciprocal and other negative powers of a positive Bernoullian variate, *Annals of Mathematical Statistics*, **16,** 50–61.

[80] Stevens, W. L. (1951). Mean and variance of an entry in a contingency table, *Biometrika*, **38,** 468–470.

[81] Stuart, A. (1963). Standard errors for percentages, *Applied Statistics*, **12,** 87–101.

[82] 'Student' (1919). An example of deviations from Poisson's law in practice, *Biometrika*, **12,** 211–213.

[83] Thionet, P. (1963). Sur le moment d'ordre (−1) de la distribution binomiale tronquée. Application a l'échantillonnage de Hajek, *Publications de l'Institut de Statistique de l'Université de Paris*, **12,** 93–102.

[84] Thomas, D. G. and Gart, J. J. (1968). The small sample performances of some estimators of the truncated binomial distribution, *American Statistical Association Meeting*, *Pittsburgh*.

[85] Uhlmann, W. (1966). Vergleich der hypergeometrischen mit der Binomial-Verteilung, *Metrika*, **10,** 145–148.

[86] Uspensky, J. V. (1931). On Ch. Jordan's series for probability, *Annals of Mathematics*, **32,** 306–312.

[87] Uspensky, J. V. (1937). *Introduction to Mathematical Probability*, New York: McGraw-Hill, Inc.

[88] Wishart, J. (1956). An approximation to the binomial distribution, *The Statistician*, **6,** 1–8.

4

Poisson Distribution

1. Definition and Genesis

A random variable $\mathbf{x}$ is said to have a Poisson distribution with parameter θ if

$$\text{(1)} \qquad \Pr[\mathbf{x} = k] = e^{-\theta}\theta^k/k! \qquad (k = 0,1,2,\ldots; \theta > 0).$$

This distribution is the limit of a sequence of binomial distributions with

$$P_{k,N} = \Pr[\mathbf{x} = k] = \begin{cases} \binom{N}{k} p^k(1-p)^{N-k} & (\text{for } k = 0,1,\ldots,N) \\ 0 & (\text{for } k > N) \end{cases}$$

in which N tends to infinity, and p tends to zero but Np remains equal to θ. It can be established by direct analysis that

$$\text{(2)} \qquad \lim_{\substack{N\to\infty \\ Np=\theta}} \sum_w P_{k,N} = \sum_w e^{-\theta}\theta^k/k!$$

where $\sum_w$ denotes summation over any (finite or infinite) subset w of the non-negative integers 0, 1, 2, ...

Figures 1a–d exhibit the approach to this limit in a pictorial form. Figures 1a–c represent binomial distributions, each with $Np = 2$, but with $N = 5, 10, 20$ respectively. Figure 1d represents a Poisson distribution with $\theta = 2$. The similarity with Figure 1c is quite clear. The stability of $\Pr[\mathbf{x} = 1]$ is noteworthy.

The Poisson distribution may also arise for events occurring 'randomly and independently' in time. If it be supposed that the future lifetime of an item of equipment is independent of its present age (Section 1, Chapter 18) the lifetime can be represented by a random variable $\mathbf{t}$ with probability density function of form

(3) $$p_t(t) = \tau^{-1} \exp(-t/\tau) \qquad (t \geq 0; \tau > 0).$$

The expected value of **t** is τ (Chapter 18, Section 4). Now imagine a situation in which each item is replaced by another item with exactly the same distribution (as defined by (3)) of lifetime. What is the distribution of the number of failures (complete lifetimes), **x**, in a period of length T? If we denote the times of successive lifetimes by $\mathbf{t}_1, \mathbf{t}_2, \ldots$ etc. then (for $k \geq 1$)

(4)
$$\Pr[\mathbf{x} = k] = \Pr[\mathbf{t}_1 + \mathbf{t}_2 + \cdots + \mathbf{t}_{k+1} > T] - \Pr[\mathbf{t}_1 + \mathbf{t}_2 + \cdots + \mathbf{t}_k > T].$$

From (3), it can be seen that $\mathbf{t}_j$ is distributed as $\frac{1}{2}\tau$ (χ^2 with 2 degrees of freedom) and so $\sum_{j=1}^{m} \mathbf{t}_j$ is distributed as $\frac{1}{2}\tau$ (χ^2 with $2m$ degrees of freedom). (Chapter 17, Section 2.) Hence, from (4)

(5)
$$\begin{aligned}\Pr[\mathbf{x} = k] &= \Pr[\chi^2_{2(k+1)} > 2T/\tau] - \Pr[\chi^2_{2k} > 2T/\tau] \\ &= \sum_{j=0}^{k} e^{-T/\tau} \frac{(T/\tau)^j}{j!} - \sum_{j=0}^{k-1} e^{-T/\tau} \frac{(T/\tau)^j}{j!} \\ &\qquad\qquad \text{(using (92) below)} \\ &= e^{-T/\tau}(T/\tau)^j/j!\end{aligned}$$

as in (1) with θ replaced by T/τ. Hence **x** has a Poisson distribution with parameter T/τ.

More general conditions under which a Poisson distribution can arise are discussed by Redheffer [79] and Walsh [105].

2. Historical Remarks

In 1837 Poisson [74] published the derivation of the distribution which bears his name (sections 73, pp. 189–190, and 81, pp. 205–207). He approached the distribution by considering limiting forms of the binomial distribution (the first of the two kinds of genesis described in Section 1). Bortkiewicz [8], in 1898, considered circumstances in which Poisson's distribution might arise. From the point of view of Poisson's own approach, these are situations where, *in addition* to the requirements of independence of trials, and constancy of probability from trial to trial, the number of trials must be very large while the probability of occurrence of the outcome under observation must be small. Although Bortkiewicz called this the 'Law of Small Numbers' there is no need for $\theta = Np$ to be small, so the actual values of the observations need not be 'small'. It is the largeness of N and the smallness of p which are important. In Bortkiewicz's work [8], one of the "outcomes" considered was the number of deaths from being kicked by mules, per annum, in Prussian Army Corps. Here was a situation where the probability of death from this cause was small while

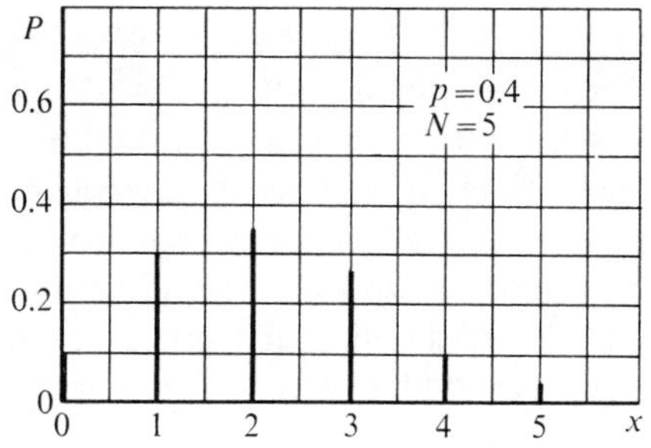

FIGURE 1a
Binomial

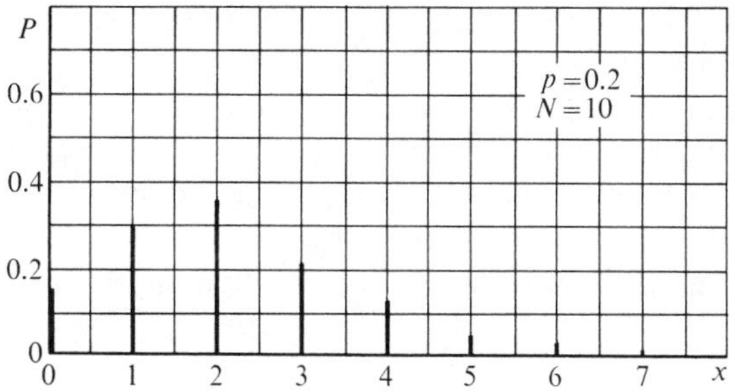

FIGURE 1b
Binomial

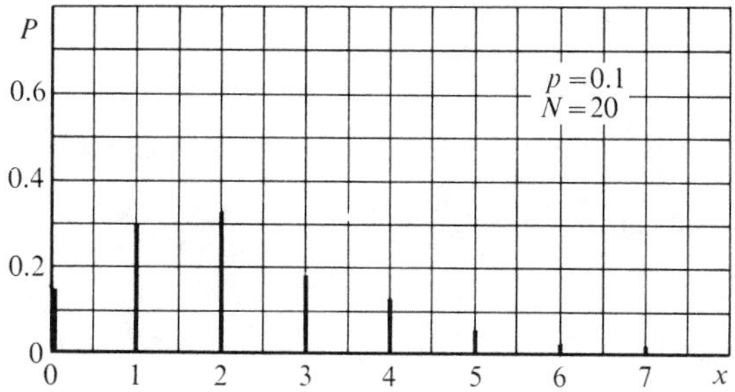

FIGURE 1c
Binomial

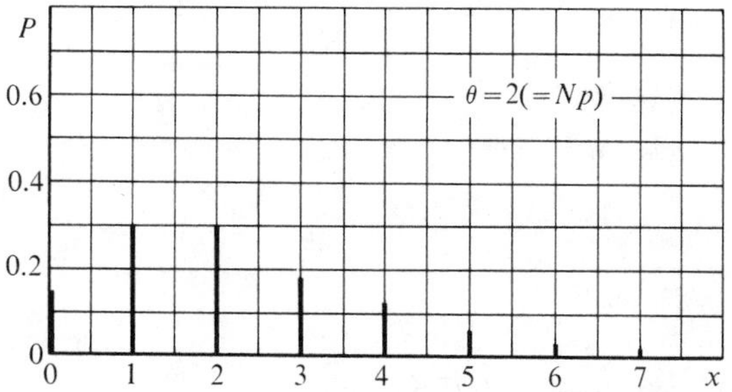

FIGURE 1d
Poisson

the number of soldiers exposed to risk (in any one Corps) is large. Whether the conditions of independence and constant probability are satisfied is doubtful. However, the data available to Bortkiewicz were quite satisfactorily fitted by a Poisson distribution, and have been very widely quoted as an example of the applicability of this distribution.

'Student' (W. S. Gosset) [95], in 1907, used the Poisson distribution to represent, to a first approximation, the number of particles falling in a small area A when a large number of such areas are spread at random over a surface large in comparison with A.

The second form of genesis described in Section 1 underlies the use of the Poisson distribution to represent variations in the number of particles ('rays') emitted by a radioactive source in forced periods of time. Rutherford and Geiger [85] gave some numerical data which were excellently fitted by a Poisson distribution (see also Rutherford et al. [84]).

In recent years the Poisson distribution has been used to a markedly increasing extent, but reasons underlying its use are, in the great majority of cases, of the kinds indicated earlier in this Section. Like the binomial distribution, the Poisson distribution often serves as a standard from which to measure departures, even when it is not itself an adequate representation of the real situation.

The handbook by Haight [43] contains much interesting information on the Poisson and related distributions, and should be consulted if further details are needed on matters discussed in this Chapter.

3. Moments and Generating Functions

For a random variable **x** with distribution (1), the rth factorial moment is

$$E[\mathbf{x}(\mathbf{x}-1)\dots(\mathbf{x}-r+1)] = \theta^r. \tag{6}$$

The rth moment about zero is

$$E(\mathbf{x}^r) = \sum_{j=0}^{r} (\theta^j/j!)\,\Delta^j 0^r. \tag{7}$$

Taking $r = 1, 2, 3, 4$:

$$E(\mathbf{x}) = \theta \tag{8.1}$$

$$\mu_2'(\mathbf{x}) = \theta + \theta^2 \tag{8.2}$$

$$\mu_3'(\mathbf{x}) = \theta + 3\theta^2 + \theta^3 \tag{8.3}$$

$$\mu_4'(\mathbf{x}) = \theta + 7\theta^2 + 6\theta^3 + \theta^4 \tag{8.4}$$

(cf. formulas (4.1)–(4.4) of Chapter 3).

On account of (8.1) **x** can be said to have a Poisson distribution with *expected value* θ (instead of *parameter* θ) if this is more convenient.

The moment generating function of $\mathbf{x}$ is

$$E(e^{t\mathbf{x}}) = e^{-\theta} \sum_{j=0}^{\infty} (\theta e^t)^j/j! = \exp[\theta(e^t - 1)]. \tag{9}$$

The cumulant generating function is

$$\theta(e^t - 1), \qquad \text{so that } \kappa_r(\mathbf{x}) = \theta \text{ for all } r \geq 2$$

whence

$$\text{var}(\mathbf{x}) = \theta \tag{10.1}$$

$$\mu_3(\mathbf{x}) = \theta \tag{10.2}$$

$$\mu_4(\mathbf{x}) = \theta + 3\theta^2. \tag{10.3}$$

These formulas could be obtained from (8.1)–(8.4).

From (10.1)–(10.3) the moment ratios $\sqrt{\beta_1}$ and β_2 are

$$\sqrt{\beta_1} = \theta^{-\frac{1}{2}} \tag{11.1}$$

$$\beta_2 = 3 + \theta^{-1}. \tag{11.2}$$

Note that for any Poisson distribution:

$$E(\mathbf{x}) = \text{var}(\mathbf{x}) \quad \text{and} \quad \beta_2 - \beta_1 - 3 = 0.$$

There is a recurrence relationship among moments ($\mu_r = E[(\mathbf{x} - \theta)^r]$) about the mean:

$$\mu_{r+1} = r\theta\mu_{r-1} + \theta \frac{\partial \mu_r}{\partial \theta}, \tag{12}$$

which should be compared with (8) of Chapter 3.

The mean deviation of the Poisson distribution is

$$\nu_1 = E[|\mathbf{x} - \theta|] = 2e^{-\theta}\theta^{[\theta]+1}/[\theta]! \tag{13}$$

The rth inverse (ascending) factorial moment is

$$\begin{aligned} \mu_{-[r]} = E[\{(\mathbf{x} + 1)^{[r]}\}^{-1}] &= e^{-\theta} \sum_{j=0}^{\infty} \{\theta^j/(j + r)!\} \\ &= \theta^{-r}\left[1 - e^{-\theta} \sum_{j=0}^{r-1} (\theta^j/j!)\right]. \end{aligned}$$

For the Poisson distribution

$$\begin{aligned} \frac{\text{mean deviation}}{\text{standard deviation}} &= \frac{2\theta^{[\theta]+\frac{1}{2}}}{[\theta]!e^\theta} \\ &\doteqdot \sqrt{\frac{2}{\pi}}\left(\frac{\theta}{[\theta]}\right)^{[\theta]+\frac{1}{2}} \exp\left[-(\theta - [\theta]) - \frac{1}{12[\theta]}\right]. \end{aligned} \tag{14}$$

As θ increases the ratio tends to $\sqrt{2/\pi} \doteqdot 0.798$ (Crow [24]).

4. Properties

From (1)

$$\Pr[\mathbf{x} = k + 1]/\Pr[\mathbf{x} = k] = \theta/(k + 1) \tag{15}$$

whence it follows that $\Pr[\mathbf{x} = k]$ increases with k to a maximum at $k = [\theta]$ (or to two equal maxima at $k = \theta - 1$ and $k = \theta$, if θ is an integer) and thereafter decreases as k increases. Some typical forms of the distribution are shown in figure 1.

Note that equation (13) can be written

$$\text{mean deviation} = 2\theta \max_k \Pr[\mathbf{x} = k]. \tag{16}$$

Hadley and Whitin [41] and Said [86] list a considerable number of formulas involving Poisson probabilities, which might be of use in engineering and operations research. A few are listed here; others appear in various parts of this Chapter. We use the notations

$$w(k,\theta) = e^{-\theta}\theta^k/k! \tag{17.1}$$

$$P(k,\theta) = \sum_{j=0}^{k} w(j,\theta) \tag{17.2}$$

$$Q(k,\theta) = 1 - P(k - 1,\theta) = \sum_{j=k}^{\infty} w(j,\theta). \tag{17.3}$$

(Particular note should be taken of (17.3). It would be more natural to define $Q(k,\theta)$ as $(1 - P(k,\theta))$, but the definition used here makes some of the formulas simpler. This notation is used in Hadley and Whitin [41] and in Haight [43].)

$$kw(k,\theta) = \theta w(k - 1,\theta) \tag{18.1}$$

$$\sum_{j=0}^{k} w(j,\theta_1)w(k-j,\theta_2) = w(k,\theta_1 + \theta_2) \tag{18.2}$$

$$\sum_{j=0}^{k} jw(j,\theta) = \theta P(k-1,\theta) \tag{18.3}$$

$$\sum_{j=k}^{\infty} jw(j,\theta) = \theta Q(k-1,\theta) \tag{18.4}$$

$$kQ(k + 1,\theta) = kQ(k,\theta) - \theta w(k-1,\theta) \tag{18.5}$$

$$\sum_{j=k}^{\infty} Q(j,\theta) = \theta Q(k - 1,\theta) + (1 - k)Q(k,\theta) \tag{18.6a}$$

$$\sum_{j=0}^{k} Q(j,\theta) = \theta[1 - Q(k,\theta)] - kQ(k + 1,\theta) \tag{18.6b}$$

$$\sum_{j=k}^{\infty} j^2 w(j,\theta) = \theta Q(k - 1,\theta) + \theta^2 Q(k - 2,\theta) \tag{18.7}$$

$$\text{(18.8a)} \qquad \sum_{j=k}^{\infty} jQ(j,\theta) = \tfrac{1}{2}\theta^2 Q(k-2)\theta) + \theta Q(k-1,\theta) - \tfrac{1}{2}k(k-1)Q(k,\theta)$$

(18.8b)

$$\sum_{j=0}^{k} jQ(j,\theta) = \tfrac{1}{2}\theta^2[1 - Q(k-1,\theta)] + \theta[1 - Q(k,\theta)] + \tfrac{1}{2}k(k+1)Q(k+1,\theta)$$

$$\text{(18.9)} \qquad \frac{\partial w(k,\theta)}{\partial\theta} = w(k-1,\theta) - w(k,\theta).$$

If $\mathbf{x}_1$ and $\mathbf{x}_2$ are independent random variables, each having a Poisson distribution, with expected values θ_1 and θ_2 respectively, then $(\mathbf{x}_1 + \mathbf{x}_2)$ has a Poisson distribution with expected value $(\theta_1 + \theta_2)$. This follows from (18.2) above.

The distribution of the difference $(\mathbf{x}_1 - \mathbf{x}_2)$ cannot be expressed in so simple a form. It will be discussed in Section 11.

The *conditional* distribution of $\mathbf{x}_1$, given $\mathbf{x}_1 + \mathbf{x}_2 = X$ is a binomial with parameters X, $\theta_1/(\theta_1 + \theta_2)$.

5. Characterizations

Raikov [76] showed that if $\mathbf{x}_1$ and $\mathbf{x}_2$ are independent random variables and $(\mathbf{x}_1 + \mathbf{x}_2)$ has a Poisson distribution, then $\mathbf{x}_1$ and $\mathbf{x}_2$ must each have Poisson distributions. (A similar property also holds for the sum of any number of independent random variables.)

Chatterji [13] has shown that if $\mathbf{x}_1$ and $\mathbf{x}_2$ are independent non-negative integer-valued random variables and if

$$\text{(19)} \qquad \Pr[\mathbf{x}_1 = X_1 \mid \mathbf{x}_1 + \mathbf{x}_2 = X] = \binom{X}{X_1} p_X^{X_1}(1 - p_X)^{X-X_1}$$

$$(X_1 = 0,1,\ldots,X)$$

then it follows that

(a) p_X does not depend on X, but equals a constant, p, for all values of X and

(b) $\mathbf{x}_1$ and $\mathbf{x}_2$ each have Poisson distributions with parameters in the ratio $p : (1 - p)$.

This characterization has been expanded by Volodin [102] to n variables $\mathbf{x}_1, \mathbf{x}_2, \ldots, \mathbf{x}_n$: The condition is that the conditional distribution of $\mathbf{x}_1, \mathbf{x}_2, \ldots, \mathbf{x}_n$ given $\sum \mathbf{x}_i$ be multinomial (Chapter 11).

Bol'shev [7] has suggested that this property might be used to generate Poisson distributed random numbers for a number of Poisson distributions, using only one original Poisson distribution with a large expected value (and so nearly normal) which is subsequently split up according to a multinomial with fixed cell probabilities.

Rao and Rubin [78] obtained the following characterization of the Poisson distribution.

If $\mathbf{x}$ is a discrete random variable taking only non-negative integer values and the conditional distribution of $\mathbf{y}$ given $\mathbf{x} = x$ is binomial with parameters x, p (p not depending on x), then the distribution of $\mathbf{x}$ is Poisson if and only if

$$\Pr[\mathbf{y} = k \mid \mathbf{y} = \mathbf{x}] = \Pr[\mathbf{y} = k \mid \mathbf{y} \neq \mathbf{x}]. \tag{20}$$

(Rao [77] gives the following physical basis for the model described above: $\mathbf{x}$ represents a 'naturally' occurring quantity, which is observed in such a way that some of the components of $\mathbf{x}$ may not be counted. $\mathbf{y}$ represents the value remaining (and actually observed) after this 'destructive process'.)

Daboni [26] has given the following characterization in terms of mixtures of binomial distributions (Chapter 3, Section 1). Suppose that $\mathbf{x}$ is a random variable distributed as a mixture of binomial distributions with parameters $\mathbf{N}$, p. Then $\mathbf{x}$ and $(\mathbf{N} - \mathbf{x})$ are independent if and only if $\mathbf{N}$ has a Poisson distribution.

6. Estimation

6.1 *Point Estimation*

There is only one parameter, θ, used in defining a Poisson distribution. Hence there is only need to estimate a single parameter, though different functions of this parameter may be estimated in different circumstances.

Given n independent random variables $\mathbf{x}_1, \mathbf{x}_2, \ldots, \mathbf{x}_n$ each with distribution (1), the maximum likelihood estimator of θ is

$$\hat{\boldsymbol{\theta}} = n^{-1} \sum_{j=1}^{n} \mathbf{x}_j . \tag{21}$$

(Since $\sum_{j=1}^{n} \mathbf{x}_j$ has a Poisson distribution with parameter $n\theta$, and $\sum_{j=1}^{n} \mathbf{x}_j$ is a complete sufficient statistic for θ we really need only consider the problem of estimating θ given a *single* observed value.)

The variance of $\hat{\boldsymbol{\theta}}$ is θ/n. This is equal to the Cramer-Rao bound and $\hat{\boldsymbol{\theta}}$ is the minimum variance unbiased estimator of θ.

A natural estimator for $e^{-\theta}$, the "probability of the zero class" is $\exp(-\hat{\boldsymbol{\theta}})$. This is a biased estimator (its expected value is $\exp[-\theta(1 - e^{-1/n})]$), the minimum variance unbiased estimator of $e^{-\theta}$ is

$$\mathbf{T} = (1 - \hat{\boldsymbol{\theta}}/n)^{n-1}. \tag{22}$$

The variance of this estimator is $e^{-2\theta}(e^{\theta/n} - 1)$. The mean square error of $\mathbf{T}$ is less than that of $\exp(-\hat{\boldsymbol{\theta}})$ for $e^{-\theta} < 0.45$ (Johnson [48]). (Of course, if there is doubt whether the distribution of each $\mathbf{x}_j$ is really Poisson, it is safer to use the proportion of $\mathbf{x}$'s which are equal to zero as estimator of the probability of the zero class.)

The estimation of the Poisson probability (1), for general k, has been considered by Barton [3] and Glasser [38]. An unbiased estimator of $e^{-\theta}\theta^k/k!$ is the random variable $\mathbf{y}$, defined by

$$\mathbf{y} = \begin{cases} 1 & \text{if } \mathbf{x}_1 = k \\ 0 & \text{if } \mathbf{x}_1 \neq k. \end{cases}$$

Then, using Blackwell's theorem (Chapter 1, Section 7.4) the minimum variance unbiased estimator of $e^{-\theta}\theta^k/k!$ is

$$\begin{aligned} (23) \qquad E[\mathbf{y} \mid \hat{\boldsymbol{\theta}}] &= \Pr[\mathbf{y} = 1 \mid \hat{\boldsymbol{\theta}}] \\ &= \Pr[\mathbf{x}_1 = k \mid \hat{\boldsymbol{\theta}}] \\ &= \Pr\left[(\mathbf{x}_1 = k) \cap \left(\sum_{j=2}^{n} \mathbf{x}_j = n\hat{\boldsymbol{\theta}} - k\right)\right] \Big/ \Pr\left[\sum_{j=1}^{n} \mathbf{x}_j = n\hat{\boldsymbol{\theta}}\right] \\ &= \frac{(e^{-\theta}\theta^k/k!)(e^{-\theta(n-1)}[(n-1)\theta]^{n\hat{\boldsymbol{\theta}}-k}/(n\hat{\boldsymbol{\theta}}-k)!)}{e^{-n\theta}(n\theta)^{n\hat{\boldsymbol{\theta}}}/(n\hat{\boldsymbol{\theta}})!} \\ &= \binom{n\hat{\boldsymbol{\theta}}}{k}\left(\frac{n-1}{n}\right)^{n\hat{\boldsymbol{\theta}}} \frac{1}{(n-1)^k}. \end{aligned}$$

The minimum variance estimator of $\sum_w (e^{-\theta}\theta^k/k!)$ is

$$(24) \qquad \sum_w (n-1)^{-k}\binom{n\hat{\boldsymbol{\theta}}}{k}(1-n^{-1})^{n\hat{\boldsymbol{\theta}}}$$

where $\sum_w$ denotes summation with respect to k over any (finite or infinite) subset of the non-negative integers.

When all data are available, the maximum likelihood estimator ($\hat{\boldsymbol{\theta}}$) of θ is so easily available that it has, in practice, no serious competitors. However, when some data are omitted, or inaccurately observed, the situation is less clear-cut. The most important special situation of this kind is that in which the zero class is not observed. The random variables so obtained have the *positive Poisson* distribution

$$(25) \qquad \Pr[\mathbf{x} = k] = \frac{e^{-\theta}\theta^k/k!}{1 - e^{-\theta}} \qquad (k = 1,2,\ldots)$$

which will be discussed in Section 9, together with appropriate methods of estimation of θ.

Estimation of the *ratio* of expected values θ_1, θ_2 of two Poisson distributions based on observed values of independent random variables which have such distributions, has been described by Chapman [12]. He showed that there is no unbiased estimator of the ratio θ_1/θ_2 with finite variance, but that the estimator

$$(26) \qquad \mathbf{x}_1/(\mathbf{x}_2 + 1)$$

(in an obvious notation) is 'almost unbiased'.

6.2 *Confidence Intervals*

Since the Poisson distribution is a discrete distribution it is not possible to construct confidence intervals for θ with an exactly specified confidence coefficient, of say, $100(1-\alpha)\%$. *Approximate* $100(1-\alpha)\%$ confidence limits for θ given an observed value of $\mathbf{x}$, where $\mathbf{x}$ has the distribution (1), are obtained by solving the equations

$$\exp(-\boldsymbol{\theta}_{\mathrm{L}}) \sum_{j=\mathbf{x}}^{\infty} (\boldsymbol{\theta}_{\mathrm{L}}^{j}/j!) = \tfrac{1}{2}\alpha \tag{27.1}$$

$$\exp(-\boldsymbol{\theta}_{\mathrm{U}}) \sum_{j=0}^{\mathbf{x}} (\boldsymbol{\theta}_{\mathrm{U}}^{j}/j!) = \tfrac{1}{2}\alpha \tag{27.2}$$

for $\boldsymbol{\theta}_{\mathrm{L}}$, $\boldsymbol{\theta}_{\mathrm{U}}$ respectively, and using the interval $(\boldsymbol{\theta}_{\mathrm{L}}, \boldsymbol{\theta}_{\mathrm{U}})$.

From the relationship between the Poisson and χ^2 distributions (Section 1), the equations (27.1) and (27.2) can be written

$$\boldsymbol{\theta}_{\mathrm{L}} = \tfrac{1}{2}\chi^2_{2\mathbf{x},\alpha/2} \tag{28.1}$$

$$\boldsymbol{\theta}_{\mathrm{U}} = \tfrac{1}{2}\chi^2_{2(\mathbf{x}+1),1-\alpha/2}. \tag{28.2}$$

So the values of $\boldsymbol{\theta}_L$ and $\boldsymbol{\theta}_U$ can be found by interpolation (with respect to the number of degrees of freedom) in tables of percentage points of the central χ^2 distribution.

If θ is expected to be fairly large, say greater than 5, a normal approximation to the Poisson distribution (see Section 7) might be used. Then

$$\Pr[|\mathbf{x}-\theta| < u_{\alpha/2}\cdot\sqrt{\theta} \mid \theta] = 1-\alpha \tag{29}$$

where $(2\pi)^{-\frac{1}{2}} \int_{u_{\alpha/2}}^{\infty} e^{-u^2/2}\,du = \tfrac{1}{2}\alpha$. From (29)

$$\Pr[\theta^2 - \theta\{2\mathbf{x}+u^2_{\alpha/2}\} + \mathbf{x}^2 < 0 \mid \theta] = 1-\alpha$$

or

$$\Pr[\mathbf{x}+\tfrac{1}{2}u^2_{\alpha/2} - u_{\alpha/2}\sqrt{\mathbf{x}+\tfrac{1}{4}u^2_{\alpha/2}} < \theta < \mathbf{x}+\tfrac{1}{2}u^2_{\alpha/2} + u_{\alpha/2}\sqrt{\mathbf{x}+\tfrac{1}{4}u^2_{\alpha/2}} \mid \theta] \doteqdot 1-\alpha.$$

The limits

$$\mathbf{x}+\tfrac{1}{2}u^2_{\alpha/2} \pm u_{\alpha/2}\sqrt{\mathbf{x}+\tfrac{1}{4}u^2_{\alpha/2}} \tag{30}$$

thus enclose a confidence interval for θ with confidence coefficient approximately equal to $100(1-\alpha)\%$.

Values of approximate 95% confidence limits for θ, given an observed value of $\mathbf{x}$, given by Mantel [57], are shown in Table 1. The 'limit factors' shown must be *multiplied* by the observed value ($\mathbf{x}$) to obtain limits for θ. There is a similar table in [90]. Extensive tables of $\boldsymbol{\theta}_L$ and $\boldsymbol{\theta}_u$ are given by Lachenbruch

TABLE 1

Factors for 95 Percent Confidence Limits for Mean of a Poisson-distributed Variable

Observed number on which the estimate is based (n)	Lower limit factor (L)	Upper limit factor (U)
1	0.0253	5.57
2	0.121	3.61
3	0.206	2.92
4	0.272	2.56
5	0.324	2.33
6	0.367	2.18
7	0.401	2.06
8	0.431	1.97
9	0.458	1.90
10	0.480	1.84
11	0.499	1.79
12	0.517	1.75
13	0.532	1.71
14	0.546	1.68
15	0.560	1.65
16	0.572	1.62
17	0.583	1.60
18	0.593	1.58
19	0.602	1.56
20	0.611	1.54
21	0.619	1.53
22	0.627	1.51
23	0.634	1.50
24	0.641	1.49
25	0.647	1.48
26	0.653	1.47
27	0.659	1.46
28	0.665	1.45
29	0.670	1.44
30	0.675	1.43

Observed number on which the estimate is based (n)	Lower limit factor (L)	Lower limit factor (U)
35	0.697	1.39
40	0.714	1.36
45	0.729	1.34
50	0.742	1.32
60	0.770	1.30
70	0.785	1.27
80	0.798	1.25
90	0.809	1.24
100	0.818	1.22
120	0.833	1.200
140	0.844	1.184
160	0.854	1.171
180	0.862	1.160
200	0.868	1.151
250	0.882	1.134
300	0.892	1.121
350	0.899	1.112
400	0.906	1.104
450	0.911	1.098
500	0.915	1.093
600	0.922	1.084
700	0.928	1.078
800	0.932	1.072
900	0.936	1.068
1000	0.939	1.064

[55]. Values are given to three decimal places for $\alpha = 0.1k^{-1}$, $0.05k^{-1}$, $0.01k^{-1}$ with $k = 1(1)10$ and $\mathbf{x} = 0(1)100$.

Charts from which $\boldsymbol{\theta}_L$ and $\boldsymbol{\theta}_U$ can be read off are contained in [70]. Crow and Gardner [25] have constructed a modified version of these charts, so that (i) the confidence belt is as narrow as possible, *measured in the direction of the observed variable*, and (ii) among such narrowest belts, it has the smallest possible upper confidence limits. Condition (ii) reduces, particularly, the width of confidence intervals corresponding to small values of the observed variable.

Let $\mathbf{x}_1$ and $\mathbf{x}_2$ be two independent Poisson random variables with parameters θ_1 and θ_2 respectively. Let $G_{m,n}(X)$ represent the distribution function of the ratio of two independent chi-square variables χ^2_{2m}/χ^2_{2n} with $2m$ and $2n$ degrees of freedom. Bol'shev [5] has shown that for any $\alpha(0 \leq \alpha < 1)$ the solution of the equation

$$G_{\mathbf{x}_1,\mathbf{x}_2+1}(\mathbf{X}) = \alpha$$

satisfies the inequality $\inf_{\theta_1,\theta_2} \Pr\left[\mathbf{X} < \frac{\theta_1}{\theta_2}\right] \geq 1 - \alpha.$

Thus, $\mathbf{X}$ can be taken to be the lower bound of a confidence interval for θ_1/θ_2 with minimal confidence coefficient $1 - \alpha$. Similarly the solution of

$$G_{\mathbf{x}_1+1,\mathbf{x}_2}(\mathbf{Y}) = 1 - \alpha$$

satisfies the inequality $\inf_{\theta_1,\theta_2} \Pr\left[\mathbf{Y} > \frac{\theta_1}{\theta_2}\right] \geq 1 - \alpha$ and thus $\mathbf{Y}$ can be taken to be the upper bound of a confidence interval for θ_1/θ_2 with minimal confidence coefficient $(1 - \alpha)$.

The results are useful for testing the hypothesis $\theta_1/\theta_2 <$ constant, and also for the hypotheses $\theta_1/\theta_2 < 1$ and $\theta_1/\theta_2 > 1$ which, however, are more easily tested using, for example, the criterion of signs, to which this procedure is equivalent.

Chapman [12] makes use of the fact that the conditional distribution of $\mathbf{x}_1$, given $\mathbf{x}_1 + \mathbf{x}_2 = X$ is binomial with parameters X, $\theta_1/(\theta_1 + \theta_2)$ (Section 4). Forming approximate confidence intervals for $\theta_1/(\theta_1 + \theta_2) = (1 + \theta_2/\theta_1)^{-1}$ is then equivalent to doing the same for θ_1/θ_2.

7. Approximations to the Poisson distribution

The relationship between the Poisson and χ^2 distributions (see Section 11, equation (93)), implies that approximations to the cumulative distribution function of (central) χ^2 distributions can also be used as approximations to Poisson probabilities (and vice versa). Thus if $\mathbf{x}$ has distribution (1) then

$$\text{(31)} \qquad \Pr[\mathbf{x} \leq X] = \Pr[\chi^2_{2(X+1)} > 2\theta].$$

Approximations to one side are also approximations to the other. Thus, the

Wilson-Hilferty approximation to the χ^2 distribution (Chapter 17) corresponds to

$$(32)\qquad \Pr[\mathbf{x} \le X] \doteqdot (2\pi)^{-\frac{1}{2}} \int_Z^\infty e^{-u^2/2}\, du$$

$$\left(Z = 3\left[\left(\frac{\theta}{X+1}\right)^{\frac{3}{2}} - 1 + \frac{1}{9(X+1)}\right](X+1)^{-\frac{1}{2}}\right).$$

As $\theta \to \infty$, the standardized Poisson distribution tends to the unit normal distribution. Modifying the crude approximation "$\mathbf{z} = (\mathbf{x} - \theta)/\sqrt{\theta}$ unit normally distributed" as in the Cornish-Fisher expansion (see Chapter 12, Section 5) leads to $\mathbf{z} - \frac{1}{3}(\mathbf{z}^2 - 1)\theta^{-\frac{1}{2}} + \frac{1}{36}(7\mathbf{z}^3 - \mathbf{z})\theta^{-1} \ldots$ unit normally distributed.

A very simple approximation is to take $2\sqrt{\mathbf{x}}$ as normally distributed with expected value zero and variance 1. An improvement, suggested by Anscombe [2], is to use $2\sqrt{\mathbf{x} + \frac{3}{8}}$. Freeman and Tukey [33] have suggested the transformed variable

$$\sqrt{\mathbf{x}} + \sqrt{\mathbf{x} + 1}.$$

Tables of this quantity, to 2 decimal places for $x = 0(1)50$, have been given by Mosteller and Youtz [65].

Individual Poisson probabilities may be approximated by the formula

$$(33)\qquad \Pr[\mathbf{x} = k] = (2\pi)^{-\frac{1}{2}} \int_{K_-}^{K_+} e^{-\frac{1}{2}u^2}\, du$$

where $K_+ = (k - \theta + \frac{1}{2})\theta^{-\frac{1}{2}}$; $K_- = (k - \theta - \frac{1}{2})\theta^{-\frac{1}{2}}$.

A modified form

$$(34)\qquad \Pr[\mathbf{x} = k] \doteqdot (2\pi\theta)^{-\frac{1}{2}} e^{-z^2/2}[1 + \tfrac{1}{6}(3z - z^3)\theta^{-\frac{1}{2}}] + R$$

where $z = (k - \theta)\theta^{-\frac{1}{2}}$ and

$$|R| < 0.0748\theta^{-\frac{3}{2}} - 0.0055\theta^{-2} + 0.3724\theta^{-\frac{5}{2}} + 0.5595\theta^{-3} + \theta^{-\frac{1}{4}}(1 + \tfrac{1}{2}\theta^{-\frac{1}{2}}) \exp[-2\theta^{\frac{1}{2}}]$$

has been proposed by Makabe and Morimura [56].

Makabe and Morimura also obtain the following approximation to sums of individual probabilities:

$$(35)\qquad \sum_{j=k_1}^{k_2} e^{-\theta}(\theta^j/j!) = (2\pi)^{-\frac{1}{2}} \int_{x_1}^{x_2} \exp(-u^2/2)\, du + \tfrac{1}{6}(2\pi\theta)^{-\frac{1}{2}}[(1 - x_2^2)\exp(-x_2^2/2) - (1 - x_1^2)\exp(-x_1^2/2)] + R'$$

TABLE 2

Principal Tables of the Poisson Distribution

Tabulated function	θ	Number of decimals or figures	Reference
Pr[x = k]	θ = 0.1(0.1)15	6D	Soper [91] Pearson [72], Table LI Pearson and Hartley [70], Table 39 Bol'shev and Smirnov [7], Table 5.3
Pr[x = k]	θ = 0.1(0.1)1(1)20	5S	Fry [34], Table VI*
Pr[x = k]	θ = 0.001(0.001)0.01(0.01)0.3,0.4 θ = 0.5(0.1)15(1)100	7D 6D	Molina [59], Table 1
Pr[x = k]	θ = 0.001(0.001)1(0.01)5 θ = 5.01(0.01)10	8D 7D	Kitagawa [53] (For some corrections, see Sexton et al. [87]).
Pr[x = k]	θ = 0.1(0.1)10.0,11(1)20		Burrington and May [10]
Pr[x = k]	θ = 0.1(0.1)15.0,16(2)30	6D	Janko [47], Table 29
Pr[x = k]	θ = 0.00000010(0.0000001)0.00000015(0.0000005) 0.000015(0.000001)0.00005(0.000005)0.0005 (0.00001)0.001(0.00005)0.005(0.0001)0.01 (0.0005)0.2(0.001)0.4(0.005)0.5(0.01)1.0(0.05)2.0 (0.1)5.0(0.5)10(1)100(5)205	8D	General Electric Company [37]

Tabulated function	θ	Number of decimals or figures	Reference
$\Pr[\mathbf{x} \geq k]$	θ = 0.1(0.1)1(1)20	5S	Fry [34], Table VII*
$\Pr[\mathbf{x} \geq k]$	θ = 0.001(0.001)0.01(0.01)0.3,0.4 θ = 0.5(0.1)15(1)100	7D 6D	Molina [59], Table 2
$\Pr[\mathbf{x} \leq k]$	θ = 0.1(0.1)10.0,11(120)		Burrington and May [10]
$\Pr[\mathbf{x} \leq k]$	θ = 0.005(0.005)0.005(0.005)0.05(0.05)1.0 1.0(0.1)5(25)10(0.5)20(1)60	5D	Pearson and Hartley [70], Table 7
$\Pr[\mathbf{x} \leq k]$	θ = 0.001,0.005(0.005)0.100(0.100)1.00 1.00(0.20)2.00(0.50)5.00(1.00)1.0.00	4D	Owen [68], pp. 260–261
$\Pr[\mathbf{x} \geq k]$	θ = 0.1(0.1)15.0,16(2)30	6D	Janko [47], Table 30
$\left.\begin{matrix}\Pr[\mathbf{x} \leq k]\\ \Pr[\mathbf{x} \geq k]\end{matrix}\right\}$	same range and same increments as for the individual terms	8D	General Electric Company [37]

*In Fry's Tables k takes all integral values for which the function tabulated is greater than 0.000001.

with

$$x_1 = (k_1 - \theta - \tfrac{1}{2})\theta^{-\frac{1}{2}},$$
$$x_2 = (k_2 - \theta + \tfrac{1}{2})\theta^{-\frac{1}{2}},$$

$$|R'| \leq 0.05440^{-1} + 0.0108\theta^{-\frac{3}{2}} + 0.2743\theta^{-2} + 0.0065\theta^{-\frac{5}{2}} + (1 + \tfrac{1}{2}\theta^{-\frac{1}{2}}) \exp(-2\theta^{\frac{1}{2}}).$$

(A similar, though not quite so precise, formula was obtained by Cheng [14].)

Coming now to bounds, which have the advantage over approximations that the sign of the error is known, we first have the inequality (Teicher [98])

$$\Pr[\mathbf{x} \leq \theta] \geq e^{-1}$$

(If θ is an integer the right hand side can be replaced by $\frac{1}{2}$.) Also, the following inequalities have been given by Bohman [4]

$$\Pr[\mathbf{x} \leq k] \leq (2\pi)^{-\frac{1}{2}} \int_{-\infty}^{(k+1-\theta)/\sqrt{\theta}} e^{-u^2/2}\, du \tag{36.1}$$

$$\Pr[\mathbf{x} \leq k] \geq [\Gamma(\theta + 1)]^{-1} \int_0^k t^{\theta} e^{-t}\, dt. \tag{36.2}$$

Kalinin [50] has obtained the asymptotic expansion (as $N \to \infty$)

$$\sum_{j=0}^{c} \binom{N}{j} \left(\frac{\theta}{N}\right)^j \left(1 - \frac{\theta}{N}\right)^{N-j} = \sum_{j=0}^{c} \frac{\theta^j e^{-\theta}}{j!} + \sum_{j=1}^{\infty} \frac{V_j(c,\theta)}{N^j} \tag{36.3}$$

where

$$V_j(c,\theta) = \sum_{i=0}^{c} \frac{\theta^i e^{-\theta}}{i!} M_j(i,\theta), \tag{36.4}$$

with $M_j(i,\theta)$ equal to the sum, over all sets of positive integers (including zeros) $\nu_1, \nu_2, \ldots, \nu_j$, satisfying $\sum_{h=1}^{j} h\nu_h = j$, of the product $\prod_{h=1}^{j} \{[L_h(c,\theta)]^{\nu_h}/\nu_h!\}$, where

$$L_h(c,\theta) = \frac{B_{h+1} - B_{h+1}(c)}{h(h+1)} + \frac{c\theta^h}{h} - \frac{\theta^{h+1}}{h+1}. \tag{36.5}$$

B_h, $B_h(\theta)$ are the Bernoulli numbers and polynomials defined in Chapter 1, Section 3.

The same author has also given an asymptotic expansion (as $\theta \to \infty$) for $\theta^j e^{-\theta}/j!$ in terms of $y = (j + \theta - \lambda)\theta^{-\frac{1}{2}}$. In these formulas θ plays the role of a continuity correction which can be assigned an arbitrary value.

8. Tables

The first detailed published tables for Poisson distribution are contained in Molina's [59] volume, where 6 place tables for the Poisson probabilities with parameter $\theta = 0.001(0.001)0.01(0.01)0.3(0.1)1(1)5$ and for the values of $k = 0(1)\infty$ are given. Cumulative probabilities are also given for the same values of k and θ.

In [70], Table 39 gives 6-place values of $\Pr[\mathbf{x} = k]$ for $\theta = 0.1(0.1)15.0$ and $k = 0(1)\infty$ and presents the probabilities $\Pr[\mathbf{x} \leq k]$ for

$$\theta = 0.0005(0.0005)0.005(0.005)0.05(0.05)1(0.1)5(0.025)10(0.5)20(1)60$$

and $k = 1(1)35$. Kitagawa's [53] tables are more detailed and present the probabilities $\Pr[\mathbf{x} = k]$ to 8 decimal places for $k = 0(1)\infty$ and $\theta = 0.001(0.001)1.000$; for the same range of k with $\theta = 1.01(0.01)5.00$ to 8 places; and for the same range of k with $\theta = 5.01(0.01)10.00$ with 7 places.

In Janko's [47] tables 6-place values of $\Pr[\mathbf{x} = k]$ are given with $\theta = 0.1(0.1)15.0$, $16(2)30$ for $k = 1(1)\infty$ and similar tables for the cumulative probabilities.

Some of the earlier tables are summarized in Eisenhart and Zelen's article [30] and reproduced herewith (supplemented with some more recent additions) for the reader's convenience. A revised and enlarged listing is available in the second edition of this Handbook.

Hald and Kousgaard [44] give values of θ satisfying the equation

$$e^{-\theta} \sum_{j=0}^{c} \{\theta^j/j!\} = P$$

to four significant figures, for $P = 0.001, 0.005, 0.01, 0.025, 0.05, 0.1, 0.2, 0.8, 0.9, 0.95, 0.975, 0.99, 0.995$, and 0.999, and $c = 0(1)50$. The *Tables of the Incomplete Gamma Function Ratio* by Khamis and Rudert (Julius von Liebig, Darmstadt, Germany, 1965) may be used to obtain Poisson probabilities, for distributions with $\theta = 0.00005(0.00005)0.0005(0.0005)0.005(0.005)0.5(0.025)3.0(0.05)8.0(0.25)33.0(0.5)83.0(1)125$. Values are given to ten decimal places.

9. Applications

The Poisson distribution is used (*a*) as an approximation to the binomial distribution, (*b*) when events occur randomly in time, or (*c*) to describe experiments in which the observed variable is a count. In this section we will be mainly concerned with type (*b*).

An important type (*b*) application is the use of Poisson processes in queueing theory. (A Poisson process is one in which the intervals between successive events have independent identical exponential distributions and, consequently, the number of events in a specified time interval has a Poisson distribution (see Section 1).) Consider, for example, the problem of finding the number of telephone channels in use (or customers waiting in line, etc.) at any one time. Suppose that there are 'infinitely many' (actually, a large number) of channels available, and that the holding time (length of call) ($\mathbf{t}$) for each call has the exponential distribution with probability density function

$$p_t(t) = \phi^{-1}e^{-t/\phi} \qquad (\phi > 0; t > 0) \tag{37}$$

and incoming calls arrive at times following a Poisson process, with the average number of calls per unit time equal to θ. Then the function $P_N(\tau)$ representing the probability that exactly N channels are being used at time τ after no channels are in use, satisfies the differential equations

$$P_0'(\tau) = -\theta P_0(\tau) + \phi^{-1}P_1(\tau) \tag{38.1}$$

$$P_N'(\tau) = -(\theta + N\phi^{-1})P_N(\tau) + \theta P_{N-1}(\tau) + (N+1)\phi^{-1}P_{N+1}(\tau) \qquad (N \geq 1). \tag{38.2}$$

The 'steady state' probabilities ($\lim_{\tau\to\infty} P_N(\tau) = P_N$) satisfy the equations

$$\theta P_0 = \phi^{-1}P_1 \tag{39.1}$$

$$(\theta + N\phi^{-1})P_N = \theta P_{N-1} + (N+1)\phi^{-1}P_{N+1}. \tag{39.2}$$

On solving these equations, it is found that

$$P_N = e^{-\theta\phi}(\theta\phi)^N/N! \qquad (N = 0,1,\ldots) \tag{40}$$

which is recognized as equivalent to (1) with θ replaced by $\theta\phi$.

Further specific examples of the use of the Poisson distribution in applied work may be found in papers by Breny [9] Duker [29], McKendrick [58] Stinson and Walsh [93] and Wallis [104].

Cohen [19] has suggested that a modification of a truncated Poisson distribution might be used in place of certain contagious distributions. This is discussed further in Chapter 8.

10. Modified Poisson Distributions

10.1 *Truncated Poisson Distributions*

The commonest form of truncation is the omission of the zero class, because the observational apparatus becomes active only when at least one event occurs. The corresponding truncated Poisson distribution is (see (25))

$$\Pr[\mathbf{x} = k] = e^{-\theta}(1 - e^{-\theta})^{-1}\theta^k/k! = (e^\theta - 1)^{-1}\theta^k/k! \qquad (k = 1,2,\ldots,). \tag{41}$$

This is usually called the *positive Poisson* distribution (see Section 6.1). Cohen [16] calls it a *conditional Poisson* distribution. The rth factorial moment of $\mathbf{x}$ is

$$E[\mathbf{x}(\mathbf{x}-1)\ldots(\mathbf{x}-r+1)] = (1 - e^{-\theta})^{-1}\theta^r. \tag{42}$$

The expected value and variance are

$$E(\mathbf{x}) = (1 - e^{-\theta})^{-1}\theta \tag{43.1}$$

$$\operatorname{var}(\mathbf{x}) = (1 - e^{-\theta})^{-1}[1 - \theta e^{-\theta}(1 - e^{-\theta})^{-1}]. \tag{43.2}$$

The expected value of $\mathbf{x}^{-1}$ is

$$e^{-\theta}(1 - e^{-\theta})^{-1} \sum_{j=1}^{\infty} \theta^j/[j!\,j]. \tag{44}$$

Roy and Tiku [83] have shown that

$$E(\mathbf{x}^{-1}) \doteqdot (\theta - 1)(1 - e^{-\theta})^{-1} \tag{45}$$

for sufficiently large θ.

Grab and Savage [40] (addenda: **49,** 906) give tables of $E(\mathbf{x}^{-1})$ to 5 decimal places for $\theta = 0.01, 0.05(0.05)1.00(0.1)2.0(0.2)5.0(0.5)7.0(1)10(2)20$.

The maximum likelihood estimator, $\hat{\boldsymbol{\theta}}$ of θ, given observed values of n independent random variables $\mathbf{x}_1, \mathbf{x}_2, \ldots, \mathbf{x}_n$ each having the distribution (41) satisfies the equation

$$\bar{\mathbf{x}} = n^{-1} \sum_{j=1}^{n} \mathbf{x}_j = \hat{\boldsymbol{\theta}}/(1 - e^{-\hat{\boldsymbol{\theta}}}). \tag{46}$$

Equation (46) may be solved numerically. The process is quite straightforward, especially if tables of the function, $\hat{\boldsymbol{\theta}}/(1 - e^{-\hat{\boldsymbol{\theta}}})$ are used (tables are in David and Johnson [27]). Also Cohen [17] gives a nomograph for solving this equation. Irwin [46] derived an explicit expression for $\hat{\boldsymbol{\theta}}$ as a function of $\bar{\mathbf{x}}$, namely:

$$\hat{\boldsymbol{\theta}} = \bar{\mathbf{x}} - \sum_{j=1}^{\infty} \frac{j^{j-1}}{j!} (\bar{\mathbf{x}}e^{-\bar{\mathbf{x}}})^j, \tag{47}$$

but the first two methods described are usually easier to apply. For ease of calculation the approximation

$$\frac{j^{j-1}}{j!} (\bar{\mathbf{x}}e^{-\mathbf{x}})^j \doteqdot \frac{1}{\sqrt{2\pi j^3}} (\bar{\mathbf{x}}e^{-(\bar{\mathbf{x}}-1)})^j \tag{48}$$

may be used for the larger values of j.

For large values of n, the variance of $\boldsymbol{\theta}$ is given by

$$\theta(1 - e^{-\theta})^2(1 - e^{-\theta} - \theta e^{-\theta})^{-1}n^{-1}. \tag{49}$$

A method of estimation of θ was proposed by McKendrick [58] as early as 1926. In this method, the available sample of size n is regarded as part of an original sample of size $n + n_0$, including n_0 observed values each equal to zero. For Poisson variables (as in (1)), the first two factorial moments are θ and θ^2.

Using this, a first estimate of $n + n_0$ is

$$n + \hat{\mathbf{n}}_0^{(1)} = \left(\sum_{j=1}^{n} \mathbf{x}_j\right)^2 \left(\sum_{j=1}^{n} \mathbf{x}_j(\mathbf{x}_j - 1)\right)^{-1}, \tag{50}$$

and a first estimate of θ is

$$\hat{\boldsymbol{\theta}}^{(1)} = \left(\sum_{j=1}^{n} \mathbf{x}_j\right)(n + \hat{\mathbf{n}}_0^{(1)})^{-1}. \tag{51}$$

The value of n_0 is now estimated as

$$\hat{\mathbf{n}}_0^{(2)} = (n + \hat{\mathbf{n}}_0^{(1)})e^{-\hat{\boldsymbol{\theta}}^{(1)}}, \tag{52}$$

and a second estimate of θ is

$$\hat{\boldsymbol{\theta}}^{(2)} = \left(\sum_{j=1}^{n} \mathbf{x}_j\right)(n + \hat{\mathbf{n}}_0^{(2)})^{-1}. \tag{53}$$

The process proceeds iteratively. This method is of considerable historical interest, and not difficult to apply, but it is not generally used at present.

David and Johnson [27] have suggested an estimator based on the sample moments

$$\mathbf{T}_r = n^{-1} \sum_{j=1}^{n} \mathbf{x}_j^r \qquad (r = 1,2,\ldots,). \tag{54}$$

This is

$$\hat{\boldsymbol{\theta}}_{(1)} = \mathbf{T}_2\mathbf{T}_1^{-1} - 1. \tag{55}$$

Its variance is approximately

$$(\theta + 2)(1 - e^{-\theta})n^{-1}. \tag{56}$$

It has an asymptotic efficiency (relative to the maximum likelihood estimator, $\hat{\boldsymbol{\theta}}$) which reaches its minimum value of about 70% between $\theta = 2.5$ and $\theta = 3$. The efficiency increases to 100% as θ increases (David and Johnson [27]).

The unbiased estimator, n^{-1} [sum of all $\mathbf{x}_j$'s for which $\mathbf{x}_j \geq 2$] has efficiency over 90% (see also (59) et seq. below).

Tate and Goen [97] have shown that the minimum variance unbiased estimator of θ is

$$\hat{\boldsymbol{\theta}}^* = n\bar{\mathbf{x}}\mathfrak{S}_{n\bar{\mathbf{x}}-1}^{(n)}/\mathfrak{S}_{n\bar{\mathbf{x}}}^{(n)} = \bar{\mathbf{x}}(1 - \mathfrak{S}_{n\bar{\mathbf{x}}-1}^{(n-1)}/\mathfrak{S}_{n\bar{\mathbf{x}}}^{(n)})$$

where $\mathfrak{S}_a^{(b)}$ is the Stirling number of the second kind, defined in Chapter 1, Section 2. There are tables of the multiplier of $\bar{\mathbf{x}}$ in [97]. For large n, the multiplier is approximately $(1 - (1 - n^{-1})^{n\bar{\mathbf{x}}-1})$.

Minimum variance unbiased estimators are also given in [97] for Poisson distributions truncated by omission of values less than or equal to $c > 0$. Maximum likelihood estimators based on censored samples from truncated

Poisson distributions have been discussed by Murakami [66].

If the first r_1 values $(0,1,\ldots,(r_1 - 1))$ are omitted, then (1) becomes a "left truncated" Poisson distribution,

$$\Pr[\mathbf{x} = k] = [e^{-\theta}\theta^k/k!]\left[1 - e^{-\theta}\sum_{j=0}^{r_1-1}(\theta^j/j!)\right]^{-1} \qquad (57)$$

$$(k = r_1, r_1 + 1,\ldots).$$

The maximum likelihood estimator, $\hat{\boldsymbol{\theta}}_{(r_1,\phi)}$ of θ, given values of n independent random variables $\mathbf{x}_1, \mathbf{x}_2, \ldots, \mathbf{x}_n$ each distributed as in (57), satisfies the equation.

$$\bar{\mathbf{x}} = \left[\hat{\boldsymbol{\theta}}_{(r_1,\phi)} - e^{-\hat{\boldsymbol{\theta}}_{(r_1,\phi)}}\sum_{j=1}^{r-1}\{\hat{\boldsymbol{\theta}}^j_{(r_1,\phi)}/(j-1)!\}\right] \qquad (58)$$

$$\times\left[1 - e^{-\hat{\boldsymbol{\theta}}_{(r_1,\phi)}}\sum_{j=0}^{r_1-1}(\hat{\boldsymbol{\theta}}^j_{(r_1,\phi)}/j!)\right]^{-1}$$

As an initial value of $\hat{\boldsymbol{\theta}}$ to use in iterative solution of (58) the simple estimator

$$n^{-1}\,[\text{sum of all } \mathbf{x}_j\text{'s for which } \mathbf{x}_j > r_1] \qquad (59)$$

(proposed by Moore [61]) might be used. For the case $r_1 = 1$ (the positive Poisson of (41)) this estimator has variance

$$\theta(1 + \theta(e^{\theta} - 1)^{-1})n^{-1}, \qquad (60)$$

(Plackett [73]) and its efficiency is quite high (over 90%). Rider [80] constructed the estimator

$$[\mathbf{T}_2 - r_1\mathbf{T}_1][\mathbf{T}_1 - n(r_1 - 1)]^{-1} \quad \text{(see (54) for definition of } \mathbf{T}_1, \mathbf{T}_2\text{)}, \qquad (61)$$

and compared it with the maximum likelihood estimator. There seemed no clear advantage to either estimator.

Rao and Rubin [78] gave a characterization of the left-truncated Poisson distribution analogous to that (Section 5) for the complete Poisson distribution.

'Right-truncation' (omission of values exceeding a specified value, r_2h) can occur if the counting mechanism is unable to deal with large numbers, though it is more usually the case that the *existence* of these high values is known, even if their exact magnitude is not. In this case, if there are n' values $>r_2$, the maximum likelihood estimator, $\hat{\boldsymbol{\theta}}$, satisfies the equation (Tippett [100]) (sum of values $\leq r_2$) $= \hat{\boldsymbol{\theta}}\left[N - n'\left[1 + (\hat{\boldsymbol{\theta}}^{r_2}/r_2!)\left\{\sum_{j=r_2+1}^{\infty}(\hat{\boldsymbol{\theta}}^j/j!)\right\}^{-1}\right]\right]$. If, however, this information is not available then it is appropriate to use the *right truncated Poisson distribution*

$$\Pr[\mathbf{x} = k] = [\theta^k/k!]\left[\sum_{j=0}^{r_2}\{\theta^j/j!\}\right]^{-1} \qquad (k = 0,1,\ldots,r_2) \qquad (62)$$

(The factor $e^{-\theta}$ is omitted from both numerator and denominator.)

If $\mathbf{x}_1, \mathbf{x}_2, \ldots, \mathbf{x}_n$ are independent random variables each with distribution (62), then the maximum likelihood estimator $\hat{\boldsymbol{\theta}}_{(0,r_2)}$ of θ satisfies the following equation

$$\sum_{j=0}^{r_2} (\bar{\mathbf{x}} - j) \frac{\hat{\boldsymbol{\theta}}^j_{(0,r_2)}}{j!} = 0. \tag{63}$$

The solution of this equation for $\hat{\boldsymbol{\theta}}_{(0,r_2)}$ is a function of $\bar{\mathbf{x}}$ alone, and not of n. Cohen [21] has provided tables from which $\hat{\boldsymbol{\theta}}_{(0,r_2)}$ can be obtained, given r_2 and $\bar{\mathbf{x}}$.

When the tables are insufficient, a solution may be obtained by interpolation. Equation (63) can be written in the form

$$\bar{\mathbf{x}} = \left[\hat{\boldsymbol{\theta}}_{(0,r_2)} \sum_{j=0}^{r_2-1} \{e^{-\hat{\boldsymbol{\theta}}_{(0,r_2)}} \hat{\boldsymbol{\theta}}^j_{(0,r_2)}/j!\}\right] \Big/ \left[\sum_{j=0}^{r_2} e^{-\hat{\boldsymbol{\theta}}_{(0,r_2)}} \hat{\boldsymbol{\theta}}^j_{(0,r_2)}/j!\right] \tag{64}$$

and appropriate tables (see Section 8) used to evaluate the sums on the right hand side of (64) for tabulated values of $\hat{\boldsymbol{\theta}}_{(0,r_2)}$.

The asymptotic variance of $\hat{\boldsymbol{\theta}}_{(0,r_2)}$ is

$$\theta \psi_{r_2}(\theta) n^{-1} \tag{65}$$

where

$$\psi_{r_2}(\theta) = \left[\sum_{j=0}^{r_2} (\theta^j/j!)\right]^2 \left[\left\{\sum_{j=0}^{r_2-1} (\theta^j/j!)\right\} \left\{\sum_{j=0}^{r_2-1} (\theta^j/j!) + \theta^{r_2+1}/r_2!\right\} - (\theta^{r_2+1}/r_2!) \sum_{j=0}^{r_2} (\theta^j/j!)\right]^{-1}.$$

Moore [61] suggests the simple estimator (analogous to (59))

$$\frac{\text{Sum of } \mathbf{x}_j\text{'s}}{\text{Number of } \mathbf{x}_j\text{'s which are less than } (r_2 - 1)}. \tag{66}$$

This is an unbiased estimator of θ.

The doubly truncated Poisson distribution is the distribution of a random variable $\mathbf{x}$ for which

$$\Pr[\mathbf{x} = k] = [\theta^k/k!] \Big/ \left[\sum_{j=r_1}^{r_2} \{\theta^j/j!\}\right] \tag{67}$$

$$(j = r_1, r_1 + 1, \ldots, r_2; r_2 \geq r_1 > 0).$$

If $\mathbf{x}_1, \mathbf{x}_2, \ldots, \mathbf{x}_n$ are independent random variables each with distribution (67), the maximum likelihood estimator, $\hat{\boldsymbol{\theta}}_{(r_1,r_2)}$ of θ satisfies the equation

$$\bar{x} = \left[\sum_{j=r_1}^{r_2} \{j\hat{\boldsymbol{\theta}}^j_{(r_1,r_2)}/j!\}\right] \Big/ \left[\sum_{j=r_1}^{r_2} \{\hat{\boldsymbol{\theta}}^j_{(r_1,r_2)}/j!\}\right]. \tag{68}$$

This can be solved by interpolation as described for (63). (The numerator and denominator of (68) can each be multiplied by $\exp(-\hat{\boldsymbol{\theta}}_{(r_1,r_2)})$, and the resulting numerator written as $\hat{\boldsymbol{\theta}}_{(r_1,r_2)}$ times the sum of Poisson probabilities between $(r_1 - 1)$ and $(r_2 - 1)$ inclusive.

Moore [61] suggested the statistic

$$(69) \qquad \frac{\text{Sum of } \mathbf{x}_j\text{'s for which } \mathbf{x}_j \geq r_1 + 1}{\text{Number of } \mathbf{x}_j\text{'s which are less than } r_2}$$

which is an unbiased estimator of θ.

Doss [28] compared the efficiency of the maximum likelihood estimators $\hat{\boldsymbol{\theta}}$ (as in (46) and (47)) and $\hat{\boldsymbol{\theta}}_{(r_1,r_2)}$, of θ for distributions (1) and (67) respectively. He showed that

$$(70) \qquad \frac{\operatorname{var}(\hat{\boldsymbol{\theta}} \mid (1))}{\operatorname{var}(\hat{\boldsymbol{\theta}}_{(r_1,r_2)} \mid (67))} \doteqdot 1 - \frac{W(r_1,r_2,\theta)}{\left[\sum_{j=r_1}^{r_2} \{\theta^j/j!\}\right]^2}$$

where

$$W(r_1,r_2,\theta) = \theta[\{\theta^{r_2-1}/(r_2-1)!\} - \{\theta^{r_1}/r_1!\}] \sum_{j=r_1-1}^{r_2-1} \{\theta^j/j!\}$$
$$- \left[\{\theta^{r_2-1}/(r_2-1)!\} \frac{\theta^{r_2-1}r_2}{(r_2-1)!} - \frac{\theta^{r_1}(r_1+1)}{r_1!} - \{\theta^{r_1}/r_1!\}\right.$$
$$\left. + \{\theta^{r_2-1}/(r_2-2)!\} - \{\theta^{r_1}/(r_1-1)!\}\right] \sum_{j=r_1}^{r_2} \{\theta^j/j!\}.$$

$W(r_1,r_2,\theta)$ is always positive so $\operatorname{var}(\hat{\boldsymbol{\theta}} \mid (1)) < \operatorname{var}(\hat{\boldsymbol{\theta}}_{(r_1,r_2)} \mid (67))$ according to approximation (70). Of course, comparisons of this kind are relevant only when it is possible to control the method of observation to produce distributions of type (1) or (67) with little difference in cost per observation. (Swamy [96] has made similar comparisons for distributions other than Poisson, and has found cases where the corresponding ratio of variances is greater than one.)

10.2 *The 'Generalized Poisson Distribution'*

This distribution attempts to take into account errors in recording a variable which in reality does have a Poisson distribution. (cf. Rao and Rubin [78], quoted in Section 5). The distribution is defined by the equations

$$(71) \qquad \begin{aligned} \Pr[\mathbf{x} = 0] &= e^{-\theta}(1 + \theta\lambda) \\ \Pr[\mathbf{x} = 1] &= \theta e^{-\theta}(1 - \lambda) \\ \Pr[\mathbf{x} = j] &= \theta^j e^{-\theta}/j! \qquad (j \geq 2). \end{aligned}$$

It corresponds to a situation in which values of a variable with distribution (1) are recorded correctly, except that, when the true value is 1, there is a probability λ that it will be recorded as zero. Cohen [18] obtained the following formulas for maximum likelihood estimators of θ and λ respectively:

(72.1) $$\tfrac{1}{2}[\bar{\mathbf{x}} - 1 + \mathbf{f}_0 + \{(\bar{\mathbf{x}} - 1 + \mathbf{f}_0)^2 + 4(\bar{\mathbf{x}} - \mathbf{f}_1)\}^{\frac{1}{2}}]$$

(72.2) $$(\mathbf{f}_0 - \mathbf{f}_1\,[\text{estimator of } \theta]^{-1})(\mathbf{f}_0 + \mathbf{f}_1)^{-1}$$

where $\bar{\mathbf{x}}$ is the sample mean, and $\mathbf{f}_0$, $\mathbf{f}_1$, are the proportions of the observed values equal to zero and one. The asymptotic variances of (72.1) and (72.2) are, respectively

(73.1) $$\theta(1 + \theta)(1 + \theta - e^{-\theta})^{-1}n^{-1}$$

(73.2) $$(1 + \lambda\theta - \lambda e^{-\theta})(1 - \lambda)(\theta e^{-\theta})^{-1}(1 + \theta - e^{-\theta})^{-1}n^{-1}$$

where n denotes the sample size.

The asymptotic correlation between the estimators is

(74) $$[(1 - \lambda)e^{-\theta}(1 - \theta)^{-1}(1 + \lambda\theta - \lambda e^{-\theta})^{-1}]^{\frac{1}{2}}.$$

Cohen [19] also considered the case when a true value of $(c + 1)$ is sometimes reported as c.

Shah and Venkataraman [88] constructed estimators of θ and λ, for distribution (71), based on sample moments. These are, respectively,

(75.1) $$[\mathbf{T}_2 - \mathbf{T}_1]^{\frac{1}{2}}$$

(75.2) $$[1 - \mathbf{T}_1(\mathbf{T}_2 - \mathbf{T}_1)^{-\frac{1}{2}}][\exp[\mathbf{T}_2 - \mathbf{T}_1]^{\frac{1}{2}}].$$

The asymptotic variances of (75.1) and (75.2) are, respectively,

(76.1) $$(\theta + \tfrac{1}{2})n^{-1}$$

(76.2) $$[A(\theta e^{-\theta})^2 + 4(B\theta)^3 + 2(B\theta)^2 - 2B\theta(\lambda\theta e^{-\theta} + 2)e^{\theta}]n^{-1}$$

where $$A = \theta(2\lambda\theta e^{-\theta} + 1) - \lambda\theta e^{-\theta}(\lambda\theta e^{-\theta} + 1)$$

$$B = \tfrac{1}{2}[\lambda\theta^{-1} + (1 - \lambda e^{-\theta})e^{\theta}\theta^{-2}].$$

These authors also give formulas for the moments of the distribution.

Another modification of the Poisson distribution is derived by recording all values of $\mathbf{x}$ that are greater than a certain value, K, as that value. The resulting distribution is

$$\Pr[\mathbf{x} = k] = e^{-\theta}\theta^k/k! \qquad (k < K)$$

$$\Pr[\mathbf{x} = K] = \sum_{j=K}^{\infty} [e^{-\theta}\theta^j/j!].$$

Newell [67] has applied this distribution to the number of hospital beds occupied when K is the total number of beds available.

10.3 *Mixtures of Poisson Distributions*

If, in (1), the parameter θ is replaced by a random variable, $\boldsymbol{\theta}$, then a new distribution, with

$$\text{(77)} \qquad \Pr[\mathbf{x} = k] = (k!)^{-1}E(\boldsymbol{\theta}^k e^{-\boldsymbol{\theta}}) \qquad (k = 0,1,2,\ldots,) \text{ is obtained.}$$

These are called *compound Poisson* distributions, and are studied in Chapter 8. Here we describe a few special cases.

If $\boldsymbol{\theta}$ has a discrete distribution taking K values $\theta_1, \theta_2, \ldots, \theta_K$ with probabilities $p_1, p_2, \ldots, p_K$ respectively $\left(\sum_{j=1}^{K} p_j = 1\right)$, then $\mathbf{x}$ has a *K-component Poisson mixture* distribution. If $K = 2$ then $p_1 + p_2 = 1$ and

$$\text{(78)} \qquad \mu_{(r)} = E(\mathbf{x}^{(r)}) = p_1\theta_1^r + p_2\theta_2^r.$$

Putting $r = 1$, 2 and 3 and solving for p_1, θ_1 and θ_2 we obtain three equations, from which p_1, θ_1 and θ_2 can be expressed in terms of $\mu_{(1)}$, $\mu_{(2)}$ and $\mu_{(3)}$.

If $\mathbf{T}_{(r)} = n^{-1} \sum_{j=1}^{n} \mathbf{x}_j(\mathbf{x}_j - 1)(\mathbf{x}_j - 2)\ldots(\mathbf{x}_j - r + 1)$ the roots (in θ) of the equation

$$\text{(79)} \qquad (\mathbf{T}_{(2)} - \mathbf{T}_{(1)}^2)\theta^2 - (\mathbf{T}_{(3)} - \mathbf{T}_{(1)}\mathbf{T}_{(2)})\theta + \mathbf{T}_{(1)}\mathbf{T}_{(3)} - \mathbf{T}_{(2)}^2 = 0$$

may be used as estimators of θ_1 and θ_2. (See Jones [49] and Rider [81].)

Cohen [22] gives a systematic description of these matters and also describes how to fit mixtures of two positive Poisson distributions. This distribution is defined by

$$\Pr[\mathbf{x} = k] = pe^{-\theta_1}(1 - e^{-\theta_1})^{-1}(\theta_1^k/k!) + (1 - p)e^{-\theta_2}(1 - e^{-\theta_2})^{-1}(\theta_2^k/k!) \qquad (k = 1,2,\ldots)$$

and now

$$\mu_{(r)} = p(1 - e^{-\theta_1})^{-1}\theta_1^r + (1 - p)(1 - e^{-\theta_2})^{-1}\theta_2^r.$$

Putting $r = 1$, 2, 3 and eliminating p we obtain

$$\text{(80.1)} \qquad \theta_1 = (\mu_{(3)} - \theta_2\mu_{(2)})/(\mu_{(2)} - \theta_2\mu_1')$$

and

$$\text{(80.2)} \qquad \frac{\mu_1' - \theta_2(1 - e^{-\theta_2})^{-1}}{\theta_1(1 - e^{-\theta_1})^{-1} - \theta_2(1 - e^{-\theta_2})^{-1}} = \frac{\mu_{(2)} - \theta_2^2(1 - e^{-\theta_2})^{-1}}{\theta_1^2(1 - e^{-\theta_1})^{-1} - \theta_2^2(1 - e^{-\theta_1})^{-1}}.$$

Replacing θ_1 in (80.2) by its expression in terms of θ_2 in (80.1) an equation in θ_2 is obtained. If μ_1', $\mu_{(2)}$ and $\mu_{(3)}$ be replaced by $\mathbf{T}_{(1)}$, $\mathbf{T}_{(2)}$, $\mathbf{T}_{(3)}$ respectively, the solution of the resulting equation is the estimator of θ_2 proposed by Cohen. Estimation of p and θ_1 then follows directly.

Cohen [22] also obtains estimators for the parameters p', θ, p of the mixture of Poisson and binomial distributions

$$\Pr[\mathbf{x} = k] = p'e^{-\theta}(\theta^k/k!) + (1 - p')\binom{N}{k} p^k(1 - p)^{N-k} \qquad (k = 0,1,\ldots).$$

(N is assumed to be known.)

For this distribution

$$\mu_{(r)} = p'\theta^r + (1 - p')N(N - 1)\ldots(N - r + 1)p^r$$

from which can be obtained the following simultaneous equations for θ and p

$$(81.1) \qquad \mu_{(2)}(\theta - Np) - Np\theta[(N - 1)p - \theta] = \mu_1'[\theta^2 - N(N - 1)p^2]$$

$$(81.2) \qquad \mu_{(3)}(\theta - Np) - Np\theta[(N - 1)(N - 2)p^2 - \theta^2] = \mu_1'[\theta^3 - N(N - 1)(N - 2)p^3].$$

If $\boldsymbol{\theta}$ has a continuous gamma distribution then $\mathbf{x}$ has a negative binomial distribution (see Chapter 5, Section 4). Other distributions for $\boldsymbol{\theta}$ can also be assumed but are not at all generally used.

The '*stuttering*' *Poisson* distribution described by Adelson [1] is not a mixture, but is the distribution of the sum $\left(\mathbf{x} = \sum_{j-1}^{m} \mathbf{x}_j\right)$ of m independent random variables, $\mathbf{x}_1, \mathbf{x}_2, \ldots, \mathbf{x}_m$, with

$$(82) \qquad \Pr[\mathbf{x}_j = jk] = e^{-\theta_j}\theta_j^k/k! \qquad (k = 0,1,2,\ldots; j = 1,\ldots)$$

Adelson [1] has shown that the probability generating function of this distribution is

$$(83) \qquad \exp\left(-\sum_{j=1}^{m} \theta_j\right) \exp\left(\sum_{j=1}^{m} \theta_j t^j\right)$$

and that, denoting $\Pr[\mathbf{x} = j]$ by P_j, there is a recurrence relation

$$(84) \qquad P_{j+1} = (j + 1)^{-1} \sum_{i=0}^{j} (i + 1)\theta_{i+1}P_{j-i}$$

with $P_0 = \exp\left(-\sum_{j=1}^{m} \theta_j\right)$. (Note that Adelson calls the distribution 'compound Poisson', but that it is not included in our definition in (77).)

The term 'stuttering Poisson' was originally used by Galliher (see Morse [64] and Galliher et al. [35]) to describe a mixture of Poisson distributions with expected values $\mathbf{n}\theta$, where $\mathbf{n}$ has a geometric distribution (see Chapter 5, Section 2). This is a generalized Poisson distributed of the kind defined in Chapter 8, Section 8. We have preferred to apply the term 'stuttering Poisson' to the class of distributions defined in [1].

10.4 *Displaced Poisson Distribution*

Staff [92] defines the *displaced Poisson distribution* as a *left-truncated* Poisson distribution which has been "displaced" by subtraction of a constant so that the lowest value that is taken by the variable is zero. Thus the left-truncated distribution

$$(85)\qquad \Pr[\mathbf{x} = k] = (\theta^k/k!)\Big/\left[\sum_{j=r+1}^{\infty} (\theta^j/j!)\right] \qquad (k = r+1, r+2,\ldots)$$

corresponds to the displaced distribution

$$(86)\qquad \Pr[\mathbf{x} = k] = (\theta^{k+r+1}/(k+r+1)!)\left[\sum_{j=r+1}^{\infty} (\theta^j/j!)\right]^{-1} \qquad (k = 0,1,2,\ldots).$$

If r is known, estimation of θ is fairly straightforward. Staff also considers the situation in which both r and θ have to be estimated. The simplest formulas he obtains are for estimators based on the sample mean ($\bar{\mathbf{x}}$), standard deviation ($\mathbf{s}$) and proportion in the zero class ($\mathbf{f}_0$). They are

$$(87)\qquad \begin{aligned} \mathbf{r}^* &= (\mathbf{s}^2 - \bar{\mathbf{x}})[1 - \mathbf{f}_0(1 + \bar{\mathbf{x}})]^{-1} \\ \boldsymbol{\lambda}^* &= \bar{\mathbf{x}} + \mathbf{r}^*(1 - \mathbf{f}_0). \end{aligned}$$

If r is less than 8, the efficiency of $\mathbf{r}^*$ (relative to the maximum likelihood estimator) exceeds 70% (for example, if $r = 2$ then the efficiency of $\mathbf{r}^*$ exceeds 89%).

10.5 *A Generalization of the Poisson Law*

In Section 1 it was shown that if the times between successive events have a common exponential distribution, then the total number of events in a fixed time T has a Poisson distribution. Morlat [63] constructed a 'generalization of the Poisson law' by supposing that the common distribution of the times (t) is a general gamma distribution with origin at zero (Chapter 17). If the common probability density function is

$$\{\Gamma(\alpha)\}^{-1} t^{\alpha-1} e^{-t} \qquad (t \geq 0)$$

then the probability of k events in time t is

$$(88)\qquad P_\alpha(k) = \{\Gamma(k\alpha)\}^{-1} \int_0^T [1 - y^\alpha \Gamma(k\alpha)/\Gamma((k+1)\alpha)] y^{k\alpha-1} e^{-y}\, dy.$$

If α is an integer, we have the remarkable formula

$$(89)\qquad P_\alpha(k) = \sum_{j=k\alpha}^{(k+1)\alpha-1} e^{-T}(T^j/j!).$$

11. Relation with Other Distributions

The limiting distribution of the standardized Poisson variable $(\mathbf{x} - \theta)\theta^{-\frac{1}{2}}$, where $\mathbf{x}$ has distribution (1), is a unit normal distribution. That is

$$(90) \qquad \lim_{\theta\to\infty} \Pr[\alpha < (\mathbf{x} - \theta)\theta^{-\frac{1}{2}} < \beta] = (2\pi)^{-\frac{1}{2}} \int_\alpha^\beta e^{-u^2/2}\, du.$$

We have already noted that the Poisson distribution is itself a limiting distribution of a sequence of binomial distributions.

Uhlmann [101] showed that, for $0 \le c < N - 1$,

$$e^{-Np} \sum_{j=0}^{c} \frac{(Np)^j}{j!} > \sum_{j=0}^{c} \binom{N}{j} p^j(1-p)^{N-j}$$

indicating that the binomial cumulative distribution is the greater, except possibly for $c = N - 1$ (as can be seen by taking $c = 1, N = 2$).

The Poisson distribution is basic in the genesis of contagious distributions (Chapter 9; see also Chapter 8 for compound Poisson distributions).

Apart from these direct probabilistic relations between the Poisson and other discrete distributions, there is an interesting formal relation between the Poisson distribution and certain gamma distributions.

If $\mathbf{y}$ has probability density function

$$(91) \qquad p_{\mathbf{y}}(y) = \frac{1}{(\alpha-1)!} y^{\alpha-1}e^{-y} \qquad (0 < y)$$

and α is a positive integer, then

$$(92) \qquad \Pr[\mathbf{y} > Y] = \frac{1}{(\alpha-1)!} \int_Y^\infty y^{\alpha-1}e^{-y}\, dy$$

$$= \frac{1}{(\alpha-1)!} Y^{\alpha-1}e^{-Y} + \frac{1}{(\alpha-2)!} \int_Y^\infty y^{\alpha-2}e^{-y}\, dy$$

(integrating by parts)

$$= \frac{1}{(\alpha-1)!} Y^{\alpha-1}e^{-Y} + \frac{1}{(\alpha-2)!} Y^{\alpha-2}e^{-Y} + \cdots$$
$$\cdots + Ye^{-Y} + e^{-Y}$$

= probability that a random variable having a Poisson distribution with expected value Y is less than α.

In particular, putting $\alpha = \frac{1}{2}\nu$ (ν is an even integer) and $Y = \frac{1}{2}x$

(93) $\qquad \Pr[\chi_\nu^2 > x] =$ probability that a random variable having a Poisson distribution with expected value $\frac{1}{2}x$ is less than $\frac{1}{2}\nu$.

The distribution of the difference between two independent random variables, each having a Poisson distribution has attracted some attention. Strackee and van der Gon [94] state "In a steady state the number of light quanta, emitted or absorbed in a definite time, is distributed according to a Poisson distribution. In view thereof, the physical limit of perceptible contrast in vision can be studied in terms of the difference between two independent variates each following a Poisson distribution." The distribution of differences may be relevant when a physical effect is estimated as the difference between two counts, one when a 'cause' is acting, and the other a 'control' to estimate 'background effect'.

Irwin [45] studied the case when the two variables, $\mathbf{x}_1$ and $\mathbf{x}_2$ each have the same expected value, θ. Evidently, for $y \geq 0$

$$(94) \qquad \begin{aligned} \Pr[\mathbf{x}_1 - \mathbf{x}_2 = y] &= e^{-2\theta} \sum_{j=y}^{\infty} \theta^{j+(j-y)}[j!(j-y)!]^{-1} \\ &= e^{-2\theta} I_{\frac{1}{2}y}(2\theta) \end{aligned}$$

where $I(\cdot)$ is a modified Bessel function. (By symmetry $\Pr[\mathbf{x}_1 - \mathbf{x}_2 = y] = \Pr[\mathbf{x}_2 - \mathbf{x}_1 = y]$.) Skellam [89] and de Castro [11] discussed the problem when $E[\mathbf{x}_1] = \theta_1 \neq E[\mathbf{x}_2] = \theta_2$. In this case, for $y > 0$

$$(95) \qquad \begin{aligned} \Pr[\mathbf{x}_1 - \mathbf{x}_2 = y] &= e^{-(\theta_1+\theta_2)} \sum_{j=y}^{\infty} \theta_1^j \theta_2^{j-y}[j!(j-y)!]^{-1} \\ &= e^{-(\theta_1-\theta_2)}(\theta_1/\theta_2)^{\frac{1}{2}y} I_{\frac{1}{2}y}(2\sqrt{\theta_1\theta_2}). \end{aligned}$$

Strackee and van der Gon [94] give tables of the cumulative probability $\Pr[\mathbf{x}_1 - \mathbf{x}_2 \leq y]$ to four decimal places for the following combinations of values of θ_1 and θ_2:

θ_1	$\frac{1}{4}$	1	4	$\frac{1}{2}$	1	2	4	8	1	2	4	2	4	8	4	8
θ_2	$\frac{1}{4}$	$\frac{1}{4}$	$\frac{1}{4}$	$\frac{1}{2}$	$\frac{1}{2}$	$\frac{1}{2}$	$\frac{1}{2}$	$\frac{1}{2}$	1	1	1	2	2	2	4	8

The differences between the normal approximation (see Fisz [31]):

$$(96) \qquad \Pr[\mathbf{x}_1 - \mathbf{x}_2 \leq y] \doteqdot [2\pi(\theta_1 + \theta_2)]^{-\frac{1}{2}} \int_{-\infty}^{y+\frac{1}{2}} \exp\left[-\frac{(\theta_1 - \theta_2 - y)^2}{2(\theta_1 + \theta_2)}\right] dy$$

and the tabled values are also shown.

Ractliffe [75] carried out a Monte Carlo experiment to assess the accuracy with which the distribution of $(\mathbf{x}_1 - \mathbf{x}_2)(\mathbf{x}_1 + \mathbf{x}_2)^{-\frac{1}{2}}$ is represented by a unit normal distribution for the special case $\theta_1 = \theta_2 = \theta$. He obtained 4,000 sample values for each of $\theta = 5, 10, 15, 30, 55, 80, 100$ and 130, and an extra 12,000 values for $\theta = 15$. For these values of θ the approximation seemed to be adequate.

The cumulants of the distribution of $(\mathbf{x}_1 - \mathbf{x}_2)$ are easily written as

$$(97) \qquad \kappa_r(\mathbf{x}_1 - \mathbf{x}_2) = \kappa_r(\mathbf{x}_1) + (-1)^r \kappa_r(\mathbf{x}_2) = \theta_1 + (-1)^r \theta_2.$$

Romani [82] has discussed the properties of the maximum likelihood estimator of $E[\mathbf{x}_1 - \mathbf{x}_2] = \theta_1 - \theta_2$.

REFERENCES

[1] Adelson, R. M. (1966). Compound Poisson distributions, *Operations Research Quarterly*, **17,** 73–75.

[2] Anscombe, F. J. (1948). The transformation of Poisson, binomial, and negative binomial data, *Biometrika*, **35,** 246–254.

[3] Barton, D. E. (1961). Unbiased estimation of a set of probabilities, *Biometrika*, **48,** 227–229.

[4] Bohman, H. (1963). Two inequalities for Poisson distributions, *Skandinavisk Aktuarietidskrift*, **46,** 47–52.

[5] Bol'shev, L. N. (1962). Comparison of parameters of Poisson distributions *Teoriya Veroyatnostei i ee Primeneniya*, **7,** 113–114. (In Russian)

[6] Bol'shev, L. N. (1965). On a characterization of the Poisson distribution *Teoriya Veroyatnostei i ee Primeneniya*, **10,** 488–499. (In Russian)

[7] Bol'shev, L. N. and Smirnov, N. V. (1965). *Tables of Mathematical Statistics*, Moscow: Nauka.

[8] Bortkiewicz, L. von (1898). *Das Gesetz der Kleinen Zahlen*, Leipzig: Teubner.

[9] Breny, H. (1953). Sur une classe de fonctions aléatoires liées à la loi de Poisson, *Bulletin de la Societé Royale de Science, Liège*, **22,** 405–416.

[10] Burrington, R. S. and May, D. C. (1953). *Handbook of Probability and Statistics with Tables*, New York: McGraw-Hill, Inc.

[11] Castro, G. de (1952). Note on differences of Bernoulli and Poisson variables, *Portugaliae Mathematica*, **11,** 173–175.

[12] Chapman, D. G. (1952). On tests and estimates for the ratio of Poisson means, *Annals of the Institute of Statistical Mathematics, Tokyo*, **4,** 45–49.

[13] Chatterji, S. D. (1963). Some elementary characterizations of the Poisson distribution, *American Mathematical Monthly*, **70,** 958–964.

[14] Cheng, Tseng-Tung (1949). The normal approximation to the Poisson distribution and a proof of a conjecture of Ramanujan, *Bulletin of the American Mathematical Society*, **55,** 396–401.

[15] Cohen, A. C. (1954). Estimation of the Poisson parameter from truncated samples and from censored samples, *Journal of the American Statistical Association*, **49,** 158–168.

[16] Cohen, A. C. (1960). Estimating the parameter in a conditional Poisson distribution, *Biometrics*, **16,** 203–211.

[17] Cohen, A. C. (1960). Estimation in truncated Poisson distribution when zeros and some ones are missing, *Journal of the American Statistical Association*, **55,** 342–348.

[18] Cohen, A. C. (1960). Estimating the parameters of a modified Poisson distribution, *Journal of the American Statistical Association*, **55,** 139–144.

[19] Cohen, A. C. (1960). Estimation in the Poisson distribution when sample values of $c + 1$ are sometimes erroneously reported as c, *Annals of the Institute of Statistical Mathematics, Tokyo*, **9,** 189–193.

[20] Cohen, A. C. (1960). An extension of a truncated Poisson distribution, *Biometrics*, **16,** 446–450.

[21] Cohen, A. C. (1961). Estimating the Poisson parameter from samples that are truncated on the right, *Technometrics*, **3,** 433–438.

[22] Cohen, A. C. (1963). Estimation in mixture of discrete distributions, *Proceedings of the International Symposium on Discrete Distributions, Montreal*, 373–378.

[23] Cox, D. R. (1953). Some simple approximate tests for Poisson variates, *Biometrika*, **40,** 354–360.

[24] Crow, E. L. (1958). The mean deviation of the Poisson distribution, *Biometrika*, **45,** 556–559.

[25] Crow, E. L. and Gardner, R. S. (1959). Confidence intervals for the expectation of a Poisson variable, *Biometrika*, **46,** 444–453.

[26] Daboni, L. (1959). A property of the Poisson distribution, *Bolletino della Unione Matematica Italiana*, **14,** 318–320.

[27] David, F. N. and Johnson, N. L. (1952). The truncated Poisson, *Biometrics*, **8,** 275–285.

[28] Doss, S. A. D. C. (1963). On the efficiency of best asymptotically normal estimates of the Poisson parameter based on singly and doubly truncated or censored samples, *Biometrics*, **19,** 588–594.

[29] Duker, S. (1955). The Poisson distribution in educational research, *Journal of Experimental Education*, **23,** 263–269.

[30] Eisenhart, C. and Zelen, M. (1958). Elements of Probability, Chapter 12, *Handbook of Physics*, New York, N.Y.: McGraw-Hill, Inc., (Second revised edition 1967).

[31] Fisz, M. (1953). The limiting distribution of the difference of two Poisson random variables, *Zastosowania Matematyki*, **1,** 41–45. (In Polish)

[32] Feller, W. (1957). *An Introduction to Probability Theory and its Applications*, Vol. 1, 413–414. New York: John Wiley & Sons, Inc.

[33] Freeman, M. F. and Tukey, J. W. (1950). Transformations related to the angular and the square root, *Annals of Mathematical Statistics*, **21,** 607–611.

[34] Fry, T. C. (1928). *Probability and Its Engineering Uses*, Princeton, N.J.: D. Van Nostrand Co., Inc.

[35] Galliher, H. P., Morse, P. M. and Simond, M. (1959). Dynamics of two classes of continuous-review inventory systems, *Operations Research*, **7,** 362–384.

[36] Garwood, F. (1936). Fiducial limits for the Poisson distribution, *Biometrika*, **28,** 437–442.

[37] General Electric Company, Defense Systems Department (1962). *Tables of the Individual and Cumulative Terms of Poisson Distribution*, Princeton, N.J.: D. Van Nostrand Co., Inc.

[38] Glasser, G. J. (1962). Minimum variance unbiased estimators for Poisson probabilities, *Technometrics*, **4,** 409–418.

[39] Govindarajulu, Z. (1960). *Central limit theorems and asymptotic efficiency for one sample non-parametric procedures.* Technical Report No. 11. University of Minnesota Department of Statistics.

[40] Grab, E. L. and Savage, I. R. (1954). Tables of the expected value of $1/x$ for positive Bernoulli and Poisson variables, *Journal of the American Statistical Association*, **49,** 169–177.

[41] Hadley, G. and Whitin, T. M. (1961). Useful properties of the Poisson distribution, *Operations Research*, **9,** 408–410.

[42] Haight, F. A. (1966). A Poisson generalization based on counts that are delayed but not lost, *Operations Research*, **14,** 943–945.

[43] Haight, F. (1967). *Handbook of the Poisson Distribution*, New York: John Wiley & Sons, Inc.

[44] Hald, A. and Kousgaard, E. (1967). A table for solving the binomial equation $B(c,n,p) = P$, *Matematisk-fysiske Skrifter, Det Kongelige Danske Videnskabernes Selskab*, **3,** No. 4 (48 pp.)

[45] Irwin, J. O. (1937). The frequency distribution of the difference between two independent variates following the same Poisson distribution, *Journal of the Royal Statistical Society, Series A*, **100,** 415–416.

[46] Irwin, J. O. (1959). On the estimation of the mean of a Poisson distribution from a sample with the zero class missing, *Biometrics*, **15,** 324–326.

[47] Janko, J. (Ed.) (1958). *Statistical Tables*, (republished by Soviet Central Statistical Bureau, 1961) Prague.

[48] Johnson, N. L. (1951). Estimators of the probability of the zero class in Poisson and certain related populations, *Annals of Mathematical Statistics*, **22,** 94–101.

[49] Jones, H. G. (1933). A note on the *n*-ages method, *Journal of the Institute of Actuaries*, **64,** 318–324.

[50] Kalinin, V. M. (1967). Convergent and asymptotic expansions of probability distributions, *Teoriya Veroyatnostei i ee Primeneniya*, **12,** 24–38.

[51] Katti, S. K. (1960). Moments of the absolute difference and absolute deviation of discrete distributions, *Annals of Mathematical Statistics*, **31,** 78–85.

[52] Katz, L. (1950). On the relative efficiencies of BAN estimates, *Annals of Mathematical Statistics*, **21,** 398–405.

[53] Kitagawa, T. (1952). *Tables of Poisson Distribution*, Tokyo: Baifukan Publishing Company.

[54] Koopman, B. O. (1950). Necessary and sufficient conditions for Poisson's distribution, *Proceedings of the American Mathematical Society*, **1,** 813–823.

[55] Lachenbruch, P. A. (1968). *Simultaneous confidence limits for the binomial and Poisson distributions*, University of North Carolina, Institute of Statistics Mimeo Series No. 596.

[56] Makabe, H. and Morimura, H. (1955). A normal approximation to the Poisson distribution, *Reports on Statistical Application Research, JUSE*, **4,** 37–46.

[57] Mantel, N. (1962). (Appendix C to Haenzel, W., Loveland, D. B., and Sorken, M. B.) Lung cancer mortality as related to residence and smoking histories. I. White males, *Journal of the National Cancer Institution*, **28,** 947–997.

[58] McKendrick, A. G. (1926). Applications of mathematics to medical problems, *Proceedings of the Edinburgh Mathematical Society*, **44,** 98–130.

[59] Molina, E. C. (1942). *Poisson's Exponential Binomial Limit*, New York: D. Van Nostrand Co., Inc.

[60] Moore, P. G. (1952). The estimation of the Poisson parameter from a truncated distribution, *Biometrika*, **39,** 247–251.

[61] Moore, P. G. (1954). A note on truncated Poisson distributions, *Biometrics*, **10,** 402–406.

[62] Moran, P. A. P. (1952). A characteristic property of the Poisson distribution, *Proceedings of the Cambridge Philosophical Society*, **48,** 206–207.

[63] Morlat, G. (1952). Sur une généralisation de la loi de Poisson, *Comptes Rendus, Académie des Sciences, Paris, Series A*, **235,** 933–935.

[64] Morse, P. M. (1966). Solutions of a class of discrete-time inventory problems, *Operations Research*, **7,** 67–78.

[65] Mosteller, F. and Youtz, C. (1961). Tables of the Freeman-Tukey transformations for the binomial and Poisson distributions, *Biometrika*, **48,** 433–440.

[66] Murakami, M. (1961). Censored sample from truncated Poisson distribution, *Journal of the College of Arts and Sciences, Chiba University*, **3,** 263–268.

[67] Newell, D. J. (1965). Unusual frequency distributions, *Biometrics*, **21,** 159–168.

[68] Owen, D. B. (1962). *Handbook of Statistical Tables*, Reading, Mass: Addison-Wesley Publishing Co., Inc.

[69] Parzen, E. (1962). *Stochastic Processes with Applications to Science and Engineering*, San Francisco: Holden Day, Inc.

[70] Pearson, E. S., and Hartley, H. O. (1958). *Biometrika Tables for Statisticians*, **1,** London: Cambridge University Press. (Second edition)

[71] Pearson, E. S. and Hartley, H. O. (1950). Tables of the χ^2 integral and of the cumulative Poisson distribution, *Biometrika*, **37,** 313–325.

[72] Pearson, K. (Ed.) (1930). *Tables for Statisticians and Biometricians*, Part I, 3d ed., London: University College.

[73] Plackett, R. L. (1953). The truncated Poisson distribution, *Biometrics*, **9,** 485–488.

[74] Poisson, Simeon Denis (1837). *Recherches sur la Probabilité des Jugements en Matière Criminelle et en Matière Civile, Précédées des Regles Générales du Calcul des Probabilitiés.* Bachelier, Imprimeur-Libraire pour les Mathematiques, la Physique, etc.: Paris.

[75] Ractliffe, J. F. (1964). The significance of the difference between two Poisson variables. An experimental investigation, *Applied Statistics*, **13,** 84–86.

[76] Raikov, D. (1938). On the decomposition of Gauss' and Poisson's laws, *Izvestia Akademie Nauk SSSR, Series A*, 91–124. (In Russian)

[77] Rao, C. R. (1963). On discrete distributions arising out of methods of ascertainment, *Proceedings of the International Symposium on Discrete Distributions, Montreal*, 320–322.

[78] Rao, C. R. and Rubin, H. (1964). On a characterization of the Poisson distribution, *Sankhyā, Series A*, **26,** 295–298.

[79] Redheffer, R. M. (1953). A note on the Poisson law, *Mathematical Magazine*, **26,** 185–188.

[80] Rider, P. R. (1953). Truncated Poisson distributions, *Journal of the American Statistical Association*, **48,** 826–830.

[81] Rider, P. R. (1961). Estimating the parameters of mixed Poisson, binomial, and Weibull distributions by the method of moments, *Bulletin of the International Statistical Institute*, **38,** 1–8.

[82] Romani, J. (1956). Distribución de la suma algebraíca de variables de Poisson, *Trabajos de Estadística*, **7**, 175–181.

[83] Roy, J. and Tiku, M. L. (1962). Laguerre series approximations to the sampling distribution of variance, *Sankhyā, Series A*, **24**, 381–384.

[84] Rutherford, E., Chadwick, J. and Ellis, C. D. (1930). *Radiation from Radioactive Substances*, London: Cambridge University Press.

[85] Rutherford, E. and Geiger, H. (1910). The probability variations in the distribution of α particles, *Philosophical Magazine, 6th Series*, **20**, 698–704 (Note by H. Bateman, 704–707).

[86] Said, A. S. (1958). Some properties of the Poisson distribution, *Journal of the American Institute of Chemical Engineering* (*A.I.Ch.E. Journal*), **4**, 290–292.

[87] Sexton, C. R., Sexton, C. A. and Sexton, J. A. (1959). Table errata No. 270 (on Kitagawa tables), *Mathematical Tables and Aids to Calculation*, 141–142.

[88] Shah, B. V. and Venkataraman, V. K. (1962). A note on modified Poisson distribution, *Metron*, **22**, No. 3/4, 27–35.

[89] Skellam, J. G. (1946). The frequency distribution of the difference between two Poisson variates belonging to different populations, *Journal of the Royal Statistical Society, Series A*, **109**, 296.

[90] Society of Actuaries (1954). *Impairment Study*, 1951.

[91] Soper, H. E. (1914–1915). Tables of Poisson's exponential binomial limit, *Biometrika*, **10**, 25–35.

[92] Staff, P. J. (1964). The displaced Poisson distribution, *Australian Journal of Statistics*, **6**, 12–20.

[93] Stinson, P. J. and Walsh, J. E. (1965). An application of the Poisson approximation to naval aviation accident data, *Bulletin of the International Statistical Institute*, **41** (1), 379–380.

[94] Strackee, J. and van der Gon, J. J. D. (1962). The frequency distribution of the difference between two Poisson variates, *Statistica Neerlandica*, **16**, 17–23.

[95] 'Student' (1907). On the error of counting with a haemacytometer, *Biometrika*, **5**, 351–360.

[96] Swamy, P. S. (1962). On the amount of information supplied by censored samples of grouped observations in the estimation of statistical parameters, *Biometrika*, **49**, 245–249.

[97] Tate, R. F. and Goen, R. L. (1958). Minimum variance unbiased estimation for the truncated Poisson distribution, *Annals of Mathematical Statistics*, **29**, 755–765.

[98] Teicher, H. (1955). An inequality on Poisson probabilities, *Annals of Mathematical Statistics*, **26**, 147–149.

[99] Tiku, M. L. (1964). A note on the negative moments of a truncated Poisson variate, *Journal of the American Statistical Association*, **59**, 1220–1224.

[100] Tippett, L. H. C. (1932). A modified method of counting particles, *Proceedings of the Royal Society of London, Series A*, **137**, 434–436.

[101] Uhlmann, W. (1966). Vergleich der hypergeometrischen mit der Binomial Verteilung, *Metrika*, **10**, 145–158.

[102] Volodin, I. N. (1965). On distinguishing between Poisson and Pólya distributions on basis of a large number of samples, *Teoriya Veroyatnostei i ee Primeneniya*, **10,** 364–367. (In Russian)

[103] Wadsworth, G. P. and Bryan, J. G. (1960). *Introduction to Probability and Random Variables*, New York: McGraw-Hill, Inc.

[104] Wallis, W. A. (1936). The Poisson distribution and the Supreme Court, *Journal of the American Statistical Association*, **31,** 376–380.

[105] Walsh, J. E. (1955). The Poisson distribution as a limit for dependent binomial events with unequal probabilities, *Operations Research*, **3,** 198–204.

5

Negative Binomial Distribution

1. Definition

Formally, the negative binomial distribution can be defined in terms of the expansion of the negative binomial expression $(Q - P)^{-N}$ where $Q - P = 1$; $P > 0$. This is exactly similar to the definition of the binomial distribution in terms of the (positive) binomial expression $(q + p)^n$, where $q + p = 1$; $p > 0$; $q > 0$ and n is a positive integer.

The $(k + 1)$-th term in the expansion of $(Q - P)^{-N}$ is

$$Q^{-N}\binom{N+k-1}{N-1}(P/Q)^k = \binom{N+k-1}{N-1}(P/Q)^k(1 - P/Q)^N.$$

The negative binomial distribution with parameters N, P is defined as the distribution of a random variable, $\mathbf{x}$, for which

$$\text{(1)} \qquad \Pr[\mathbf{x} = k] = \binom{N+k-1}{N-1}(P/Q)^k(1 - P/Q)^N \qquad (k = 0,1,2,\ldots).$$

The parameter $M = NP$ (the expected value, see Section 4 below) is often used instead of P, giving the form

$$\text{(2)} \qquad \Pr[\mathbf{x} = k] = \binom{N+k-1}{N-1}(M/N)^k(1 + M/N)^{-(N+k)} \qquad (k = 0,1,2,\ldots).$$

Note that there is a non-zero probability for $\mathbf{x}$ taking any specified non-negative integer value, as in the Poisson distribution, but unlike the binomial distribution. N need not be an integer. When N is an integer, the distribution is sometimes called the *Pascal* distribution (Pascal [43]).

Other names for the negative binomial distribution are the 'binomial waiting-time distribution' and the 'Pólya distribution'.

2. Geometric Distribution

In the special case $N = 1$ the distribution is

$$\text{Pr}[\mathbf{x} = k] = Q^{-1}(P/Q)^k \qquad (k = 0,1,2,\ldots). \tag{3}$$

These values are in geometric progression, with common ratio (P/Q) and this distribution is called a *geometric distribution* (sometimes *Furry* [27] distribution). It is a discrete analogue of the exponential distribution (see Chapter 18). In fact, putting $P = \theta\delta$ (and $Q = 1 + \theta\delta$) then (keeping θ fixed)

$$\lim_{\delta\to 0} \sum_{k\geq T/\delta} \text{Pr}[\mathbf{x} = k] = e^{-\theta T}.$$

The geometric distribution possesses a property similar to the "non-ageing" (or "Markovian") property [(2), Section 2, Chapter 18] of the exponential distribution. This is

$$\begin{aligned}\text{Pr}[\mathbf{x} = X + k \mid \mathbf{x} \geq k] = \frac{Q^{-1}(P/Q)^{X+k}}{(P/Q)^k} &= Q^{-1}(P/Q)^X \\ &= \text{Pr}[\mathbf{x} = X].\end{aligned} \tag{4}$$

This property characterizes the geometric distribution (among all distributions restricted to the non-negative integers) just as the corresponding property characterizes the exponential distribution.

It is also possible to characterize any geometric distribution by the distribution of the absolute value of the difference between two independent random variables having the same geometric distribution (Puri [48]). Puri also showed that if the common distribution has parameter P, then the absolute difference distribution can be constructed as the distribution of the sum of two independent random variables, one distributed binomially with parameters $N = 1$, $p = P(1 + 2P)^{-1}$; and the other distributed geometrically with parameter P.

The geometric distribution may be extended to cover the case of a variable taking values $\theta_0, \theta_0 + \delta, \theta_0 + 2\delta, \ldots$ $(\delta > 0)$. Then, in place of (3), we have

$$\text{Pr}[\mathbf{x} = \theta_0 + k\delta] = Q^{-1}(P/Q)^k. \tag{5}$$

The characterization summarized in (4) also applies to this distribution with X replaced by $(\theta_0 + X\delta)$ and k replaced by $(\theta_0 + k\delta)$. A different characterization, demonstrated by Ferguson [24], shows that if $\mathbf{x}_1$, and $\mathbf{x}_2$ are independent, discrete random variables (each with positive standard deviation) then $\min(\mathbf{x}_1,\mathbf{x}_2)$ and $(\mathbf{x}_1 - \mathbf{x}_2)$ are mutually independent if and only if $\mathbf{x}_1$ and $\mathbf{x}_2$ both have geometric distributions with the same values of θ_0 and δ (but not necessarily the same values of P). Srivastava [60] has obtained a modified form of this characterization, applicable when n independent discrete random variables $\mathbf{x}_1, \mathbf{x}_2, \ldots, \mathbf{x}_n$ are available. The essential condition is that $\min(\mathbf{x}_1,\mathbf{x}_2,\ldots,\mathbf{x}_n)$

and $\sum_{i=1}^{n} [\mathbf{x}_i - \min(\mathbf{x}_1,\mathbf{x}_2,\ldots\mathbf{x}_n)] = n[\bar{\mathbf{x}} - \min(\mathbf{x}_1,\mathbf{x}_2,\ldots,\mathbf{x}_n)]$ be mutually independent.

Among special properties of the geometric distributions, it is of particular interest that if a mixture of negative binomial distributions (as in (1)) is formed by supposing **N** to have the geometric distribution

$$\text{(6)} \qquad \Pr[\mathbf{N} = k] = Q'^{-1}(P'/Q')^{k-1} \qquad (k = 1,2,\ldots)$$

(i.e. as in (5), with $\theta_0 = \delta = 1$) then the mixture distribution is also a geometric distribution of form (3) with Q replaced by $(QQ' - P')$ (Magistad [37]).

Margolin and Winokur [38] have given tables of the mean and variance of certain order statistics for random samples of sizes 5, 10, 15 and 20 from geometric distributions. The figures are given to two decimal places.

3. Historical Remarks

Special forms of the negative binomial distribution were discussed by Pascal [43] and Fermat. There is a derivation by Montmort in [39], published in 1714.

'Student' [61] in 1907, used the distribution as an alternative to the Poisson distribution in describing counts on the plates of a haemacytometer.

Greenwood and Yule [30] in 1920 obtained the distribution as a consequence of certain simple assumptions in accident proneness models, while Eggenberger and Pólya [23], in 1923, obtained the distribution as a limiting case of an 'urn-scheme'. (See also Section 4.)

Since this time there has been an increasing number of applications of the negative binomial, and an associated development of statistical techniques based on this distribution.

4. Genesis

(*a*) A simple model, leading to the negative binomial distribution, is that representing the number of independent trials necessary to obtain m occurrences of an event which has constant probability p of occurring at each trial.

If **x** be the random variable representing the necessary number of trials, then

$$\text{(7)} \qquad \Pr[\mathbf{x} = m + k] = \binom{m+k-1}{m-1} p^m(1-p)^k \qquad (k = 0,1,2,\ldots).$$

So **x** has a negative binomial distribution with $N = m$; $P = (1 - p)/p$.

From this approach it follows that if $\mathbf{x}_1, \mathbf{x}_2$ are independent negative binomial random variables with respective parameters (N_1,P), (N_2,P), N_1 and N_2 integers, the sum $(\mathbf{x}_1 + \mathbf{x}_2)$ is a negative binomial variable with parameters $(N_1 + N_2,P)$. This remains true, even when N_1 and N_2 are not integers.

(*b*) Suppose we have a mixture of Poisson distributions, such that the expected values, $\boldsymbol{\theta}$, of the Poisson distributions vary according to a Type III (gamma) distribution with probability density function

$$(8) \qquad p_\theta(\theta) = [\beta^\alpha\Gamma(\alpha)]^{-1}\theta^{\alpha-1}\exp[-\theta/\beta] \qquad (\theta > 0; \alpha > 0; \beta > 0).$$

Then

$$(9) \qquad \Pr[\mathbf{x} = k] = [\beta^\alpha\Gamma(\alpha)]^{-1}\int_0^\infty \theta^{\alpha-1}e^{-\theta/\beta}[\theta^k e^{-\theta}/k!]\,d\theta$$

$$= [\beta^\alpha\Gamma(\alpha)]^{-1}\int_0^\infty \theta^{\alpha+k-1}\exp[-\theta(\beta^{-1}+1)]\,d\theta$$

$$= \binom{\alpha+k-1}{\alpha-1}\left(\frac{\beta}{\beta+1}\right)^k\left(\frac{1}{\beta+1}\right)^\alpha.$$

So $\mathbf{x}$ has a negative binomial distribution with parameters (α,β).

This type of model was used to represent 'accident-proneness' by Greenwood and Yule [30]. The parameter θ then represents the expected number of accidents for an individual. This is assumed to vary from individual to individual.

This kind of model can be used generally when mixtures of random sources are to be represented. Wise [65] describes an application of this kind of model to represent the distribution of number of pinholes per 'unit' (actually 50-yard) length of enamelled wire.

The negative binomial distribution can also arise as the distribution of the sum of $\mathbf{n}$ independent variables each having the same logarithmic series distribution (Chapter 7), when $\mathbf{n}$ has a Poisson distribution (Quenouille [49]).

(*c*) The negative binomial distribution is also obtained as a limiting case of the Pólya-Eggenberger distribution (see Chapter 9, Section 4). Thompson [63] has shown that a negative binomial distribution can be obtained (approximately) from a modified form of Neyman's contagious distribution model (Chapter 9).

The negative binomial distribution is very often a first choice as alternative when it is felt that a Poisson distribution might be inadequate. While the negative binomial does not have the same flexibility as certain contagious distributions (with more than two assignable parameters) it often gives an adequate representation when the strict randomness requirements for the Poisson distribution are not approximated sufficiently closely.

5. Moments and Mode

The moment generating function is

$$(10) \qquad E(e^{\mathbf{x}t}) = (Q - Pe^t)^{-N}.$$

The probability generating function is

$$(11) \qquad (Q - Pt)^{-N}.$$

The cumulant generating function is

$$-N \log (Q - Pe^t). \tag{12}$$

The jth moment about zero is

$$\mu'_j = \sum_{k=1}^{\infty} k^j \binom{N+k-1}{N-1} (P/Q)^k (1 - P/Q)^N.$$

This can be expressed in terms of differences of zero as

$$\mu'_j = \sum_{k=1}^{j} \binom{N+k-1}{N-1} P^k \Delta^k 0^j. \tag{13}$$

(Formally $\mu'_j = (1 - P\Delta)^{-N} 0^j$.)

The jth factorial moment is

(14)

$$\begin{aligned}\mu'_{(j)} = E(\mathbf{x}^{(j)}) &= \sum_{k=j}^{\infty} k^{(j)} \binom{N+k-1}{N-1} (P/Q)^k (1 - P/Q)^N \\ &= (N+j-1)^{(j)} (P/Q)^j Q^{-N} \sum_{k=j}^{\infty} \binom{N+j+k-j-1}{N+j-1} (P/Q)^{k-j} \\ &= (N+j-1)^{(j)} P^j.\end{aligned}$$

From these formulas we find

$$\begin{cases} E(\mathbf{x}) = NP; \quad \operatorname{var}(\mathbf{x}) = NPQ = NP(1+P) \\ \sqrt{\beta_1(\mathbf{x})} = \dfrac{Q+P}{(NPQ)^{\frac{1}{2}}}; \quad \beta_2(\mathbf{x}) = 3 + \dfrac{1+6PQ}{NPQ}. \end{cases} \tag{15}$$

The variance is greater than the mean, and β_2 is greater than 3.

Note the recurrence relation among the cumulants:

$$\kappa_{r+1} = PQ \frac{\partial \kappa_r}{\partial Q} \qquad (r \geq 1). \tag{16}$$

The mean deviation is:

$$2m \binom{N+m-1}{N-1} (P/Q)^m Q^{-(N-1)} \tag{17}$$

where m is the greatest integer not greater than NP.

From the relation

$$\Pr[\mathbf{x} = k+1]/\Pr[\mathbf{x} = k] = [(N+k)P]/[(k+1)Q] \tag{18}$$

it can be seen that $\Pr[\mathbf{x} = k+1] \gtreqless \Pr[\mathbf{x} = k]$

according as $\qquad k \lesseqgtr NP - Q.$

So, if $NP > Q$ there is a mode at the least integer not less than $NP - Q$ (two equal modes if $NP = Q$). If $NP < Q$ the mode is at $k = 0$.

The information generating function ($(u - 1)$-th frequency moment) of the *geometric* distribution is

$$T_{\mathbf{x}}(u) = Q^{-u} \sum_{k=0}^{\infty} (P/Q)^{ku} = (Q^u - P^u)^{-1}. \tag{19}$$

The entropy is

$$-T'_{\mathbf{x}}(1) = P \log P - Q \log Q. \tag{20}$$

As N tends to infinity and P to zero, with NP remaining fixed ($NP = \theta$), the right hand side of (1) tends to the value $e^{-\theta}\theta^k/k!$, corresponding to a Poisson distribution with expected value θ. Figures 1a–d represent this process. Note, the increasing importance of the positive 'tail' as N decreases.

6. Computing Formulas and Approximations

The sum of a number of negative binomial terms can be expressed in terms of an incomplete beta function ratio (and so as a sum of binomial terms). In fact, in the notation of Section 1,

$$\Pr[\mathbf{x} \geq X] = I_{Q^{-1}}(N, X + 1) \qquad (X \text{ integral})$$

or equivalently, $\Pr[\mathbf{x} \leq X] = \Pr[\mathbf{y} \geq N]$ where $\mathbf{y}$ is a binomial variable with parameters $(N + X)$, Q^{-1}.

This formula has been rediscovered on many occasions. Patil [44] gives a list of references, the earliest of which is to Fieller (quoted by Pearson [45]). An additional, more recent reference is to Morris [40].

Computation of negative binomial probabilities (for integer values of N) can thus be reduced to calculation of the corresponding binomial probabilities. For similar reasons, it is not necessary to discuss approximations to negative binomial distributions separately. The approximations to binomial distributions, already discussed in Chapter 3, Section 8, can be applied to the corresponding negative binomial distributions.

However, an approximately normalizing and variance equalizing transformation does merit special discussion. From the formulas $E(\mathbf{x}) = NP$; $\text{var}(\mathbf{x}) = NP(1 + P)$, the transformation

$$\mathbf{y}' = \sqrt{N} \sinh^{-1} \sqrt{\mathbf{x}/N} \tag{21}$$

with $\mathbf{y}'$ approximately distributed as a standard normal variable, is suggested.

This transformation was applied by Beall [7] to some entomological data. More detailed investigations by Anscombe [1] indicated that the transformation

$$\mathbf{y}'' = \sqrt{N - \tfrac{1}{2}} \sinh^{-1} \sqrt{(\mathbf{x} + \tfrac{3}{8})/(N - \tfrac{3}{4})} \tag{22}$$

would be preferable.

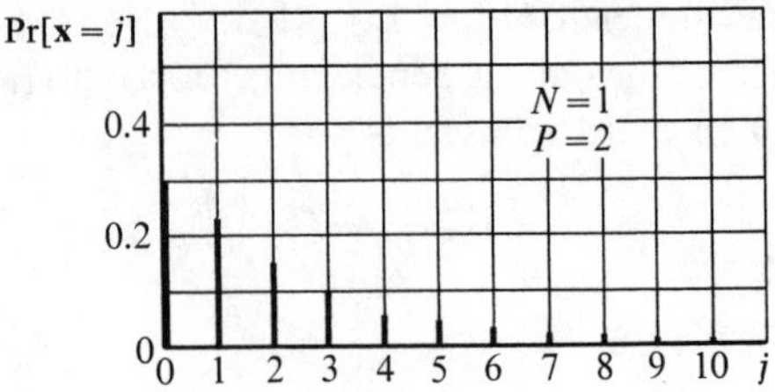

FIGURE 1a

Negative Binomial

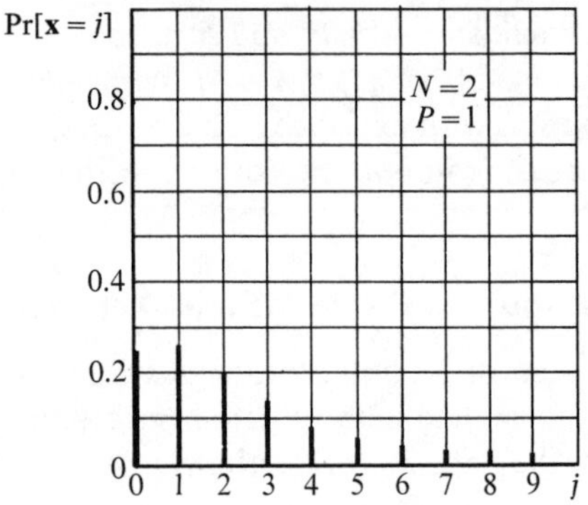

FIGURE 1b

Negative Binomial

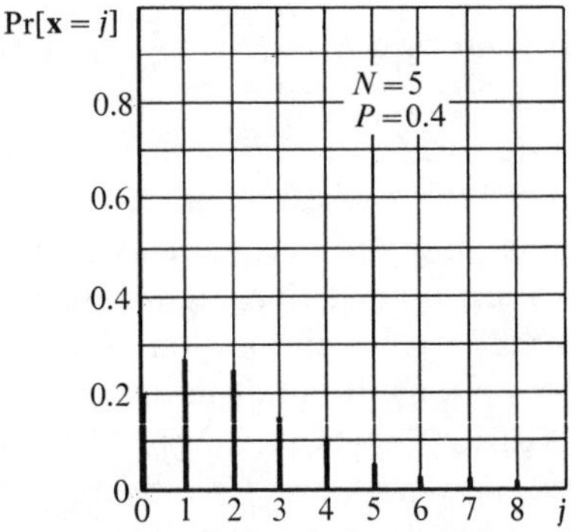

FIGURE 1c

Negative Binomial

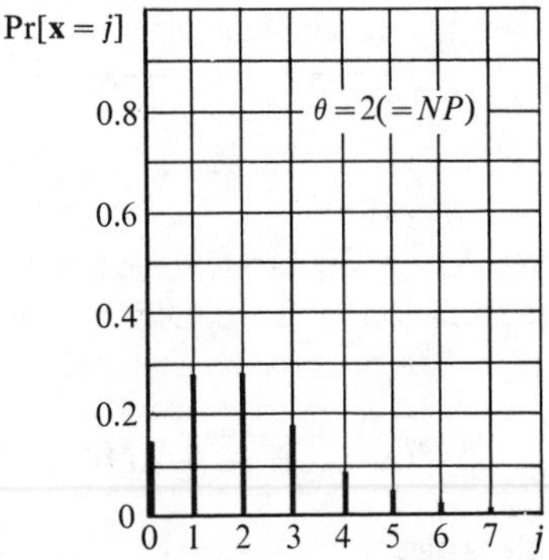

FIGURE 1d

Poisson

For occasional calculations, the recurrence relation (see Section 6)

$$\text{(23)} \qquad \Pr[\mathbf{x} = k + 1] = (N + k)(k + 1)^{-1}(P/Q) \Pr[\mathbf{x} = k]$$

is useful.

Bartko [6] has compared a number of approximations to the negative binomial distribution. The accuracy is similar to that of approximation to the binomial distribution. The two most useful approximations, in Bartko's opinion, are

(*a*) A corrected (Gram-Charlier) Poisson approximation

$$\text{(24)} \qquad \Pr[\mathbf{x} \le k] = e^{-NP} \sum_{j=0}^{k} \frac{(NP)^j}{j!} - \frac{k - NP}{2(1 + P)} e^{-NP} \frac{(NP)^k}{k!}.$$

(*b*) The Camp-Paulson approximation (see also Chapters 3 and 26)

$$\text{(25)} \qquad \Pr[\mathbf{x} \le k] = \frac{1}{\sqrt{2\pi}} \int_{-\infty}^{K} e^{-\frac{1}{2}u^2}\, du \qquad \text{with}$$

$$K = \frac{1}{3}\left[\frac{9k + 8}{k + 1} - \frac{(9N - 1)\{NP/(k + 1)\}^{\frac{1}{3}}}{N}\right]\left[\frac{\{NP/(k + 1)\}^{\frac{2}{3}}}{N} - \frac{1}{k + 1}\right]^{-\frac{1}{2}}.$$

Of these, (25) is remarkably accurate, but much more complicated than (24). Tables 1a and 1b, taken from Bartko's paper, show the maximum absolute error in individual probabilities of (24) and (25) respectively.

TABLE 1a

Maximum Absolute Errors of Approximation (24)

	N				
$(1 + P)^{-1}$	5	10	25	50	100
.05	.309				
.10	.232	.221			
.20	.146	.138	.131		
.30	.096	.090	.085	.082	
.40	.062	.058	.054	.053	
.50	.039	.036	.034	.033	.032
.60	.023	.022	.020	.019	.018
.70	.013	.011	.010	.010	.010
.80	.005	.005	.004	.004	.004
.90	.001	.001	.001	.001	.001
.95	.0001	.0002	.0003	.0002	.0003

TABLE 1b

Maximum Absolute Error of Approximation (25)

$(1 + P)^{-1}$	N = 5	10	25	50	100
.05	.001	.001	.0002		
.10	.001	.001	.0002	.0001	
.20	.001	.001	.0002	.0001	
.30	.001	.001	.0001	.0001	
.40	.001	.001	.0001	.0001	.00002
.50	.001	.001	.0001	.0001	.00003
.60	.001	.001	.0002	.0001	.00004
.70	.001	.001	.0003	.0001	.0001
.80	.002	.002	.0006	.0003	.0001
.90	.004	.002	.002	.0007	.0004
.95	.003	.005	.003	.002	.001

7. Tables

It is unnecessary to have separate tables for the negative binomial distribution, for integer values of N. However, for fractional values of N, and for convenience in looking up sequences of values, direct tables are useful. Values of $\Pr[\mathbf{x} = k]$ for:

$$Q^{-1} = 0.05;\ N = 0.1(0.1)0.5:$$
$$Q^{-1} = 0.10;\ N = 0.1(0.1)1.0:$$
$$Q^{-1} = 0.12(0.02)0.20;\ N = 0.1(0.1)2.5:$$
$$Q^{-1} = 0.22(0.02)0.40;\ N = 0.1(0.1)2.5(0.5)5.0:$$
$$Q^{-1} = 0.42(0.02)0.60;\ N = 0.1(0.1)2.5(0.5)10.0:$$
$$Q^{-1} = 0.62(0.02)0.80;\ N = 0.2(0.2)5.0(1)20,$$

and a few other pairs of values of Q^{-1} and N are given (to six decimal places) in the tables of Williamson and Bretherton [64]. The choice of values of Q^{-1} and N in these tables is such that a range of values of the expected value (NP) up to or over 5.0 is covered.

Grimm [31] gives values of individual probabilities and of the cumulative distribution function to five decimal places for

$$NP = 0.1(0.1)1.0(0.2)4.0(0.5)10.0$$
$$Q = 1.2, 1.5, 2.0(1)5.$$

Brown [16] gives values (to four decimal places) of the same quantities for

$$NP = 0.25(0.25)1.00(1)10$$
$$Q = 1.5(0.5)5.0(1)7.$$

There are also tables for $N = 1(1)15$; $Q^{-1} = 0.1(0.1)0.9(0.01)0.99$ by Berndt and Brocky [9] though these may now be out of print.

Short tables of negative binomial probabilities, calculated for particular purposes (such as fitting observed distributions) have appeared in a number of papers (e.g. Wise [65]).

Taguti [62] gives minimum values of k for which

$$\sum_{j=0}^{k} (n!)^{-1} h(h + d) \ldots (h + (j - 1)d)(1 + d)^{-(h/d)-j} \geq \alpha$$

for $\alpha = 0.95, 0.99$. These are (approximate) percentage points of negative binomial distributions with $N = h/d$, $P = d$.

8. Estimation of Parameters

Method (1). The simplest way of estimating the parameters N and P is by equating the sample mean and variance to the corresponding population values. Thus, if $\mathbf{x}_1, \mathbf{x}_2, \ldots \mathbf{x}_n$ be n observed values (supposed independent), we calculate the solutions $\hat{\mathbf{N}}$, $\hat{\mathbf{P}}$ of the equations

$$\hat{\mathbf{N}}\hat{\mathbf{P}} = \bar{\mathbf{x}}; \ \hat{\mathbf{N}}\hat{\mathbf{P}}(1 + \hat{\mathbf{P}}) = \mathbf{s}^2 \tag{26}$$

where $\bar{\mathbf{x}} = n^{-1} \sum_{i=1}^{n} \mathbf{x}_i$; $\mathbf{s}^2 = (n - 1)^{-1} \sum_{i=1}^{n} (\mathbf{x}_i - \bar{\mathbf{x}})^2$. This gives

$$\hat{\mathbf{P}} = \mathbf{s}^2/\bar{\mathbf{x}} - 1; \ \hat{\mathbf{N}} = \bar{\mathbf{x}}^2/(\mathbf{s}^2 - \bar{\mathbf{x}}^2). \tag{27}$$

Note that $\hat{\mathbf{P}}$ is negative if $\mathbf{s}^2 < \bar{\mathbf{x}}$. If this should happen it indicates that a negative binomial distribution may not be appropriate.

Method (2). In place of the second equation $[\hat{\mathbf{N}}\hat{\mathbf{P}}(1 + \hat{\mathbf{P}}) = \mathbf{s}^2]$ an equation obtained by equating the observed and expected numbers of zero values among the $\mathbf{x}$'s may be used. This equation is

$$\mathbf{f}_0 = (1 + \hat{\mathbf{P}})^{-\hat{\mathbf{N}}} \tag{28}$$

where $\mathbf{f}_0$ = proportion of zero $\mathbf{x}$'s $= n^{-1}$ [number of zero $\mathbf{x}$'s]. Combining this equation with the equation $\hat{\mathbf{N}}\hat{\mathbf{P}} = \bar{\mathbf{x}}$ gives

$$\frac{\hat{\mathbf{P}}}{\log(1 + \hat{\mathbf{P}})} = \frac{\bar{\mathbf{x}}}{-\log \mathbf{f}_0}. \tag{29}$$

This equation has to be solved for $\hat{\mathbf{P}}$. Provided $\bar{\mathbf{x}} > -\log \mathbf{f}_0$, there is always a unique solution for $\hat{\mathbf{P}}$.

Method (3). The *maximum likelihood* estimators (Fisher [25], [26], Wise [65]) satisfy the equations

$$(30)\qquad \hat{\mathbf{N}}\hat{\mathbf{P}} = \bar{\mathbf{x}};\quad \log(1+\hat{\mathbf{P}}) = \sum_{j=1}^{\infty}(\hat{\mathbf{N}}+j-1)^{-1}\mathbf{F}_j$$

where $\mathbf{F}_j = \sum_{i=j}^{\infty}\mathbf{f}_i$ = proportion of $\mathbf{x}$'s which are greater than or equal to j. From these two equations we obtain

$$(31.1)\qquad \log(1+\hat{\mathbf{P}}) = \sum_{j=1}^{\infty}(\bar{\mathbf{x}}\hat{\mathbf{P}}^{-1}+j-1)^{-1}\mathbf{F}_j \qquad \text{or}$$

$$(31.2)\qquad \log(1+\bar{\mathbf{x}}\hat{\mathbf{N}}^{-1}) = \sum_{j=1}^{\infty}(\hat{\mathbf{N}}+j-1)^{-1}\mathbf{F}_j.$$

In the second of these equations, the left hand side tends to zero as $\hat{\mathbf{N}}$ tends to zero, while (unless $\mathbf{f}_0 = 1$) the right-hand side tends to a positive limit. For large $\hat{\mathbf{N}}$, the left hand side is greater than

$$\bar{\mathbf{x}}\hat{\mathbf{N}}^{-1} - \tfrac{1}{2}(\bar{\mathbf{x}}\hat{\mathbf{N}}^{-1})^2$$

while the right hand side can be written as

(32)

$$\hat{\mathbf{N}}^{-1}\sum_{j=1}^{\infty}[1+(j-1)\hat{\mathbf{N}}^{-1}]\mathbf{F}_j < \hat{\mathbf{N}}^{-1}\sum_{j=1}^{\infty}[1-(j-1)\hat{\mathbf{N}}^{-1}]\mathbf{F}_j + C\hat{\mathbf{N}}^{-3}$$
$$< \hat{\mathbf{N}}^{-1}(\bar{\mathbf{x}} - \tfrac{1}{2}(\bar{\mathbf{x}}^2+\mathbf{s}^2-\bar{\mathbf{x}})\hat{\mathbf{N}}^{-1}) + C\hat{\mathbf{N}}^{-3}$$

where $\mathbf{s}^2 = \sum_{j=1}^{\infty} j^2\mathbf{f}_j - \left(\sum_{j=1}^{\infty} j\mathbf{f}_j\right)^2$, and C is a constant depending on the $\mathbf{f}_j$'s but not on $\hat{\mathbf{N}}$.

Hence if $\mathbf{s}^2 > \bar{\mathbf{x}}$, the right hand side will be less than the left hand side for $\hat{\mathbf{N}}$ sufficiently large, and so there must be at least one solution with $\hat{\mathbf{N}} > 0$. If $\mathbf{s}^2 \leq \bar{\mathbf{x}}$, a negative binomial distribution may not be appropriate. For N and NP both large, the following method (4) was suggested by Anscombe.

Method (4). This is an iterative method using the transformation

$$(33)\qquad \mathbf{y} = 2\sinh^{-1}\sqrt{\frac{\mathbf{x}+\frac{3}{8}}{N-\frac{3}{4}}}. \qquad \text{(see Section 6)}$$

The variance of $\mathbf{y}$ is approximately $\psi^{(1)}(N)$. Starting with a trial value, $\hat{\mathbf{N}}_0$, of N (possibly obtained by method (1) or (2)), values of $\mathbf{y}$ corresponding to the observed values of $\mathbf{x}$ are calculated. The variance, $s_{\mathbf{y}}^2$, of these $\mathbf{y}$'s is then calculated, and a new value $\hat{\mathbf{N}}_1$ obtained by solving the equation

$$(34)\qquad \mathbf{s}_{\mathbf{y}}^2 = \psi^{(1)}(\hat{\mathbf{N}}_1).$$

The process is continued until $\hat{\mathbf{N}}_{i+1} \doteqdot \hat{\mathbf{N}}_i = \hat{\mathbf{N}}$ within a sufficient degree of approximation.

The variance of $\hat{\mathbf{N}}$ determined by this method is approximately:

$$(35) \qquad [\psi^{(3)}(N) + 2\{\psi^{(1)}(N)\}^2]\left[\psi^{(2)}(N) + \frac{1}{(N-1)^2P}\right]^{-1} N^{-1}.$$

Anscombe suggests this formula is sufficiently accurate for $NP > 50$, $N > 5$.

The variance of $\bar{\mathbf{x}}$ is $n^{-1}NP(1 + P)$. This is the (exact) variance of $\widehat{\mathbf{NP}}$ for each of the first three methods described here. Anscombe [3] gives the following *approximate* formulas for the variance of $\hat{\mathbf{N}}$:

Method (1): $2N(N + 1)(Q/P)^2n^{-1}$

Method (2): $[Q^N - 1 - NP/Q][\log Q - P/Q]^{-2}n^{-1}$

Method (3): $$\frac{2N(N+1)(Q/P)^2}{1 + \sum_{j=2}^{\infty} \frac{2}{j+1} \cdot \frac{j!}{(N+2)(N+3)\cdots(N+j)}(P/Q)^j} n^{-1}.$$

Shenton [55] notes that if NP is large the variance for Method (1) is approximately $[\{\psi^{(1)}(N) - N^{-1}\}n]^{-1}$ and lies between

$$(1 + P^{-1})^2[\{\psi^{(1)}(N) - N^{-1}\}n]^{-1} \qquad \text{and} \qquad 2N(N + 1)(1 + P^{-1})^2n^{-1}.$$

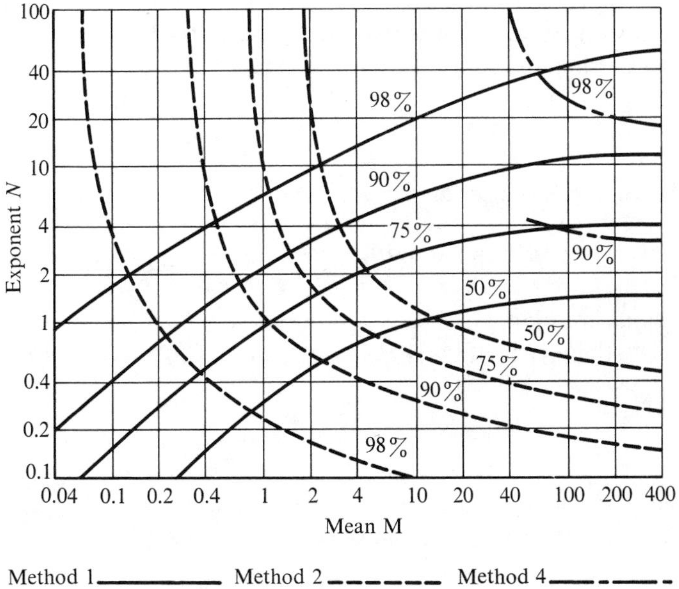

FIGURE 2

Large Sample Efficiencies of Estimators of N

Anscombe [3, p. 371] gives a graph from which the efficiencies of (1) and (2), relative to (3), as estimators of N, can be judged.

From this graph, it appears that method (1) is useful when N is large and NP is not large, while (2) is useful when neither N nor NP are large.

In fact, the expected value and also, of course, higher moments of the moment estimator of N (Method (1)) are infinite, but approximate values for moments can be obtained by neglecting the possibility that $\mathbf{s}^2 < \bar{\mathbf{x}}$. Similarly the maximum likelihood equations (31) may not be soluble if $\bar{\mathbf{x}} > \mathbf{s}^2$ (see [13]), but useful approximations can be obtained by neglecting this possibility. Bowman and Shenton [13] have given asymptotic formulas for the variances and co-variances of the estimations obtained by Methods (1) and (3), up to terms in n^{-2}. Later the same authors [14] gave similar formulas for biases of the moment estimators.

Shenton and Myers [56], in an exhaustive survey of methods of estimating the parameters of negative binomial distributions, obtained a series expansion in n^{-1} for the bias of the moment estimator (see (27)) of N, and gave tables for calculating the first four terms. Their paper also contains a table contrasting various sets of parameters which may be used in defining the distribution.

Maximum likelihood fitting of negative binomial to coarsely grouped data has been described by O'Carroll [42].

Anscombe [3] also discusses briefly, estimation of the (assumed) common value of N from data for a number of negative binomial distributions. A much more detailed discussion is given by Bliss and Owen [11], with a useful bibliography. Supposing that in a sequence of k samples the observed means and standard deviations are denoted by $\bar{\mathbf{x}}_i$, $\mathbf{s}_i^2$ respectively. $(i = 1,2,\ldots,k)$ then the moment estimators (Method (1)) of N are

$$\hat{\mathbf{N}}_i = \bar{\mathbf{x}}_i^2/(\mathbf{s}_i^2 - \bar{\mathbf{x}}_i) \qquad (i = 1,2,\ldots,k). \tag{36}$$

The approximate variances are

$$2N(N+1)(Q_i/P_i)^2 n_i^{-1} \qquad (i = 1,2,\ldots,k) \tag{37}$$

in an obvious notation.

So as a first approximation, weights $\mathbf{w}_i = n_i(\mathbf{P}_i/\hat{\mathbf{Q}})^2$ may be used and

$$\hat{\mathbf{N}}^{(1)} = \sum_{i=1}^{k} \mathbf{w}_i\hat{\mathbf{N}}_i \Big/ \sum_{i=1}^{k} \mathbf{w}_i \tag{38}$$

calculated. Using this value of $\hat{\mathbf{N}}^{(1)}$, new estimates $\hat{\mathbf{P}}_i^{(1)} = \bar{\mathbf{x}}_i/\hat{\mathbf{N}}^{(1)}$ of P_i $(i = 1,2,\ldots,k)$ are calculated, and the corresponding weights

$$\mathbf{w}_i^{(1)} = n_i(\hat{\mathbf{P}}^{(1)}/\hat{\mathbf{Q}}^{(1)})^2$$

used in calculating

$$\hat{\mathbf{N}}^{(2)} = \sum_{i=1}^{k} \mathbf{w}_i^{(1)}\hat{\mathbf{N}}_i \Big/ \sum_{i=1}^{k} \mathbf{w}_i^{(1)}. \tag{39}$$

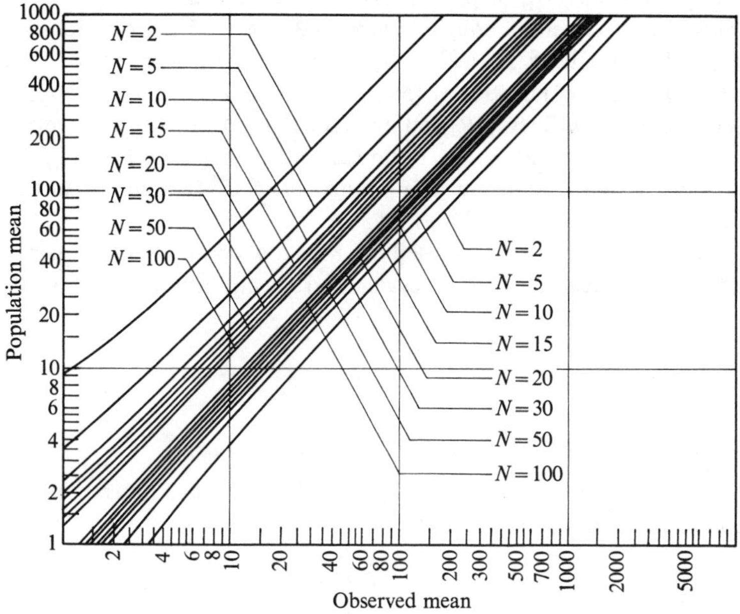

FIGURE 3

Central 90% confidence belt for the mean of a geometric distribution based on a sample of size N.

If the distribution is geometric, then N is known to be 1, and it is only necessary to estimate P. If $\mathbf{x}_1, \mathbf{x}_2, \ldots, \mathbf{x}_n$ are random variables each with the geometric distribution $\Pr[\mathbf{x}_i = k] = Q^{-1}(P/Q)^k$ $(k = 0,1,2,\ldots)$ then (see Section 4(a)) $\sum_{i=1}^{n} \mathbf{x}_i$ is a negative binomial variable with parameters $n. P$. Using this fact, Clemans [20] has constructed charts from which confidence intervals for P, given $\hat{\mathbf{M}} = n^{-1} \sum_{i=1}^{n} \mathbf{x}_i$, can be read off. One such chart, from which 90% intervals can be obtained, is shown in Figure 3.

9. Applications

It has been mentioned, in Section 4, that the negative binomial distribution is frequently used as a substitute for the Poisson distribution when it is doubtful whether the strict requirements, particularly independence, for a Poisson distribution will be satisfied. The occurrence of negative binomial distributions as mixtures of Poisson distributions has already been noted in Section 4(b).

Among specific fields where negative binomial distributions have been found to provide useful representations may be mentioned accident statistics (Arbous and Kerrich [4], Greenwood and Yule [30]) in birth-and-death processes

(Furry [27], Kendall [36]) in psychological data (Sichel [58]), in demand (by households) for 'frequently bought products', observed distributions of consumer expenditure (Chatfield et al. [18]), and as weights (lag distributions) for times series in economics (Solow [59]). Medical and military applications have been described by Chew [19] and by Bennett and Birch [8]. However, there are cases where a good fit is not obtained with a negative binomial, and in such cases it is usual to consider the possibility of using a compound or contagious distribution (Chapters 8 and 9).

The geometric distribution is used in Markov chain models, particularly meteorological models of weather cycles and precipitation amounts (Gabriel and Neumann [28]).

10. Truncated Negative Binomial Distribution

In the most common form of truncation, the zeros are not recorded. The corresponding truncated distribution is

$$\text{(40)} \qquad \Pr[\mathbf{x} = k] = (1 - Q^{-N})^{-1} \binom{N + k - 1}{N - 1} (P/Q)^k (1 - P/Q)^N \qquad (k = 1,2, \ldots).$$

Examples of this kind of truncated negative binomial distribution have been given by Sampford [53] and Brass [15].

The moments about zero of this distribution are easily calculated as

$(1 - Q^{-N})^{-1} \times$ (corresponding moment of (untruncated) negative binomial distribution).

Thus

$$\text{(41.1)} \qquad E(\mathbf{x}) = NP(1 - Q^{-N})^{-1}$$

$$E(\mathbf{x}^2) = [NPQ + (NP)^2](1 - Q^{-N})^{-1}$$

whence

$$\text{(41.2)} \qquad \operatorname{var}(\mathbf{x}) = NPQ(1 - Q^{-N})^{-1}[1 - NPQ^{-1}\{(1 - Q^{-N})^{-1} - 1\}].$$

Rider [51] has given tables of $E(\mathbf{x}^{-1})$ and of $\operatorname{var}(\mathbf{x}^{-1})$ to 5 decimal places for $P = 0.01, 0.05(0.05)1.00(1)5$; $N = 1(1)10$. Govindarajulu [29] proposes the approximations

$$\text{(42.1)} \qquad E(\mathbf{x}^{-1}) \doteqdot (NP - Q)^{-1}$$

$$\text{(42.2)} \qquad \operatorname{var}(\mathbf{x}^{-1}) \doteqdot Q(NP - Q)^{-2}(NP - 2Q)^{-1}.$$

Equating sample and expected values of mean and variance gives two simultaneous equations for N and P, whence estimates of these parameters can be obtained. The equations do not have simple explicit solutions, and David and

Johnson [22] considered introducing a third equation (equating sample and observed third moments), with which simple explicit solutions could be obtained. The estimators so obtained are, however, very inefficient, and these authors recommended using maximum likelihood methods. Rider [50] also obtained estimators based on the moments and Shah [54] gave formulas for their asymptotic variances and covariances.

Sampford [53], however, proposed a trial-and-error method of solving the equations

$$\bar{\mathbf{x}} = \hat{\mathbf{N}}\hat{\mathbf{P}}\hat{\mathbf{Q}}(1 - \mathbf{Q}^{-\hat{\mathbf{N}}})^{-1} \tag{43}$$

$$\mathbf{s}_x^2 = \hat{\mathbf{N}}\hat{\mathbf{P}}\hat{\mathbf{Q}}(1 - \mathbf{Q}^{-\hat{\mathbf{N}}})^{-1}[1 - \hat{\mathbf{N}}\hat{\mathbf{P}}\hat{\mathbf{Q}}^{-(\hat{\mathbf{N}}+1)}(1 - \hat{\mathbf{Q}}^{-\hat{\mathbf{N}}})^{-1}].$$

He showed that the efficiency of these estimators (relative to maximum likelihood estimators as ratio of variance-covariance determinants) is high for N large and $M = NP$ small, and not very low even for other values of the parameters. (For all combinations of $N = 0.5,1(1)5$ and $M = 0.5, 1, 2, 5$ the lowest value is 55.9%, for $N = 0.5$, $M = 5$; the value is above 85% for all cases when $N \geq 3$, above 91% when $N = 5$.) This method is analogous to Method (1) of Section 8.

Brass [15] proposed using $\mathbf{f}_1$, the observed proportion of unit values, together with $\bar{\mathbf{x}}$ and $\mathbf{s}^2$. This method is analogous to Method (2) of Section 8, $\mathbf{f}_1$ being used instead of $\mathbf{f}_0$. From the equations

$$\begin{aligned} \bar{\mathbf{x}} &= \hat{\mathbf{N}}\hat{\mathbf{P}}(1 - \hat{\mathbf{Q}}^{-\hat{\mathbf{N}}})^{-1} \\ \mathbf{s}^2 &= \hat{\mathbf{N}}\hat{\mathbf{P}}\hat{\mathbf{Q}}(1 - \hat{\mathbf{Q}}^{-\hat{\mathbf{N}}})^{-1}[1 - \hat{\mathbf{N}}\hat{\mathbf{P}}\hat{\mathbf{Q}}^{\hat{\mathbf{N}}+1}(1 - \hat{\mathbf{Q}}^{-\hat{\mathbf{N}}})^{-1}] \\ \mathbf{f}_1 &= \hat{\mathbf{N}}\hat{\mathbf{P}}(1 - \hat{\mathbf{Q}}^{-\hat{\mathbf{N}}})^{-1}\hat{\mathbf{Q}}^{-\hat{\mathbf{N}}+1} \end{aligned} \tag{44}$$

the formulas

$$\hat{\mathbf{Q}} = \mathbf{s}^2\bar{\mathbf{x}}^{-1}(1 - \mathbf{f}_1)^{-1}; \ \hat{\mathbf{N}} = (\bar{\mathbf{x}} - \hat{\mathbf{Q}}\mathbf{f}_1)(\hat{\mathbf{Q}} - 1)^{-1} \tag{45}$$

are obtained. Brass found that the efficiency (as measured by ratio of variance-covariance determinants) of his method was greater than that of Sampford's moment method when $N \leq 5$, and not much less when $N > 5$. Estimation of M is more accurate than by Sampford's method (and in all cases calculated, asymptotic efficiency was over 90% relative to maximum likelihood).

Brass also suggested a modification of the maximum likelihood equations, replacing $\hat{\mathbf{N}}\hat{\mathbf{P}}(1 - \hat{\mathbf{Q}}^{-\hat{\mathbf{N}}})^{-1}\hat{\mathbf{Q}}^{-(\hat{\mathbf{N}}+1)}$ by $\mathbf{f}_1$, leading to the equations

$$\begin{aligned} \hat{\mathbf{Q}} &= (\hat{\mathbf{N}} + \bar{\mathbf{x}})(\hat{\mathbf{N}} + \mathbf{f}_1)^{-1} \\ &- \frac{\bar{\mathbf{x}}(\hat{\mathbf{N}} + \mathbf{f}_1)}{\hat{\mathbf{N}}(\bar{\mathbf{x}} - \mathbf{f}_1)} \log_e \hat{\mathbf{Q}} + \sum_{j=1}^{\infty} (\hat{\mathbf{N}} + j - 1)^{-1}\mathbf{F}_j = 0 \end{aligned} \tag{46}$$

where $\mathbf{F}_j = \sum_{i \geq j} \mathbf{f}_i$.

This gives very little reduction in efficiency, except for small N (<1).

11. Distributions Related to the Negative Binomial

The negative binomial distribution is a limiting form of the Pólya-Eggenberger distribution (see Section 4(c)). In turn the logarithmic distribution (Chapter 7) is a limiting form of the negative binomial distribution.

Relationships among the negative binomial, Poisson and binomial distributions have been mentioned in Chapters 2, 3 and 4. It is convenient to make some final remarks on these relationships here.

Each of these three distributions can be regarded as being generated from the expansion of

$$[(1 + \omega) - \omega]^{-m}.$$

For the negative binomial, $\omega > 0$, $m > 0$; for the binomial $-1 < \omega < 0$, $m < 0$. The Poisson distribution corresponds to the limiting intermediate case, where $\omega \to 0$, $m \to 0$, with $m\omega = \theta$.

The most obvious distinction among the three types of distribution is in the values of the ratio of variance to mean. This is less than 1 for the binomial, equal to 1 for the Poisson, and greater than 1 for the negative binomial distribution. Sampling variation may yield values of this ratio which do not accord with the underlying type of distribution, but the value of (sample variance)/(sample mean) is a good guide as to which of these three types of distribution is to be used, *given that one of them is to be used.*

To decide whether to restrict attention to these three types of distribution, use of the "approximation coefficient" suggested by Guldberg [32], and discussed in Chapter 2.

It is interesting to note that for all Poisson distributions, the point (β_1,β_2) lies on the straight line $\beta_2 - \beta_1 - 3 = 0$, while for binomial distributions $\beta_2 - \beta_1 - 3 = -2n^{-1}$, and for negative binomial distributions

$$\beta_2 - \beta_1 - 3 = 2N^{-1}.$$

The relationship between the geometric and the exponential distributions has been described, in Section 2. Pessin [46] noted the following property of the negative binomial distribution:

"As $Q \to \infty$ with N constant the standardized negative binomial distribution tends to a gamma distribution."

Compound distributions formed from the negative binomial will be discussed in Chapter 8. The most important of these is the *Poisson-Pascal* (Chapter 8, Section 2(xix)) obtained by assigning a Poisson distribution to N/K (K being a constant). If Q^{-1} has a Beta distribution (Chapter 24) (range 0 to 1) a generalized hypergeometric distribution of Type IV (see Chapter 6, Section 8) is obtained. In the special case $N = 1$, this is sometimes called a *Miller* distribution (Haight [34]). Mixtures of two negative binomial distributions have been discussed by Chahine [17]. Conditions under which a distribution can be a mixture of geometric distributions have been discussed by Daniels [21].

REFERENCES

[1] Anscombe, F. J. (1948). The transformation of Poisson, binomial and negative binomial data, *Biometrika*, **35**, 246–254.

[2] Anscombe, F. J. (1949). The statistical analysis of insect counts based on the negative binomial distribution, *Biometrics*, **5**, 165–173.

[3] Anscombe, F. J. (1950). Sampling theory of the negative binomial and logarithmic series distributions, *Biometrika*, **37**, 358–382.

[4] Arbous, A. G. and Kerrich, J. E. (1951). Accident statistics and the concept of accident proneness, *Biometrics*, **7**, 340–432.

[5] Bartko, J. J. (1961). The negative binomial distribution: A review of properties and applications, *Virginia Journal of Science* (*New Series*), **12**, 18–37.

[6] Bartko, J. J. (1966). Approximating the negative binomial, *Biometrics*, **8**, 340–342.

[7] Beall, G. (1942). The transformation of data from entomological field experiments so that the analysis of variance becomes applicable, *Biometrika*, **32**, 243–262.

[8] Bennett, B. M. and Birch, B. (1964). Sampling inspection tables for comparison of two groups using the negative binomial model, *Trabajos de Estadistica*, **15**, 1–12.

[9] Berndt, G. D. and Brocky, S. J. (1964). [*Tables of the negative binomial distribution*, Operations Analysis Hq. SAC Offutt AFB, Nebraska (out of print).]

[10] Bliss, C. I. (1953). Fitting the negative binomial distribution to biological data, *Biometrics*, **9**, 176–196.

[11] Bliss, C. I. and Owen, A. R. C. (1958). Negative binomial distributions with a common k, *Biometrika*, **45**, 37–58.

[12] Blom, G. (1954). Transformations of the binomial, negative binomial, Poisson and χ^2 distributions, *Biometrika*, **41**, 302–316, (Correction in *Biometrika*, **43**, (1956), 235).

[13] Bowman, K. O. and Shenton, L. R. (1965). *Asymptotic covariance for the maximum likelihood estimators of the parameters of a negative binomial distribution*, Report No. K–1643, Union Carbide Corporation, Oak Ridge, Tennessee.

[14] Bowman, K. O. and Shenton, L. R. (1966). *Biases of estimators for the negative binomial distribution*, Report No. ORNL-4005, Union Carbide Corporation, Oak Ridge, Tennessee.

[15] Brass, W. (1958). Simplified methods of fitting the truncated negative binomial distribution, *Biometrika*, **45**, 59–68.

[16] Brown, Bernice (1965). *Some tables of the negative binomial distribution and their use*. RAND Memorandum RM-4577-PR, RAND Corporation, Santa Monica, California.

[17] Chahine, J. (1965). Une généralisation de la loi binomiale négative, *Revue de Statistique Appliquée*, **13**(4), 33–43.

[18] Chatfield, C. Ehrenberg, A. S. C. and Goodhardt, G. J. (1966). Progress on a simplified model of stationary purchasing behaviour, *Journal of the Royal Statistical Society, Series A*, **129**, 317–360. (discussion 360–367)

[19] Chew, V. (1964). Application of the negative binomial distribution with probability of misclassification, *Virginia Journal of Science (New Series)*, **15,** 34–40.

[20] Clemans, K. G. (1959). Confidence limits in the case of the geometric distribution, *Biometrika*, **46,** 260–264.

[21] Daniels, H. E. (1961). Mixtures of geometric distributions, *Journal of the Royal Statistical Society, Series B*, **23,** 409–413.

[22] David, F. N. and Johnson, N. L. (1952). The truncated Poisson, *Biometrics*, **8,** 275–285.

[23] Eggenberger, F. and Pólya, G. (1923). Über die Statistik verketteter Vorgänge, *Zeitschrift für angewandte Mathematik und Mechanik*, **1,** 279–289.

[24] Ferguson, T. S. (1965). A characterization of the geometric distribution, *American Mathematical Monthly*, **72,** 256–260.

[25] Fisher, R. A. (1941). The negative binomial distribution, *Annals of Eugenics, London*, **11,** 182–187.

[26] Fisher, R. A. (1953). Note on the efficient fitting of the negative binomial, *Biometrics*, **9,** 197–200.

[27] Furry, W. H. (1939). On fluctuation phenomena in the passage of high energy electrons through lead, *Physical Review*, **52,** 569–581.

[28] Gabriel, K. R. and Neumann, J. (1957), (1962). On a distribution of weather cycles by length, *Quarterly Journal of the Royal Meteorological Society*, **83,** 375–380, *ibid.* **88,** 90–95.

[29] Govindarajulu, Z. (1962). The reciprocal of the decapitated negative binomial variable, *Journal of the American Statistical Association*, **57,** 906–913.

[30] Greenwood, M. and Yule, G. U. (1920). An enquiry into the nature of frequency distributions of multiple happenings, with particular reference to the occurence of multiple attacks of disease or repeated accidents, *Journal of the Royal Statistical Society, Series A*, **83,** 255–279.

[31] Grimm, H. (1962). Tafeln der negativen Binomialverteilung, *Biometrische Zeitschrift*, **4,** 239–262.

[32] Guldberg, A. (1931). On discontinuous frequency functions and statistical series, *Skandinavisk Aktuarietidskrift*, **14,** 167–187.

[33] Gurland, J. (1957). Some applications of the negative binomial and other contagious distributions, *American Journal of Public Health*, **49,** 1388–1399.

[34] Haight, F. A. (1961). Index to the distributions of mathematical statistics, *Journal of Research, National Bureau of Standards*, **65B,** 23–60.

[35] Katti, S. K. and Gurland, J. (1962). Efficiency of certain methods of estimation for the negative binomial and Neyman type A distributions, *Biometrika*, **49,** 215–226.

[36] Kendall, D. G. (1949). Stochastic processes and population growth, *Journal of the Royal Statistical Society, Series B*, **11,** 230–264.

[37] Magistad, J. G. (1961). Some discrete distributions associated with life testing, *Proceedings of the 7th National Symposium on Reliability and Quality Control*, 1–11.

[38] Margolin, B. H. and Winokur, H. S. (1967). Exact moments of the order statistics of the geometric distribution and their relation to inverse sampling and reliability of redundant systems, *Journal of the American Statistical Association*, **62,** 915–925.

[39] Montmort, P. R. (1714). *Essai d'analyse sur les jeux de hasards*, Paris.

[40] Morris, K. W. (1963). A note on direct and inverse binomial sampling, *Biometrika*, **50,** 544–545.

[41] Murty, V. N. (1956). A note on Bhattacharyya bounds for the negative binomial distribution, *Annals of Mathematical Statistics*, **27,** 1182–1183.

[42] O'Carroll, F. M. (1962). Fitting a negative binomial distribution to coarsely grouped data by maximum likelihood, *Applied Statistics*, **11,** 196–201.

[43] Pascal, B. (1679). *Varia opera mathematica D. Petri de Fermat* (Tolossae).

[44] Patil, G. P. (1960). On the evaluation of the negative binomial distribution with examples, *Technometrics*, **2,** 501–505.

[45] Pearson, K. (1933). On the applications of the double Bessel function $K_{\tau_1,\tau_2}(x)$ to statistical problems, *Biometrika*, **25,** 158–178.

[46] Pessin, V. (1961). Some asymptotic properties of the negative binomial distribution, *Annals of Mathematical Statistics*, **32,** 922–923 (abstract).

[47] Pessin, V. (1963). Some discrete distribution limit theorems using a new derivative, *Proceedings of the International Symposium on Discrete Distributions, Montreal*, 109–122.

[48] Puri P. S. (1966). Probability generating functions of absolute difference of two random variables. *Proceedings of the National Academy of Science*, **56,** 1059–1061.

[49] Quenouille, M. H. (1949). A relation between the logarithmic, Poisson, and negative binomial series, *Biometrics*, **5,** 162–164.

[50] Rider, P. R. (1955). Truncated binomial and negative binomial distributions, *Journal of the American Statistical Association*, **50,** 877–883.

[51] Rider, P. R. (1962). The negative binomial distribution and the incomplete beta function, *American Mathematical Monthly*, **69,** 302–304.

[52] Rider, P. R. (1962). Expected values and standard deviations of the reciprocal of a variable from a decapitated negative binomial distribution, *Journal of the American Statistical Association*, **57,** 439–445.

[53] Sampford, M. R. (1955). The truncated negative binomial distribution, *Biometrika*, **42,** 58–69.

[54] Shah, S. M. (1961). The asymptotic variances of method of moments estimates of the parameters of the truncated binomial and negative binomial distributions, *Journal of the American Statistical Association*, **56,** 880–994.

[55] Shenton, L. R. (1963). A note on bounds for the asymptotic sampling variance of the maximum likelihood estimator of a parameter in the negative binomial distribution, *Annals of the Institute of Statistical Mathematics, Tokyo*, **15,** 145–151.

[56] Shenton, L. R. and Myers, R. (1963). Comments on estimation for the negative binomial distribution, *Proceedings of the International Symposium on Discrete Distributions, Montreal*, 241–262.

[57] Shenton, L. R. and Wallington, P. A. (1962). The bias of moment estimators with an application to the negative binomial distribution, *Biometrika*, **49,** 193–204.

[58] Sichel, H. S. (1951). The estimation of the parameters of a negative binomial distribution with special reference to psychological data, *Psychometrika*, **16,** 107–127.

[59] Solow, R. M. (1960). On a family of lag distributions, *Econometrica*, **28**, 392–406.

[60] Srivastava, M. S. (1965). Characterization theorems for some distributions, *Annals of Mathematical Statistics*, **36**, 361 (Abstract).

[61] 'Student' (1907). On the error of counting with a haemacytometer, *Biometrika*, **5**, 351–360.

[62] Taguti, G. (1952). Tables of 5% and 1% points for the Polya-Eggenberger distribution function, *Reports of Statistical Application Research, JUSE*, **2**, 27–32.

[63] Thompson, H. R. (1954). A note on contagious distributions, *Biometrika*, **41**, 268–271.

[64] Williamson, E. and Bretherton, M. H. (1963). *Tables of the Negative Binomial Probability Distribution*, New York: John Wiley & Sons, Inc.

[65] Wise, M. E. (1946). The use of the negative binomial distribution in an industrial sampling problem, *Journal of the Royal Statistical Society, Series B*, **8**, 202–211.

6

Hypergeometric Distribution

1. Genesis and Definition

We first consider the classical situation in which a hypergeometric distribution arises naturally. Suppose an urn contains N balls, of which X are white and $(N - X)$ are black. If a sample of n balls is drawn at random from the urn (without replacing any balls in the urn at any stage)* then the probability of the number, x, of white balls among the n balls chosen in the sample being equal to k is

$$\text{(1)} \qquad \Pr[\mathbf{x} = k] = \frac{\binom{X}{k}\binom{N-X}{n-k}}{\binom{N}{n}}$$

$$\text{for } \max(0, n - N + X) \leq k \leq \min(n, X).$$

This defines a *hypergeometric distribution* with parameters n, X, N. Note that the same distribution is obtained if n and X are interchanged.

The reason for the name 'hypergeometric' is that the quantities on the right hand side are successive terms in the expansion of

$$\frac{(N-n)!(N-X)!}{N!(N-X-n)!} F(-n, -X; N - X - n + 1; 1)$$

where $F(\alpha,\beta;\gamma;z) = 1 + \frac{\alpha\beta}{\gamma}\cdot\frac{z}{1!} + \frac{\alpha(\alpha+1)\beta(\beta+1)}{\gamma(\gamma+1)}\cdot\frac{z^2}{2!} + \cdots$ is a hypergeometric function (Chapter 1, Section 3).

*If each ball were replaced after drawing, the probability of drawing a white ball would be X/N for each individual drawing and the distribution of x would be binomial with parameters n, X/N.

Another natural way in which the hypergeometric distribution arises is in the *theory of exceedances*. Consider two independent random samples of sizes n_1 and n_2, drawn from a population in which a measured character has a continuous distribution. The number of *exceedances* ${}^m\mathcal{E}_{2,1}$ is defined as the number (out of n_2) of observed values in the second sample exceeding at least $(n_1 - m + 1)$ of the values in the first sample. Then

$$\Pr[{}^m\mathcal{E}_{2,1} = k] = \binom{n_1 + n_2 - m - k}{n_1 - m}\binom{k + m - 1}{m - 1}\Big/\binom{n_1 + n_2}{n_1}$$

within appropriate limits for k (Sarkadi [42], Gumbel and von Schelling [17]). This corresponds to a hypergeometric distribution with parameters $(n_1 - 1)$, $(n_1 + n_2 - 1)$, $(k + m - 1)$.

2. Moments and Generating Functions

The rth factorial moment of a random variable $\mathbf{x}$ with distribution (1) is

$$\mu_{(r)} = E(\mathbf{x}^{(r)}) = n^{(r)}X^{(r)}/N^{(r)}. \tag{2}$$

Putting $r = 1$ and $r = 2$, we find

$$E(\mathbf{x}) = nX/N \tag{3.1}$$

$$\text{var}(\mathbf{x}) = [(N - n)/(N - 1)]n(X/N)(1 - X/N). \tag{3.2}$$

The first two moment-ratios are

$$\sqrt{\beta_1} = \frac{(1 - X/N) - X/N}{\left[\left(\frac{N - n}{N - 1}\right) n \cdot \frac{X}{N}\left(1 - \frac{X}{N}\right)\right]^{\frac{1}{2}}} \cdot \frac{N - 2n}{N - 2} \tag{4.1}$$

$$\beta_2 = \frac{3(N - 1)(N + 6)}{(N - 2)(N - 3)} + \frac{(N - 1)N(N + 1)}{(N - n)(N - 2)(N - 3)} \times \left[1 - \frac{6N}{N + 1}\left(pq + \frac{n(N - n)}{N^2}\right)\right] \cdot \frac{1}{npq}. \tag{4.2}$$

(These equations may be compared with the corresponding formulas for the binomial distribution [see Chapter 3, Section 4].) Pearson [34] obtained formulas for these ratios, and higher moments in 1899.

The moment-generating function is

$$M_{\mathbf{x}}(t) = E(e^{t\mathbf{x}}) = \frac{(N - n)!(N - X)!}{N!} F(-n, -X; N - X - n + 1; e^t). \tag{5}$$

The function $M_{\mathbf{x}}(t)$ satisfies the differential equation

$$(1 - e^t)\left\{\frac{d^2M}{dt^2} - (n + X)\frac{dM}{dt} + nXM\right\} - nXM + N\frac{dM}{dt} = 0. \tag{6}$$

The following finite difference relation holds among the central moments $\{\mu_j\}$:

(7) $$N\mu_{r+1} = \{(1+E)^r - E^r\}[\mu_2 + \alpha\mu_1 + \beta\mu_0]$$

where E is the displacement operator $(E^p\mu_s \equiv \mu_{s+p})$; $\alpha = -(X + n - 2Xn/N)$; $\beta = (X/N)(1 - X/N)N(N - n)$; and $\mu_0 = 1, \mu_1 = 0$ (Pearson [36]). A similar relationship holds among the incomplete central moments:

(7.1) $$\mu_{r(\rho)} = \sum_{j\geq\rho}(j - nX/N)^r \binom{X}{j}\binom{N-X}{n-j}\Big/\binom{N}{n}$$

except that μ_{r+1} on the left side of (7) is replaced by $\mu_{r+1(\rho)} - (\rho - X/N)^r\mu_{1(\rho)}$, and not by $\mu_{r+1(\rho)}$ alone. (For more details about incomplete moments, see Ayyangar [1].)

The mean deviation is

(8)
$$\nu_1 = E[|\mathbf{x} - nX/N|] = 2mN^{-1}(N - X - n + m)\binom{X}{m}\binom{N-X}{n-m}\Big/\binom{N}{n}$$

where m is the greatest integer not exceeding $1 + nX/N$.

Denoting the rth inverse (ascending) factorial moment by

$$\mu_{-[r]} = E[\{(\mathbf{x}+1)^{[r]}\}^{-1}]$$

we have

(9) $$E[(\mathbf{x}+1)^{-1}] = \frac{N+1}{(n+1)(X+1)}\left[1 - \frac{(N-X)^{(n+1)}}{(N+1)^{(n+1)}}\right] \text{ if } N > n + X$$

$$= (N+1)(n+1)^{-1}(X+1)^{-1} \text{ if } N \leq n \quad X$$

and the recurrence relation:

(10) $$\mu_{-[r]} = \frac{1 - \sum_{k=0}^{r-1} P(k \mid N+r, n+r, X+r)}{1 - \sum_{k=0}^{r-1} P(k \mid N+r-1, n+r-1, X+r-1)} \cdot \frac{N+r}{(n+r)(X+r)} \cdot \mu_{-[r-1]} \qquad (r > 1)$$

(Stephan [47]). From (9) and (10) all values $\mu_{-[r]}$ can be calculated.

3. Properties

Denoting the expression on the right hand side of (1) by $P(k \mid X,n,N)$, the following recurrence relations hold:

(11.1) $$P(k+1 \mid X,n,N) = \frac{(X-k)(n-k)}{(k+1)(N-X-n+k+1)}P(k \mid X,n,N)$$

$$(11.2)\qquad P(k \mid X+1,n,N) = \frac{(X+1)(N-X-n+k)}{(N-X)(X+1-k)} P(k \mid X,n,N)$$

$$(11.3)\qquad P(k \mid X,n+1,N) = \frac{(N-X-n+k)(n+1)}{(n+1-k)(N-n)} P(k \mid X,n,N)$$

$$(11.4)\qquad P(k \mid X,n,N+1) = \frac{(N+1-n)(N+1-X)}{(N+1-n-X+k)(N+1)} P(k \mid X,n,N)$$

(see, e.g. Guldberg [16]).

From (11.1) it can be seen that $P(k+1 \mid X,n,N) \gtrless P(k \mid X,n,N)$ according as

$$k \lessgtr \frac{(n+1)(X+1)}{N+2} - 1.$$

Hence $P(k \mid X,n,N)$ increases with k, and reaches a maximum value at the greatest integer which does not exceed $(n+1)(X+1)/(N+2)$. If $(n+1)(X+1)/(N+2)$ is an integer, there are two equal maximum values at $k = (n+1)(X+1)/(N+2) - 1$ and $k = (n+1)(X+1)/(N+2)$. (Note that if N and X are large the maximum is near the expected value nX/N.)

4. Estimation

There has been little systematic development of methods to estimate parameters of hypergeometric populations.

In the most common situation N (also n) is known, and X has to be estimated. The maximum likelihood estimator $\hat{X}$ is the integer maximizing $\binom{\hat{X}}{\mathbf{x}}\binom{N-\hat{X}}{n-\mathbf{x}}$ for the observed value $\mathbf{x}$. From (11.2), $P(\mathbf{x} \mid X+1,n,N) \gtrless P(\mathbf{x} \mid X,n,N)$ according as $X \lessgtr n^{-1}\mathbf{x}(N+1) - 1$. Hence $\hat{X}$ is the greatest integer not exceeding $n^{-1}\mathbf{x}(N+1)$. (If $n^{-1}\mathbf{x}(N+1)$ is an integer, then either $n^{-1}\mathbf{x}(N+1) - 1$ or $n^{-1}\mathbf{x}(N+1)$ is a maximum likelihood estimator.) We may note that the variance of $n^{-1}\mathbf{x}(N+1)$ is (from (3.2)):

$$n^{-1}(N+1)^2(N-n)(N-1)^{-1}(X/N)(1-X/N).$$

Occasionally, we want to estimate N, with X and n known (see Section 7). Then, from (11.4), $P(\mathbf{x} \mid X,n,N+1) \gtrless P(\mathbf{x} \mid X,n,N)$ according as $N \lessgtr nX/\mathbf{x} - 1$. Hence the maximum likelihood estimator, N, of N is the greatest integer not exceeding $nX/\mathbf{x}$ (if $nX/\mathbf{x}$ is an integer, $nX/\mathbf{x} - 1$ and $nX/\mathbf{x}$ are both maximum likelihood estimators). The variance of $nX/\mathbf{x}$ is infinite if (as is usual) n is not greater than $(N-X)$. However, some idea of variability can be gained by considering the *positive hypergeometric distribution* (see Section 8).

Alternately some arbitrary value (e.g., $2nX$) may be assigned to the estimator if $\mathbf{x}$ is zero.

Chapman [6] noted that in equation (9) "the second term in square brackets is negligible for n/N and X/N sufficiently large", and suggested using the estimator

$$\mathbf{N}^* = \frac{(n+1)(X+1)}{\mathbf{x}+1} - 1 \tag{12}$$

since $E[\mathbf{N}^*] \doteqdot N$. He gave the table shown below, from which can be found the minimum sample size needed to make $|E[\mathbf{N}^*] - N| < 1$.

TABLE 1

Minimum Values of nX/N to make $|E[\mathbf{N}^*] - N| < 1$

N	10^4	10^5	10^6	10^7	10^8	10^9
nX/N	9.2	11.5	13.8	16.1	18.4	20.7

Using Waring's formula

$$x^{-1} = \sum_{j=0}^{m-1} [j!/(x+1)^{[j+1]}] + R_m$$

where $$R_m = m!/x^{[m+1]},$$

Chapman obtained the inequality (for $nX/N > 10$)

$$E[nX/\mathbf{x}] \geq N[1 + N/(nX) + 2(N/(nX))^2] \tag{13}$$

showing that the 'natural' estimator $nX/\mathbf{x}$ has a noticeable positive bias even when (according to Table 1) $\mathbf{N}^*$ is nearly unbiased.

The variance of $\mathbf{N}^*$ is approximately

$$N^2[(nX/N)^{-1} + 2(nX/N)^{-2} + 6(nX/N)^{-3}] \tag{14}$$

and this is also approximately the mean square error $E[(\mathbf{N}^* - N)^2]$ if n is large enough. This compares very favorably with the mean square error of $\hat{\mathbf{N}}$ (with $\hat{\mathbf{N}} = 2nX$ if $\mathbf{x} = 0$) which is approximately

$$N^2[3(nX/N)^{-1} + 6(nX/N)^{-2}]. \tag{15}$$

For the estimator

$$\mathbf{N}^{**} = (n+2)(X+2)/(\mathbf{x}+2), \text{ with } n \text{ sufficiently large,}$$

$$E[\mathbf{N}^{**}] \doteqdot N(1 - N/(nX)) \tag{16.1}$$

$$\text{var}[\mathbf{N}^{**}] \doteqdot N^2[(nX/N)^{-1} - (nX/N)^{-2} - (nX/N)^{-3}] \tag{16.2}$$

which indicates that this should be at least as good as $\mathbf{N}^*$.

Pearson [34] obtained the following formula for N in terms of the moments of the distribution:

$$N = \frac{\mu_5\mu_3 - 10\mu_4\mu_2^2 + 6\mu_3^3 + 2\mu_2^4}{\mu_5\mu_3 - 4\mu_4\mu_2^2 + 3\mu_3^3 + 2\mu_2^4}.$$

He suggested that this formula might be used for estimating N, but it has not been so used, and would probably give very inaccurate estimates.

5. Approximations

Unlike the situation of estimation, there are a considerable variety of approximations to the individual probabilities, and also to cumulative sums of these probabilities, for hypergeometric distributions. Many of these are based on the approximation of the hypergeometric distribution (1) by a binomial distribution with parameters n, X/N. More precisely, Śròdka [46] has obtained the following bounds on $P(k \mid X,n,N)$:

$$(17) \quad \binom{n}{k}\left(\frac{X-k}{N}\right)^k\left(\frac{(N-X)-(n-k)}{N}\right)^{n-k}\left(1+\frac{6n^2-6n-1}{12N}\right)$$
$$< P(k \mid X,n,N) < \binom{n}{k}\left(\frac{X}{N}\right)^k\left(1-\frac{X}{N}\right)^{n-k}\left(1+\frac{6n^2+6n-1}{12N}\right)^{-N}.$$

For sufficiently large N this can be written in the simpler form:

$$(18) \quad \binom{n}{k}\left(\frac{X-k}{N}\right)^k\left(\frac{(N-X)-(n-k)}{N}\right)^{n-k}$$
$$< P(k \mid X,n,N) < \binom{n}{k}\left(\frac{X}{N}\right)^k\left(1-\frac{X}{N}\right)^{n-k}\left(1-\frac{n}{N}\right)^{-n}.$$

It is usually adequate to use the simple binomial approximation

$$\binom{n}{k}(X/N)^k(1-X/N)^{n-k}$$

when $n < 0.1N$.

If n, X/N be replaced by n^*, p^* so that n^*p^*, $n^*p^*(1-p^*)$ are equal to the right-hand sides of (3.1), (3.2) respectively, there is marked improvement (Sandiford [40]), even if it is necessary to round n^* to the nearest integer. Greater accuracy may be obtained (Ord [32]) by multiplying by the corrective factor $\{1 + b_1(k)N^{-1}\}$ where

$$b_1(k) = [2(X/N)(1-X/N)]^{-1}[k(1-2X/N) + n(X/N)^2 - (k - nX/N)^2].$$

An approximation of similar form has been constructed by Bennett [2]. This is (for $X \geq n$)

$$\sum_{j=0}^{K} P(j \mid X,n,N) \doteqdot \sum_{j=0}^{K}\binom{n}{j}(X/N)^j(1-X/N)^{n-j}$$

$$-\frac{1}{2}\cdot\frac{n(n-1)(X/N)(1-X/N)}{N-1}\left\{\binom{n}{K+2}(X/N)^{K+2}(1-X/N)^{n-K-2}\right.$$
$$\left.-\binom{n}{K+1}(X/N)^{K+1}(1-X/N)^{n-K-1}\right\}$$

(By the remark following equation (1) we can always arrange that $X \geq n$.)

The following approximation was obtained by Wise [51]:

$$\sum_{k=0}^{K} P(k \mid X,n,N) \doteq \sum_{j=0}^{K} \binom{n}{j} w^j(1-w)^{n-j}$$

where $w = (X - \frac{1}{2}K)/(N - \frac{1}{2}n + \frac{1}{2})$, i.e. the distribution (1) is approximated by a binomial distribution with parameters n, $(X - \frac{1}{2}K)/(N - \frac{1}{2}n + \frac{1}{2})$.

If the binomial approximation to the hypergeometric distribution may itself be approximated by a Poisson or normal approximation (see Chapter 3), then there is a corresponding Poisson or normal approximation to the hypergeometric. Thus when X/N is small but n is large, the Poisson approximation

$$P(k \mid X,n,N) = \exp[-nX/N](nX/N)^k/k! \tag{19}$$

may be used. If n is large, but X/N not small then (using a continuity correction) the cumulative probability

$$\Pr[\mathrm{x} \leq K] = \sum_{k \leq K} P(k \mid X,n,N)$$

is approximated by

$$(2\pi)^{-\frac{1}{2}}\int_{-\infty}^{Y} \exp(-\tfrac{1}{2}u^2)\,du \qquad \text{with}$$

$$Y = (K - nXN^{-1} + \tfrac{1}{2})/\sqrt{(N-n)(N-1)^{-1}n(X/N)(1-X/N)}\,. \tag{20}$$

Hemelrijk [20] has reported that, unless the tail probability is less than about 0.07 *and* $X + n \leq \frac{1}{2}N$, some improvement is effected by replacing $(N-1)^{-1}$ by N^{-1} under the square root sign. A more refined normal approximation suggested by Feller [12] and improved by Nicholson [31], gives bounds on the cumulative probability. We first define:

$$p_1 = (X+1)/(N+2);\ q_1 = 1 - p_1;\ s = (n+1)/(N+2);\ t = 1 - s$$
$$a = \tfrac{1}{6}(p_1 - q_1)(t - s);\ \sigma = \sqrt{(N+1)tpq}$$
$$x_j = [j + \tfrac{1}{2} - (n+1)p_1]\sigma^{-1}.$$

Then for

$$\begin{cases} \sigma > 3 \\ K_1 \geq (N+1)p_1 \\ K_2 + \frac{1}{2} \leq (N+1)p_1 + \frac{2}{3}\sigma^2 \\ n - K_1 \geq 4 \quad \text{and} \quad X - K_2 \geq 4 \end{cases}$$

we have

$$\left(\frac{N+1}{N+2}\right) e^R \frac{1}{\sqrt{2\pi}} \int_{Y_{K_1}}^{Y_{K_2+1}} \exp\left(-\tfrac{1}{2}u^2\right) du \le \sum_{j=K_1}^{K_2} P(k \mid X,n,N)$$

$$\le \left(\frac{N+1}{N+2}\right) e^R \frac{1}{\sqrt{2\pi}} \int_{Y_{K_1}}^{Y'_{K_2+1}} \exp\left(-\tfrac{1}{2}u^2\right) du$$

where

$$R = \frac{5(1 - p_1q_1)(1 - st)}{36\sigma^2} + \frac{2}{3(N+2)}$$

$$(21) \qquad Y_j = \frac{j - (N+1)p_1}{\sigma} + \frac{a}{\sigma}\left[\frac{j - (N+1)p_1}{\sigma}\right]^2 + \frac{2a}{\sigma} + \frac{x_j^3}{\sigma} + \frac{1}{7\sigma}$$

$$Y'_j = \frac{j - (N+1)p_1}{\sigma} + \frac{a}{\sigma}\left[\frac{j - (N+1)p_1}{\sigma}\right]^2 + \frac{2a}{\sigma} - \frac{1}{2\sigma^2}.$$

This approximation is rather complicated. It is good when the distribution is highly skewed ($|a|$ large). For the symmetrical case ($a = 0$) there is little advantage to be gained by using this formula.

Since, as already noted, the hypergeometric distribution is unchanged by interchanging n and X it is clear that a binomial with parameters X, n/N has a claim equal to that of a binomial with parameters n, X/N as an approximating distribution for (1). In addition the distribution of $(n - x)$ might be approximated by a binomial with parameters $N - X$, n/N or that of $(X - x)$ by a binomial with parameters $N - n$, X/N.

Brunk et al. [4] have compared these approximations. Their investigations support (for $X < n$) the opinion of Lieberman and Owen [27] that it is best to use the binomial with smallest first parameter (i.e. min $(n,X,N - X,N - n)$).

Pearson [35] has approximated hypergeometric distributions by (continuous) Pearson Type distributions. This work was continued by Davies [10] who showed, incidentally, that for the hypergeometric distribution $\beta_2 - \beta_1 > 3$, while $\beta_1(\beta_2 + 3)^2 > 4(2\beta_2 - 3\beta_1 - 6)(4\beta_2 - 3\beta_1)$. The Pearson curves which fit hypergeometric distributions are thus mostly of Types VI or III. (See Figure 1 of Chapter 2.) Bol'shev [3] also proposed an approximation of this kind which gives good results for $N \ge 25$. The cumulative distribution function

$$\sum_{k=0}^{K} P(k \mid X,n,N)$$

is approximated by the incomplete beta function ratio

$$(22) \qquad I_{1-\xi}(\lambda - K + c, K - c + 1)$$

with

$$\xi = N^{-1}(N-2)^{-1}[N(n+X-1)-2nX]$$
$$c = n(n-1)X(X-1)(N-1)^{-1}[(N-n)(N-X)+nX-N]^{-1}$$
$$\lambda = (N-2)^2 nX(N-n)(N-X)(N-1)^{-1}[(N-n)(N-X)+nX-N]^{-1} \cdot [N(n+X-1)-2nX]^{-1}.$$

In the limit as $N \to \infty$, the approximation becomes more exact.

For the case $N = 20$, $X = n = 10$ (a symmetric case, which has a close simple normal approximation), approximation (22) is compared with the simple normal approximation (20) in Table 2.

TABLE 2

Comparison of Beta and Normal Approximations to the Hypergeometric Distribution

$P(x \mid 10,10,20)$	$x = 1$	$x = 2$	$x = 3$	$x = 4$
Exact	0.0005	0.0115	0.0894	0.3281
Normal Approximation	0.0001	0.0146	0.0955	0.3315
Beta Approximation	0.0000	0.0094	0.0938	0.3342

Uhlmann [49] has made a systematic comparison between the hypergeometric distribution (with parameters n, Np, N $(0 < p < 1)$) and the binomial distribution (with parameters n, p). Denoting $\Pr[\mathbf{x} \le c]$ for the two distributions by $L_{N,n,c}(p)$ and $L_{n,c}(p)$ respectively, he shows that in general:

(23)

$$L_{N,n,c}(p) - L_{n,c}(p) \begin{cases} =0 & \text{for } p = 0 \\ >0 & \text{for } 0 < p \le c(n-1)^{-1}N(N+1)^{-1} \\ <0 & \text{for } c(n-1)^{-1}N(N+1)^{-1} + (N+1)^{-1} \le p < 1 \\ =0 & \text{for } p = 1. \end{cases}$$

If n is *odd* and $c = \frac{1}{2}(n-1)$ it is possible to eliminate the range of values,

$$c(n-1)^{-1}N(N+1)^{-1} < p < c(n-1)^{-1}N(N+1)^{-1} + (N+1)^{-1},$$

where the sign of the difference is indeterminate, and replace (23) by:

$$L_{N,n,(n-1)/2}(p) - L_{n,(n-1)/2}(p) \begin{cases} >0 & \text{for } 0 < p < \frac{1}{2} \\ =0 & \text{for } p = \frac{1}{2} \\ <0 & \text{for } \frac{1}{2} < p < 1. \end{cases} \tag{24}$$

6. Tables

An extensive set of tables of individual and cumulative probabilities for hypergeometric distribution has been prepared by Lieberman and Owen [27]. They give values for individual and cumulative probabilities to six decimal places for

(i) $N = 2(1)100$; $n = 1(1)50$ for all possible values of X

(ii) $N = 1000$; $n = 500$ for all possible values of X

(iii) $N = 100(100)2000$; $n = \frac{1}{2}N$; $X = n - 1$ and $X = n$.

Less extensive tables were published earlier by DeLury and Chung [11]. There are tables by Neild [30], however they are not as yet easily accessible.

Graphs based on evaluation of hypergeometric probabilities are given by Clark and Koopmans [8].

7. Applications

The hypergeometric distribution frequently replaces a binomial distribution when it is desirable to make allowance for finiteness of sample size.

An interesting application is in the estimation of the size of animal populations from "capture-recapture" data. This kind of application dates back at least to 1896 (Petersen [37], quoted by Chapman [7]). Consider, for example, estimation of the number (N) of fish in a pond. First, a known number (X) of fish are netted, marked ('tagged') and returned to the pond. A short time later, long enough to ensure (hopefully) random dispersion of the tagged fish, but not long enough for natural changes to affect the population size too much, a sample of size n is taken from the pond, and the number (x) of tagged fish in the sample is observed. The likelihood (to be regarded as a function of N) is then $P(\mathbf{x} \mid X,n,N)$ and the maximum likelihood estimator of N is approximately $nX/\mathbf{x}$ (see Section 4). The positive moments of this statistic are infinite unless n exceeds $(N - X)$, which is unlikely to be the case. However, as pointed out in Section 4, the estimators $\frac{(n+1)(X+1)}{(\mathbf{x}+1)} - 1$ or $\frac{(n+2)(X+2)}{(\mathbf{x}+2)}$ appear to be more suitable and do not have infinite moments.

There are many variations in the conditions under which this problem is encountered (see for example, Freeman [13] and Darroch [9]

The following is an application of this distribution to linguistic problems:

In primitive Indo-European, the parent language of English, German, Greek, Latin, Russian and many other languages, all words can be expressed as *PRSE* where *P* denotes 0, or 1, or 2, or . . . prefixes; *R* denotes a root; *S* denotes 0, or 1, or 2, or . . . suffixes; and *E* denotes 0 or 1 ending. One way of assessing the closeness of two "branches" (i.e. languages) of this Indoeuropean tree is as follows: construct a table whereby a column is reserved for each branch of the Indoeuropean "tree," and a row is reserved for each "attested" Indoeuropean root. An "attested" root is defined as one which appears in at least two languages.

Root Number	Italic	Greek	Germanic
1	*	*	
2		*	*
3	*	*	*

For each row, corresponding to a given root, indicate by some symbol, say *, all the columns (languages) which contain this root. The number of common *'s will then be a measure of the closeness of any two languages. If the number of common roots is R, then one could compute the probability of obtaining at least R common *'s if the *'s were arranged in the two columns in a purely random fashion, and if this probability were small, then conclude that the two languages are related. Of course this requires a fixed 'a priori' definition of what will be considered 'small enough' to reach this conclusion. Consider a case where there are N attested roots with n_A and n_B entries in the two columns and R is the number of common roots.

	Root	Language A	Language B	Common
	1	*		
	2			
	3	*	*	✓
	4		*	
	⋮			
	N	*	*	✓
Total	N	n_A	n_B	

If $N \geq (n_A + n_B)$ we can consider the probability of obtaining exactly R common *'s as the hypergeometric probability of R white balls in a sample of n_A balls drawn without replacement from an urn containing N balls, n_B of which are white (n_A and n_B can be interchanged as shown in the section defining the hypergeometric distribution). Thus, the probability of obtaining at least R common *'s is the upper hypergeometric cumulative probability $\overline{P}(R \mid n_B,n_A,N)$; that is

$$\begin{aligned} P(R) &= P\{\text{at least } R \text{ common } *\text{'s}\} \\ &= \overline{P}(R \mid n_B,n_A,N) \\ &= \frac{\binom{N-n_B}{n_A}}{\binom{N}{n_A}} F_R(-n_B,n_A;N-n_A-n_B+1;1) \end{aligned}$$

where

$$F_R(-\alpha,-\beta;\gamma;z)$$

$$= \sum_{S=R}^{\min(\alpha,\beta)} \frac{\alpha(\alpha-1)\ldots(\alpha-S+1)\beta(\beta-1)\ldots(\beta-S+1)}{\gamma(\gamma+1)\ldots(\gamma+S-1)} \cdot \frac{z^S}{S!}.$$

If $N < (n_A + n_B)$, we can consider the probability of obtaining exactly R common $*$'s as the hypergeometric probability of obtaining $N + R - n_A - n_B$ white balls in a sample of size $N - n_A$ drawn without replacement from an urn containing N balls, $N - n_B$ of which are white (n_A and n_B can again be interchanged). Hence the probability of at least R common $*$'s becomes the upper hypergeometric cumulative probability

$$\bar{P}(N + R - n_A - n_B \mid N - n_B, N - n_A, N)$$

that is

$$\begin{aligned} P(R) &= P\{\text{at least } R \text{ common } *\text{'s}\} \\ &= \bar{P}(N + R - n_A - n_B \mid N - n_B, N - n_A, N) \\ &= \frac{\dbinom{N - (N - n_B)}{N - n_A}}{\dbinom{N}{N - n_A}} \\ &\quad \times F_{N+R-n_A-n_B}(-N + n_B, -N + n_A; N - n_A - n_B + 1; 1). \end{aligned}$$

Some data collected by Ross [38] are shown below.

TABLE 3

*Number of Roots Common to Certain Indo-European Branches**

	n_1	n_2	R
Ce It — Gr	1,184	1,165	783
— Ar	1,184	442	333
— Ir Sk	1,184	1,016	694
— Sl Ba	1,184	1,213	777
— Ge	1,184	1,256	865
— Al	1,184	290	236
Gr — Ar	1,165	442	333
— Ir Sk	1,165	1,016	694
— Sl Ba	1,165	1,213	753
— Ge	1,165	1,256	763
— Al	1,165	290	242

*Abbreviations as in Tables 4 and 5.

Ross [38] points out that this method of analysis does not take into account that the presence of a root in *only* X and Y languages is stronger evidence for a relationship between them, than merely X and Y having a common root. The statistical results must be judged accordingly.

Kroeber and Chrétien [26] have considered different criteria for judging the relationship between two languages. They use a selection of 74 "philological" features and construct a table showing the presence or absence of such features as shown below. The analysis of this data was carried out by Ross [38] using the method outlined previously. Some of these results are given below. (The original table contained 74 rows.)

TABLE 4

Showing whether Certain Philological Features are Present (+) *or Absent* (−) *in Certain Indo-European Branches**

Number of feature	Ce	It	Gr	Ar	Ir	Sk	Sl	Ba	Ge
1	−	−	−	−	+	+	−	−	−
2	−	−	−	−	+	+	−	−	−
3	−	−	−	−	+	+	−	−	−
4	+	−	−	+	+	+	−	−	−
5	−	−	−	−	+	+	−	−	(+)
6	−	−	−	−	+	+	−	−	−
7	−	−	+	+	−	−	−	+	+
8	+	+	−	+	−	−	−	+	+
9	+	+	−	−	−	−	−	−	−
10	+	+	−	−	−	−	−	−	−
37	−	−	+	(−)	+	−	−	−	+

**Abbreviations:* Ce = Celtic, It = Italic, Gr = Greek, Ar = Armenian, Ir = Iranian, Sk = Sanskrit, Sl = Slavonic, Ba = Baltic, Ge = Germanic. Only the entries + or (+) are reckoned as showing the presence of a feature in a branch.

For more details of this linguistic problem and interesting discussions, see the paper by Ross [38].

A further application of the hypergeometric distribution is in a model by Irwin [22] for the number of children attacked by an infectious disease, when a fixed number (N) are exposed to it.

8. Related Distributions

8.1 *Positive Hypergeometric Distribution*

This distribution is formed by omitting the zero observed value. (If n exceeds $(N - X)$ there is no zero observed value.) If $n \leq N - X$ then the distribution

TABLE 5

*Tests of Significance of the Data of Table 4**

	n_1	n_2	R	$P(R)$
Ce — It	25	29	21	$.0^7 19$
— Gr	25	23	5	.94
— Ar	25	22	3	.99
— Ir	25	31	7	.97
— Sk	25	27	6	.97
— Sl	25	27	7	.91
— Ba	25	28	6	.99
— Ge	25	25	7	.84
Gr — Ar	23	22	9	.18
— Ir	23	31	12	.31
— Sk	23	27	10	.94
— Sl	23	27	6	.47
— Ba	23	28	7	.89
— Ge	23	25	6	.91

*Abbreviations as for Table 4. Results to two significant figures, subject to the convention that, if ϵ is positive, $(0.99 + \epsilon)$ is entered as .99 whatever be the value of ϵ.

is

$$\Pr[\mathbf{x} = k] = \binom{X}{k}\binom{N-X}{n-k}\Bigg/\left[\binom{N}{n} - \binom{N-X}{n}\right] \qquad (25)$$

$$[k = 1, \ldots \min(n, X)].$$

(Reference has already been made to this distribution in Sections 4 and 7.)

Govindarajulu [15] has made a detailed study of the inverse moments of this distribution. He has given values of $E(\mathbf{x}^{-r})$ for $r = 1, 2$ and

$N = 1(1)20;\ X = 1(1)N;\ n = 1(1)X$

$N = 25(5)50;\ X/N = 5\%(5\%)100\%;\ n = 1(1)X$

$N = 55(5)100(10)140;\ X/N = 5\%(5\%)100\%;\ n/N(\leq X/N) = 5\%(5\%)100\%.$

8.2 *Compound Hypergeometric Distributions*

The most common compound distributions are obtained by supposing that **X** is a random variable. Hald [18] and Horsnell [21] considered various prior distributions of **X**. Distributions of this kind are discussed in Chapter 8.

8.3 *Negative Hypergeometric Distribution*

If sampling without replacement (as described in Section 1) is continued until a white balls (or, alternatively, b black balls) are obtained ($0 < a \leq X$; $0 < b \leq N - X$), then the distribution of the number of draws needed is a *negative hypergeometric distribution.* (The definition is analagous to that of the negative binomial distribution, but the range of values of a negative hypergeometric distribution is finite.)

For this distribution

$$(26) \qquad \Pr[\mathbf{x} = k] = \frac{\binom{X}{a-1}\binom{N-X}{k-a}}{\binom{N}{k-1}} \cdot \frac{X - a + 1}{N - k + 1}$$

($a \leq k \leq N - X + a$). Gart [14] and Skellam [45] have obtained a special case of the negative hypergeometric distribution as a compound binomial distribution (see Chapter 8). The distribution and its properties are discussed in detail by Bol'shev [3]. Extension to multivariate case (e.g. Keats [23]) is discussed in Chapter 11.

The rth *ascending* factorial moment is

$$(27) \qquad E[\mathbf{x}^{[r]}] = E[\mathbf{x}(\mathbf{x} + 1) \ldots (\mathbf{x} + r - 1)] = a^{[r]}(N + 1)^{[r]}/(X + 1)^{[r]}.$$

In particular, the expected value is

$$a(N + 1)/(X + 1)$$

and the variance is

$$a(N + 1)(N - X)(X + 1 - a)(X + 1)^{-2}(X + 2)^{-1}$$

(see Matuszewski [28] and Chahine [5]).

As N and X tend to infinity with X/N tending to a fixed limit (between 0 and 1), this distribution tends to a negative binomial distribution.

8.4 *Noncentral Hypergeometric Distribution*

This is the name given by Wallenius [50] to a distribution constructed by supposing that in sampling without replacement (as in Section 1) the probability of drawing a white ball, given that there are X' white and $(N' - X')$ black balls is not X'/N', but $X'[X' + \theta(N' - X')]^{-1}$ with $\theta \neq 1$. The mathematical analysis following from this assumption is rather involved. Starting from the recurrence formula

$$\Pr[\mathbf{x} = k \mid X,n,N] = [X + \theta(N - X)]^{-1}[X\Pr(\mathbf{x} = k - 1 \mid X - 1,n - 1,N - 1) + \theta(N - X)\Pr(\mathbf{x} = k \mid X,n - 1,N - 1)],$$

Wallenius obtains the formula

$$\text{(28)} \qquad \Pr[\mathbf{x} = k] = \binom{X}{k}\binom{N - X}{n - k}\int_0^1 (1 - t^c)^k(1 - t^{\theta c})^{n-k}\,dt$$

with $c = [X - k + \theta(N - X - n + k)]^{-1}$; and the bounds (according as $\theta \gtrless 1$)

$$\text{(29)} \qquad \binom{n}{k}\frac{X^{(k)}(N - X)^{(n-k)}}{[X + \theta(N - X)]^{(k)}[N - X + (x - k)/\theta]^{(n-k)}} \lesseqgtr \Pr[\mathbf{x} = k]$$
$$\lesseqgtr \binom{n}{k}\frac{X^{(k)}(N - X)^{(n-k)}}{[X + \theta(N - X - n + k)]^{(k)}[N - X + X/\theta]^{(n-k)}}.$$

For n small compared with X and $N - X$, $\mathbf{x}$ is approximately distributed binomially with parameters n, $[1 + \theta X/(N - X)]^{-1}$.

8.5 *Generalized Hypergeometric Distributions*

In formula (1), it is not essential that all the parameters n, X and N be positive. In fact, we can take any two of them to be negative, and the remaining one to be positive, and still obtain positive values for the probabilities calculated formally from (1). Such distributions are termed *generalized hypergeometric distributions*. Approximations of such distributions by Pearson Type distributions were studied by Davies [10], who showed that they were adequate for many practical purposes.

Kemp and Kemp [24] have studied conditions under which the quantities

$$\text{(30)} \qquad P_r = \binom{a}{r}\binom{b}{n - r}\Big/\binom{a + b}{n} \qquad (r = 0,1,2,\ldots)$$

can represent a distribution, with a, b and n possibly being allowed to take any real values. It is clear that among the class of distributions included in (30) will be some for which $P_r = 0$ if r exceeds an integer, R. This will be so if a or n is a positive integer. Kemp and Kemp introduce the convention that $P_r = 0$ for all $r \geq R + 1$ if $P_{R+1} = 0$. They also define, for $\alpha < 0$ and $\beta < 0$ with β an integer

$$\frac{\alpha!}{(\alpha + \beta)!} = \frac{(-1)^\beta(-\alpha - \beta - 1)!}{(-\alpha - 1)!} = (-1)^\beta\frac{\Gamma(-\alpha - \beta)}{\Gamma(-\alpha)}.$$

With these conventions, they distinguished four main types of distribution included in (30), divided into sub-types according to the following scheme.

(Here J denotes some non-negative integer — the same for any one type)

Type IA(i):	$n - b - 1 < 0$; n integral; $0 \le n - 1 < a$	$(r = 0,1,\ldots,n)$
Type IA(ii):	$n - b - 1 < 0$; a integral; $0 \le a - 1 < n$	$(r = 0,1,\ldots,a)$
Type IB:	$n - b - 1 < 0; J < a < J + 1; J < n < J + 1$	
Type IIA:	$a < 0$; $n > 0$; n integral; $b < 0$; $b \neq -1$	$(r = 0,1,\ldots,n)$
Type IIB:	$a < 0 < a + b + 1; J < n < J + 1; J < n - b - 1 < J + 1$	
Type IIIA:	$n < 0 < a; b < n - a; b \neq n - a - 1$; a integral	$(r = 0,1,\ldots,a)$
Type IIIB:	$n < 0 < a + b + 1; J < a < J + 1; J < n - b - 1 < J + 1$	
Type IV:	$a < 0, n < 0; 0 < a + b + 1$	

(where no limits are shown for r, it can take any non-negative integral value).

The general formula for the rth factorial moment (*if it exists*) for this distribution is:

$$\mu_{(r)} = \frac{a!n!(a + b - r)!}{(a - r)!(n - r)!(a + b)!}. \tag{31}$$

Moments exist for:

Type IA(i)	$r \le n$
Type IA(ii)	$r \le a$
Type IB	$r \le a + b + 1$
Type IIA	Always (zero if $r > n$)
Type IIB	Never
Type IIIA	Always (zero if $r > a$)
Type IIIB	Never
Type IV	$r < a + b + 1$

In other words

$\mu_{(r)}$ is finite,

for all r,	for Types IIA and IIIA
for all $r < a + b + 1$	for Types IB and IV
for $r \le n$	for Type IA(i)
for $r \le a$	for Type IA(ii)

The 'ordinary' hypergeometric distribution belongs to Type IA(i) or IA(ii) with n, a and b all integers. The negative hypergeometric distribution (Section 8.3) belongs to Type IIIA, with n, a, b all integers. A sort of dualism can be noted between Types IA(i) and IA(ii); IIA and IIIA; and IIB and IIIB (making the substitutions $a \leftrightarrow n$; $a + b - n \rightarrow b$).

Type IIA are *Polya-Eggenberger* distributions, which will be discussed in Chapter 9, Section 4. This type can arise as a mixture of binomial distributions (see Chapter 8, Section 2(x)); Skellam [45] obtained the distribution in this way. Kemp and Kemp [24] also note another derivation by Irwin [22] which will be described in Chapter 9. Maximum likelihood equations for estimators $\hat{\mathfrak{a}}$, $\hat{\mathfrak{b}}$ of parameters a, b of Type IIA distributions have been given by Kemp and Kemp in [25].

Given N independent observations, the equations are

$$(32.1)\qquad \sum_{j=0}^{n-1} \mathbf{A}_j(j - \hat{\mathfrak{a}})^{-1} - \mathbf{A}_{-1} \sum_{j=1}^{n-1} (j - \hat{\mathfrak{a}} - \hat{\mathfrak{b}})^{-1} = 0$$

$$(32.2)\qquad \sum_{j=0}^{n-1} \mathbf{B}_j(j - \hat{\mathfrak{b}})^{-1} - \mathbf{A}_{-1} \sum_{j=1}^{n-1} (j - \hat{\mathfrak{a}} - \hat{\mathfrak{b}})^{-1} = 0$$

where $\mathbf{A}_j = \text{number of observations} \geq j + 1$

$\mathbf{B}_j = \text{number of observations} \leq n - j - 1$

The equations may be solved for $\hat{\mathfrak{a}}$ and $\hat{\mathfrak{b}}$ iteratively. If $\hat{\mathfrak{a}}_1$, $\hat{\mathfrak{b}}_1$ are trial values of $\hat{\mathfrak{a}}$, $\hat{\mathfrak{b}}$ respectively making the left hand sides of (32.1) and (32.2) equal to F_1, F_2 then improved values are $\hat{\mathfrak{a}} - (\delta\hat{\mathfrak{a}})$, $\hat{\mathfrak{b}} - (\delta\hat{\mathfrak{b}})$ where

(33.1)

$$F_1 = (\delta\hat{\mathfrak{a}}) \sum_{j=0}^{n-1} \mathbf{A}_j(j - \hat{\mathfrak{a}}_1)^{-2} - \{(\delta\hat{\mathfrak{a}}) + (\delta\hat{\mathfrak{b}})\}\mathbf{A}_{-1} \sum_{j=0}^{n-1} (j - \hat{\mathfrak{a}}_1 - \hat{\mathfrak{b}}_1)^{-2}$$

(33.2)

$$F_2 = (\delta\hat{\mathfrak{b}}) \sum_{j=0}^{n-1} \mathbf{B}_j(j - \hat{\mathfrak{b}}_1)^{-2} - \{(\delta\hat{\mathfrak{a}}) + (\delta\hat{\mathfrak{b}})\}\mathbf{A}_{-1} \sum_{j=0}^{n-1} (j - \hat{\mathfrak{a}}_1 - \hat{\mathfrak{b}}_1)^{-2}$$

Initial values for $\hat{\mathfrak{a}}_1$, $\hat{\mathfrak{b}}_1$ can be obtained by equating sample and population mean and variance. Kemp and Kemp [25] report that about five iterations usually suffice to give satisfactory results.

Sarkadi [41] pointed out that Types IA, IIA and IIIA are all forms of Pólya distribution (see Chapters 8–9). He also extended the class of distribution corresponding to (30) by including cases when the lowest value of r is greater than zero ($r \geq n - b$, $n - b$ integral) and also the cases $b = -1$, $n - a - 1$ that were excluded from Types IIA and IIIA, respectively, by Kemp and Kemp. He pointed out that the sum of P_r over the ranges $0 \leq r \leq n$, and $0 \leq r \leq a$ respectively is equal to 1 in these cases, and so formula (30) defines a proper distribution. In the special case $a = b = -1$ the *discrete rectangular* distribution

$$P_r = (n + 1)^{-1} \qquad (r = 0,1,\ldots n)$$

is obtained. This will be discussed in Chapter 10.

A comparison between generalized hypergeometric, binomial and negative binomial distributions (from Kemp and Kemp [24]) is presented in Table 6.

8.6 '*Extended*' *Hypergeometric Distributions*

This is the name given by Harkness [19] to the conditional distribution of one of two independent binomial variables given their sum. If $\mathbf{x}_j$ has parameters $n_j, p_j (j = 1,2)$ then

TABLE 6

Comparison of Generalized Hypergeometric, Binomial and Negative Binomial Distributions†

$\Pr[\mathbf{x} = r]$

r	IA(ii)	Bin.	IIA	IIIA	Neg. Bin.	IV
0	0.076	0.107	0.137	0.123	0.162	0.197
1	0.265	0.269	0.266	0.265	0.269	0.267
2	0.348	0.302	0.270	0.284	0.247	0.220
3	0.222	0.201	0.184	0.195	0.164	0.144
4	0.075	0.088	0.093	0.093	0.089	0.083
5	0.013	0.026	0.036	0.032	0.042	0.044
6	0.001	0.006	0.011	0.007	0.017	0.023
7	0.000	0.001	0.003	0.001	0.007	0.011
8	0.000	0.000	0.000	0.000	0.002	0.005
9	—‡	0.000‡	0.000	—	0.001	0.003
10	—	0.000	0.000	—	0.000	0.001
11	—	—	—	—	0.000	0.001
≥12	—	—	—	—	0.000	0.001

†For each distribution, expected value is 2 and n (or N) is 10.
‡A dash, —, means probability is zero; 0.000 means $0 <$ probability < 0.0005.

$$\text{(34)} \quad \Pr[\mathbf{x}_1 = k \mid \mathbf{x}_1 + \mathbf{x}_2 = m] = \binom{n_1}{k}\binom{n_2}{m-k}\left(\frac{p_1q_2}{p_2q_1}\right)^k \times \left[\sum_j \binom{n_1}{j}\binom{n_2}{n-j}\left(\frac{p_1q_2}{p_2q_1}\right)^j\right]^{-1}$$

with $q_j = 1 - p_j$, for $\max(0,m - n_2) \leq k \leq \min(n_1,m)$; the same limits apply to j in the summation.

An alternative form to (34) is

$$\text{(34)}' \quad \Pr[\mathbf{x}_1 = k \mid \mathbf{x}_1 + \mathbf{x}_2 = m] = \frac{\left[\binom{n_1}{k}\binom{n_2}{m-k}\left(\frac{p_1q_2}{p_2q_1}\right)^k \Big/ \binom{n_1+n_2}{m}\right]}{F(-n_1,-m;-m+n_2+1;p_1q_2/(p_2q_1))}$$

The rth factorial moment of the distribution is

$$\text{(35)} \quad \mu'_{(r)} = \frac{F(r-n_1,r-m;r-m+n_2+1;p_1q_2/(p_2q_1))}{F(-n_1,-m;-m-n_2+1;p_1q_2/(p_2q_1))}$$

There are no simple explicit expressions for the moments although it is possible to obtain recurrence relations, such as (putting $t = p_1q_2/(p_2q_1)$):

$$2(1 - t)\mu_2 = n_1mt - \{n_1 + n_2 - (n_1 + m)(1 - t)\}\mu_1'$$

Harkness [19] has obtained limits for the maximum likelihood estimator, $\hat{\mathbf{t}}$, of t, based on n independent random variables $\mathbf{y}_1, \mathbf{y}_2, \ldots, \mathbf{y}_n$ each having distribution (34). These are:

(*a*) for $\bar{\mathbf{y}} = n^{-1} \sum_{j=1}^{n} \mathbf{y}_j \le mn_1/(n_1 + n_2)$

$$\bar{\mathbf{t}} \le \hat{\mathbf{t}} \le \bar{\mathbf{t}} + \frac{\bar{\mathbf{y}}(mn_1 - (n_1 + n_2)\bar{\mathbf{y}})}{mn_1(n_1 - \mathbf{y})(\overline{m} - \bar{\mathbf{y}})} \tag{36.1}$$

where $\bar{\mathbf{t}}$ is the "natural" estimator

$$\bar{\mathbf{t}} = \frac{\bar{\mathbf{y}}(n_2 - m + \bar{\mathbf{y}})}{(n_1 - \bar{\mathbf{y}})(m - \bar{\mathbf{y}})}$$

(*b*) for $\bar{\mathbf{y}} \ge mn_1/(n_1 + n_2)$

$$\bar{\mathbf{t}}\left[1 + \frac{(n_1 - \bar{\mathbf{y}})(n_1 + n_2 - n_1m - \bar{\mathbf{y}})}{n_1(n_1 + n_2 - m)(n_1 - \bar{\mathbf{y}})(m - \bar{\mathbf{y}})}\right]^{-1} \le \hat{\mathbf{t}} \le \bar{\mathbf{t}} \tag{36.2}$$

REFERENCES

[1] Ayyangar, A. A. K. (1934). A note on the incomplete moments of the hypergeometrical series, *Biometrika*, **26**, 264–265.

[2] Bennett, W. S. (1965). *A new binomial approximation for cumulative hypergeometric probabilities*, Unpublished Ph.D. thesis, American University, Washington, D.C.

[3] Bol'shev, L. N. (1964). Distributions related to the hypergeometric distribution *Teoriya Veroyatnostei i ee Primeneniya*, **9**, 687–692. (In Russian)

[4] Brunk, H. D., Holstein, J. E. and Williams, F. (1968). A comparison of binomial approximations to the hypergeometric distribution, *American Statistician*, **22**, 24–26.

[5] Chahine, J. (1965). Une généralization de la loi binomiale négative, *Revue Statistique Appliquée*, **13**, 33–43.

[6] Chapman, D. G. (1951). Some properties of the hypergeometric distribution with applications to zoological sample censuses, *University of California Publications in Statistics*, **1**, 131–159.

[7] Chapman, D. G. (1952). Inverse, multiple and sequential sample censuses, *Biometrics*, **8**, 286–306.

[8] Clark, C. R. and Koopmans, L. H. (1959). *Graphs of the hypergeometric O.C. and A.O.Q. functions for lot sizes 10 to 225*, Sandia Corporation Monograph, SCR-121.

[9] Darroch, J. N. (1964). On the distribution of the number of successes in independent trials, *Annals of Mathematical Statistics*, **35**, 1317–1321.

[10] Davies, O. L. (1933, 1934). On asymptotic formulae for the hypergeometric series, Part I, *Biometrika*, **25**, 295–322; Part III, **26**, 59–107.

[11] DeLury, D. B. and Chung, J. H. (1950). *Confidence Limits for the Hypergeometric Distribution*, Toronto: University of Toronto Press.

[12] Feller, W. (1957). *An Introduction to Probability Theory and Its Applications*, Vol. I. New York: John Wiley & Sons, Inc.

[13] Freeman, H. (1963). *Introduction to Statistical Inference*, Reading, Mass.: Addison-Wesley Publishing Company, Inc.

[14] Gart, J. J. (1963). A median test with sequential application, *Biometrika*, **50**, 55–62.

[15] Govindarajulu, Z. (1962). The first two moments of the reciprocal of the positive hypergeometric variable, *Case Institute of Technology Statistical Laboratory Report*, No. 1061.

[16] Guldberg, A. (1931). On discontinuous frequency-functions and statistical series, *Skandinavisk Aktuarietidskrift*, **14**, 167–187.

[17] Gumbel, E. J. and von Schelling, H. (1950). The distribution of the number of exceedances, *Annals of Mathematical Statistics*, **21**, 247–262.

[18] Hald, A. (1960). The compound hypergeometric distribution and a system of single sampling inspection plans based on prior distributions and costs, *Technometrics*, **2**, 275–340.

[19] Harkness, W. L. (1965). Properties of the extended hypergeometric distribution, *Annals of Mathematical Statistics*, **36**, 938–945.

[20] Hemelrijk, J. (1967). The hypergeometric, the normal and chi-squared, *Statistica Neerlandica*, **21**, 225–229.

[21] Horsnell, G. (1957). Economical acceptance sampling schemes, *Journal of the Royal Statistical Society, Series A*, **120**, 148–191.

[22] Irwin, J. O. (1954). A distribution arising in the study of infectious diseases, *Biometrika*, **41**, 266–268.

[23] Keats, J. A. (1964). Some generalizations of a theoretical distribution of mental test scores, *Psychometrika*, **29**, 215–231.

[24] Kemp, C. D. and Kemp, A. W. (1956). Generalized hypergeometric distributions, *Journal of the Royal Statistical Society, Series B*, **18**, 202–211.

[25] Kemp, C. D. and Kemp, A. W. (1956). The analysis of point quadrat data, *Australian Journal of Botany*, **4**, 167–174.

[26] Kroeber, A. L. and Chrétien, C. D. (1937). Quantitative classification of Indo-European languages, *Language*, **13**, 83–103.

[27] Lieberman, G. J. and Owen, D. B. (1961). *Tables of the Hypergeometric Probability Distribution*, Stanford: Stanford University Press.

[28] Matuszewski, T. I. (1962). Some properties of Pascal distribution for finite population, *Journal of the American Statistical Association*, **57**, 172–174 (Correction: **57**, 919).

[29] Mood, A. M. (1943). On the dependence of sampling inspection plans upon population distributions, *Annals of Mathematical Statistics*, **14**, 415–425.

[30] Neild, E. F. (1960). Hypergeometric distribution, *I–IX*, Thesis (S.B.), Massachusetts Institute of Technology.

[31] Nicholson, W. L. (1956). On the normal approximation to the hypergeometric distribution, *Annals of Mathematical Statistics*, **27**, 471–483.

[32] Ord, J. K. (1968). Approximations to distribution functions which are hypergeometric series, *Biometrika*, **55**, 243–248.

[33] Pearson, K. (1895). Contributions to the mathematical theory of evolution: I. Skew variation in homogeneous material, *Philosophical Transactions of the Royal Society of London*, **186**, 343–414.

[34] Pearson, K. (1899). On certain properties of the hypergeometrical series, and on the fitting of such series to observation polygons in the theory of chance, *Philosophical Magazine, 5th series*, **47**, 236–246.

[35] Pearson, K. (1906). On the curves which are most suitable for describing the frequency of random samples of a population, *Biometrika*, **5**, 172–175.

[36] Pearson, K. (1924). On the moments of the hypergeometrical series, *Biometrika*, **16**, 157–162.

[37] Petersen, G. G. J. (1896). The yearly immigration of young plaice into the Linfjord from the German Sea, etc. *Report of the Danish Biological Station*, **6**, 1–48.

[38] Ross, A. S. C. (1950). Philological probability problems, *Journal of the Royal Statistical Society, Series. B*, **12**, 19–41.

[39] Sandelius, M. (1951). Unbiased estimation based on inverse hypergeometric sampling, *Kungl. Lantbrukshögskolans Annaler*, **18**, 123–127.

[40] Sandiford, P. J. (1960). A new binomial approximation for use in sampling from finite populations, *Journal of the American Statistical Association*, **55,** 718–722.

[41] Sarkadi, K. (1957). Generalized hypergeometric distributions, *Magyar Tudományos Akadémia Matematikai Kutató Intézetének Közleményei*, **2,** 59–68.

[42] Sarkadi, K. (1957). On the distribution of the number of exceedances, *Annals of Mathematical Statistics*, **28,** 1021–1023.

[43] Sarkadi, K. (1960). On the median of the distribution of exceedances, *Annals of Mathematical Statistics*, **31,** 225–226.

[44] Sawkins, D. T. (1947). A new method of approximating the binomial and hypergeometric probabilities, *Proceedings of the Royal Society of New South Wales*, **81,** 38–46.

[45] Skellam, J. G. (1948). A probability distribution derived from the binomial distribution by regarding the probability of success as variable between the sets of trials, *Journal of the Royal Statistical Society, Series B*, **10,** 257–261.

[46] Śródka, T. (1963). On approximation of hypergeometric distribution, *Zeszyty Naukowe Politechniki Lódzkiej*, **53,** 5–17.

[47] Stephan, F. F. (1945). The expected value and variance of the reciprocal and other negative powers of a positive Bernoullian variate, *Annals of Mathematical Statistics*, **16,** 50–61.

[48] Stevens, W. L. (1938). Distributions of groups in a sequence of alternatives, *Annals of Eugenics, London*, **9,** 10–17.

[49] Uhlmann, W. (1966). Vergleich der hypergeometrischen mit der Binomial-Verteilung, *Metrika*, **10,** 145–158.

[50] Wallenius, K. T. (1963). *Biased sampling; The noncentral hypergeometric probability distribution*, Ph.D. thesis, Stanford University.

[51] Wise, M. E. (1954). A quickly convergent expansion for cumulative hypergeometric probabilities, direct and inverse, *Biometrika*, **41,** 317–329.

7

Logarithmic Series Distribution

1. Definition and Properties

The random variable **x** has a logarithmic series distribution if

(1) $$\Pr[\mathbf{x} = k] = \alpha\theta^k/k \qquad (k = 1,2,\ldots;\ 0 < \theta < 1)$$

where $\alpha = -[\log(1 - \theta)]^{-1}$. The probabilities are the terms in the series expansion of $-\alpha \log (1 - \theta)$.

The moment generating function of **x** is

(2) $$E(e^{t\mathbf{x}}) = [\log(1 - \theta e^t)]/[\log(1 - \theta)].$$

The probability generating function is

(3) $$\alpha \sum_{j=1}^{\infty} (\theta^j/j)t^j = [\log(1 - \theta t)]/[\log(1 - \theta)].$$

The rth factorial moment is

(4) $$\begin{aligned}\mu_{(r)} = E(\mathbf{x}^{(r)}) &= \alpha\theta^r \sum_{k=r}^{\infty} (k-1)(k-2)\ldots(k-r+1)\theta^{k-r} \\ &= \alpha\theta^r \frac{d^{r-1}}{d\theta^{r-1}}\left[\sum_{k=1}^{\infty} \theta^{k-1}\right] \\ &= \alpha\theta^r(r-1)!(1-\theta)^{-r}.\end{aligned}$$

The first four moments about the origin are

(5.1) $$\mu_1' = \alpha\theta(1 - \theta)^{-1}$$

$$\mu_2' = \alpha\theta^2(1-\theta)^{-2} + \alpha\theta(1-\theta)^{-1} = \alpha\theta(1-\theta)^{-2} \tag{5.2}$$

$$\mu_3' = \alpha\theta(1+\theta)(1-\theta)^{-3} \tag{5.3}$$

$$\mu_4' = \alpha\theta(1 + 4\theta + \theta^2)(1-\theta)^{-4}. \tag{5.4}$$

Central moments are

$$\text{var}(\mathbf{x}) = \alpha\theta(1-\alpha\theta)(1-\theta)^{-2} = \mu_1'[(1-\theta)^{-1} - \mu_1'] \tag{6.1}$$

$$\mu_3 = \alpha\theta(1 + \theta - 3\alpha\theta + 2\alpha^2\theta^2)(1-\theta)^{-3} \tag{6.2}$$

$$\mu_4 = \alpha\theta\{1 + 4\theta + \theta^2 - 4\alpha\theta(1+\theta) + 6\alpha^2\theta^2 - 3\alpha^3\theta^3\}(1-\theta)^{-4}. \tag{6.3}$$

The moment ratios $\beta_1 = \mu_3^2/\mu_2^3$ and $\beta_2 = \mu_4/\mu_2^2$ both tend to ∞ as θ tends to 0 or as θ tends to 1, with

$$\lim_{\theta\to 0} (\beta_2/\beta_1) = 1; \quad \lim_{\theta\to 1} (\beta_2/\beta_1) = \tfrac{3}{2}.$$

(β_1,β_2) points of logarithmic series distributions are presented in Figure 1, and Table 1 gives these values in a tabular form. (Type III (Gamma) line is presented in Figure 1 for comparison.)

TABLE 1

(β_1,β_2)-Points of Logarithmic Series Distributions

θ	β_1	β_2	θ	β_1	β_2
0.05	43.456	49.858	0.55	9.009	16.463
0.1	23.587	30.062	0.6	9.029	16.656
0.15	17.061	23.613	0.65	9.144	16.972
0.2	13.877	20.512	0.7	9.362	17.424
0.25	12.038	18.762	0.75	9.700	18.044
0.3	10.879	17.699	0.8	10.196	18.891
0.35	10.116	17.040	0.85	10.928	20.085+
0.4	9.608	16.646	0.9	12.078	21.906
0.45	9.168	16.363	0.95	14.254	25.282
0.5	9.086	16.388			

The moments about zero satisfy the relation

$$\mu_{r+1}' = \theta\frac{d\mu_r'}{d\theta} + \frac{\alpha\theta}{1-\theta}\mu_r'. \tag{7.1}$$

The central moments satisfy the relation

$$\mu_{r+1} = \theta\frac{d\mu_r}{d\theta} + r\mu_2\mu_{r-1}. \tag{7.2}$$

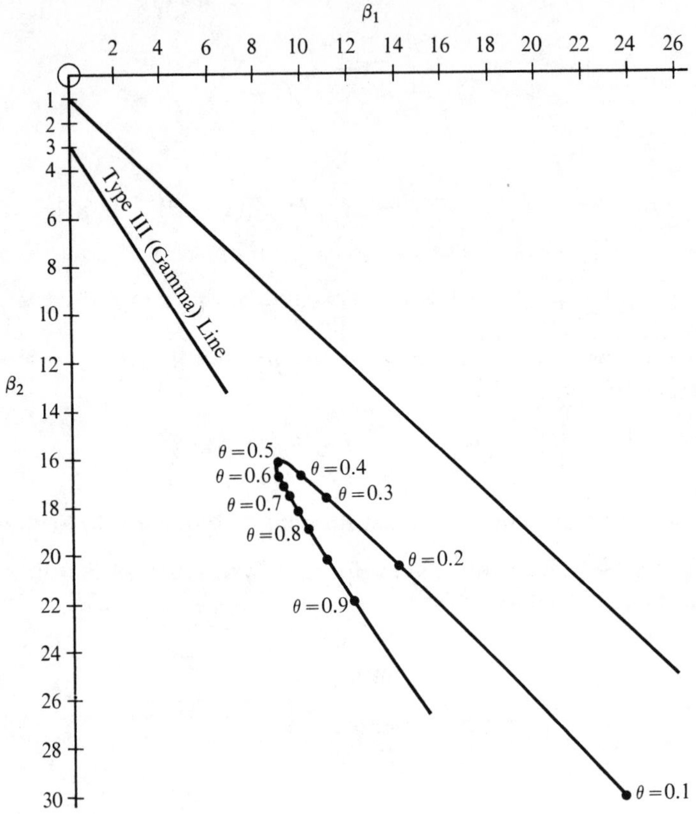

FIGURE 1

(β_1,β_2) Points of Logarithmic Series Distributions

Since the logarithmic series is a power series distribution (see Noack [15], Patil [19] and Chapter 2, Section 4) it follows that the cumulants satisfy the recurrence relations (Khatri [13])

$$\kappa_r = \theta \frac{d\kappa_{r-1}}{d\theta} \tag{7.3}$$

and the factorial cumulants satisfy

$$\kappa_{(r)} = \theta \frac{d\kappa_{(r-1)}}{d\theta} - (r - 1)\kappa_{(r-1)}. \tag{7.4}$$

The entropy of the distribution is

$$\alpha \sum_{j=1}^{\infty} \theta^j j^{-1} \log j - \alpha\theta(1 - \theta)^{-1} \log \theta - \log \alpha.$$

Siromoney [28] has shown that this is an increasing function of θ.

From (1)

$$\text{(8)} \qquad \frac{\Pr[\mathbf{x} = k+1]}{\Pr[\mathbf{x} = k]} = \frac{k\theta}{k+1}.$$

This ratio is less than 1 for *all* values of k $(=1,2,\ldots)$, since $\theta < 1$. Hence the maximum value of $\Pr[\mathbf{x} = k]$ is at the initial value $k = 1$; and the value of $\Pr[\mathbf{x} = k]$ decreases as k increases.

The mean deviation of $\mathbf{x}$ is

$$\text{(9)} \qquad \nu_1 = \alpha \sum_{k=1}^{[\mu_1']} (\mu_1' - k)\theta^k/k = \mu_1' \Pr[\mathbf{x} \leq [\mu_1']] - \frac{\alpha\theta(1 - \theta^{[\mu_1']})}{1-\theta}$$

where μ'_1 is given by (5.1).

Methods of approximating $\Pr[\mathbf{x} \leq [\mu'_1]]$ will be described in Section 3. Using one of these approximations (21), the following approximate formula for the median of the distribution is obtained:

$$\text{(10)} \qquad \text{median} \doteqdot e^{-\gamma}(1-\theta)^{-\frac{1}{2}} = 0.56146(1-\theta)^{-\frac{1}{2}}$$

(where γ is Euler's constant). This is a good approximation if θ is not too small. It can be improved by adding $\frac{1}{2} + e^{-2\gamma}(=0.81524)$ as suggested by Grundy (reported in [10]).

The distribution has a rather long positive tail. For large values of k the *shape* of the tail is similar to that of a geometric distribution (Chapter 5, Section 2) with parameter θ, for which

$$\Pr[\mathbf{x} = k+1]/\Pr[\mathbf{x} = k] = \theta^{k+1}/\theta^k = \theta$$

as compared with (8) for the logarithmic distribution.

Graphs comparing a logarithmic series distribution with two other distributions which have the same expected values are exhibited in Figures 1a–c of Chapter 10.

Anscombe [2] describes the logarithmic series distribution differently. He says: "It is a multivariate distribution, consisting of a set of independent Poisson distributions with mean values αX, $\frac{1}{2}\alpha X^2$, $\frac{1}{3}\alpha X^3$, ... A 'sample' comprises one reading from each Poisson distribution." We will not make use of Anscombe's definition.

2. Historical Remarks and Genesis

The logarithmic series was first used systematically in a paper published in 1942 (Fisher et al. [9]). It was, on this occasion, applied to the results of sampling butterflies (Corbet's data), and also to data obtained in connection with the collection of moths by means of a light-trap (Williams' data). In these experiments it was found that if the number of species represented by exactly one individual in the sample is n_1, then the number of species represented by two, three, etc. . . . individuals are approximately

$$(n_1/\theta)\theta^2/2,\ (n_1/\theta)\theta^3/3,\ \text{etc.} \ldots$$

respectively, where θ is a positive number less than 1. The total number of species is (approximately)

$$S = \sum_{k=1}^{\infty} (n_1/\theta)(\theta^k/k) = -(n_1/\theta)\log(1-\theta) \tag{11}$$

and the total number of individuals collected is (again approximately)

$$M = \sum_{k=1}^{\infty} (n_1/\theta)\theta^k = n_1/(1-\theta). \tag{12}$$

The quantity n_1/θ is sometimes called the 'index of diversity'. Of recent years the *truncated discrete lognormal distribution* has been an effective competitor with the logarithmic series distribution in describing this kind of data (Preston [25], Bliss [6]). A brief description of the newer distribution is given in Section 9. It is discussed in detail in Chapter 8.

A theoretical analysis was given by Fisher [9] leading to the logarithmic series distribution: He supposed that the number of individuals from a given species, say the ith, to be caught in a specified trap can be represented by a Poisson variable with expected value λ_i. If the λ's are chosen randomly from a gamma distribution (with origin at zero) then the expected total numbers of species represented in a given catch by $k = 1, 2, 3, \ldots$ individuals would be proportional to terms in a negative binomial expansion, truncated by exclusion of the first term, corresponding to $k = 0$. As the exponent $(-N)$ of the negative binomial approaches zero (corresponding to increasing variability among the λ's) then the probability of taking the value k (see Chapter 5) tends to

$$\lim_{N\to 0} \frac{\Gamma(N+k)}{k!\Gamma(N+1)} \frac{N}{Q^N - 1}\left(\frac{P}{Q}\right)^k = \alpha\theta^k/k \tag{13}$$

where $\theta = P/Q$ (for $k = 1,2,\ldots$).

Notice that the Poisson means could have come from some other distribution. The truncated discrete lognormal distribution, referred to above, can be derived in this way (see Section 9 and Chapter 8).

Some further details on possible modes of genesis of logarithmic series distributions are given in Nelson and David [14].

3. Applications

In this section we shall consider a few of the many applications of the logarithmic series distribution.

3.1 *Sampling of quadrats for plant species*

Blackman [5] gives the average number of different plant species found on quadrats of various sizes in a grassland formation. As the size of quadrat increases the number (M) of plants observed may be expected to increase proportionately. Supposing the 'index of diversity', δ, to remain constant then M and the number of species (S) observed would be expected [see (11) and (12)] to be related by the formula

$$e^{S/\delta} = 1 + M/\delta. \tag{14}$$

If M/δ is large (as is often the case), then

$$e^{S/\delta} \doteqdot M/\delta \tag{15}$$

and if M be doubled, S/δ would be expected to increase by $\log_e 2$: that is S would increase by the constant amount $\delta \log_e 2$. The last column of Table 2 shows that there is, in fact, a roughly constant increase (of approximately 2) for each doubling of quadrat size. This may be taken as being broadly in agreement with the theory, and indicating a value for δ of approximately $2/\log_e 2 \doteqdot 2.9$.

TABLE 2

Quadrat size in in^2.	Average number of species of plants	Average additional number obtained by doubling area
2	5.9	—
4	7.8	1.9
8	9.5	1.7
16	11.1	1.6
32	13.6	2.5
64	16.1	2.5
128	18.2	2.1

Average = 2.05.

3.2 *Distribution of animal species*

Table 3 gives the numbers (N) of mosquitoes caught in light traps in various cities in Iowa (Rowe [27]), together with the numbers of species (S) observed. The last column gives the estimated value of δ obtained by solving equation (14). Again there seems sound evidence of a constant value of δ (this time about 2).

TABLE 3

City	No. of individuals	No. of species	δ
Ruthven	20,239	18	1.95
Des Moines	17,077	20	2.24
Davenport	15,280	18	2.02
Ames	12,504	16	1.81
Muscatine	6,426	16	1.98
Dubuque	6,128	14	1.71
Lansing	5,564	16	2.03
Bluffs	1,756	13	1.90
Sioux City	661	12	2.08
Burlington	595	12	2.13

Similar data have been analysed by Williams [31] for observations on the distribution of hosts according to the number of head-lice on each (Buxton [7]) and the distribution of species of British nesting birds by numbers of individuals (Witherby et al. [35]) inter alia. (See also Blackman [5], and Williams [30], [32], and [33]).

3.3 *Population growth*

Kendall [12] has represented a form of population growth by a discrete Markov process leading to a negative binomial distribution, and as its limiting form, a logarithmic series distribution. With $P_n(t)$ representing the probability that at time t the population size is n, and assuming that there are constant instantaneous relative rates of mortality (μ) and birth (β) with net immigration at a constant absolute rate K, the difference-differential equations

$$(16.1) \qquad \frac{dP_n(t)}{dt} = (n+1)\mu P_{n+1}(t) - \{n(\mu+\beta)+K\}P_n(t) + \{(n-1)\beta+K\}P_{n-1}(t) \qquad (n > 0)$$

$$(16.2) \qquad \frac{dP_0(t)}{dt} = \mu P_1(t) - KP_0(t)$$

are obtained.

We will consider two problems, corresponding to different initial-value conditions. In the first (I) we assume that the population size is zero at time $t = -T$, in the second (II) that population size is 1 at this instant.

The system of equations (16.1)–(16.2) can be solved by introducing the population size probability generating function at time t

$$\phi(z,t) = \sum_{j=-\infty}^{\infty} z^j P_j(t) \tag{17}$$

with boundary conditions

$$\text{(I)} \qquad \phi(z,-T) = 1$$

or

$$\text{(II)} \qquad \phi(z,-T) = z.$$

The solution for $\phi(z,0)$ may be expanded in powers of z. The coefficient of z^j is then $P_j(0)$, giving the distribution of population-size at time zero (i.e. after a time T has elapsed). It is found that, if $\beta \neq \mu$, for (I)

$$\phi(z,0) = \left(\frac{\beta - \mu}{\beta\Lambda - K}\right)^{K/\beta} \left[1 - \frac{\beta(\Lambda - 1)}{\beta\Lambda - \mu} z\right]^{-K/\beta} \tag{18}$$

where $\Lambda = \exp[(\beta - \mu)T]$, while for (II)

(19)

$$\phi(z,0) = \frac{(\beta - \mu)^{K/\beta}}{(\beta\Lambda - \mu)^{1+K/\beta}} \{\mu(\Lambda - 1) - (\mu\Lambda - \beta)z\} \left[1 - \frac{\beta(\Lambda - 1)}{\beta\Lambda - \mu} z\right]^{-1-K/\beta}.$$

(18) is the probability generating function of a negative binomial distribution. (19) is the probability generating function of a negative binomial distribution 'with zeroes' (See Chapter 8, Section 4)

The logarithmic series distribution is approached as a limiting case if $K/\beta \to 0$ in case (I) (with population size >0 at time T).

3.4 *Economic applications*

Chatfield et al. [8] have used logarithmic distributions to represent the distribution of numbers of items of a product purchased by a buyer in a specified period of time. They point out that the logarithmic series distribution is likely to be useful when a negative binomial distribution with a low value of N (less than 0.1, say) is obtained. The logarthmic series has the advantage of depending on only one parameter θ, instead of two (N and P) for the negative binomial distribution.

4. Approximations

The cumulative probability

$$\Pr[x \leq X] = \alpha \sum_{j=1}^{X} (\theta^j/j)$$

can be approximated using the formulas

$$\sum_{j=1}^{X} (\theta^j/j) = -\log(1-\theta) - \sum_{j=X+1}^{\infty} (\theta^j/j)$$

and

(20)

$$\begin{aligned}\sum_{j=X+1}^{\infty} (\theta^j/j) &= -\log(1-\theta) - \sum_{j=1}^{X} j^{-1} + \int_0^{1-\theta} \phi^{-1}\{1-(1-\phi)^X\}\, d\phi \\ &\doteqdot -\log(1-\theta) - (\gamma + \tfrac{1}{2}X^{-1} - \tfrac{1}{12}X^{-2} + \log X) \\ &\quad + [-Ei(-X(1-\theta)) + \gamma + \log\{X(1-\theta)\}] \\ &\doteqdot -Ei[-X(1-\theta)] - \tfrac{1}{2}X^{-1} + \tfrac{1}{12}X^{-2}\end{aligned}$$

with an error less than $\frac{1}{4}X(1-\theta)^{2}$*. (The function $Ei(u) = \int_{-\infty}^{u} z^{-1}e^{z}\,dz$ is tabulated in Siromoney [28].)

If $X(1-\theta)$ is small then the approximation

(21) $$\sum_{j=X+1}^{\infty} (\theta^j/j) \doteqdot -\log[X(1-\theta)] + X(1-\theta) - \gamma$$

has an error less than $\frac{1}{4}[X(1-\theta)]^2$. This approximation can be used to obtain an approximation to the median of the distribution (see Section 1).

These formulas are given by Gower [10]. He also refers to the following formulas obtained by Grundy:

(22) $$\sum_{j=X+1}^{\infty} (\theta^j/j) \doteqdot Ei(X\log\theta) + \tfrac{1}{2}\theta^X/X.$$

Owen [16] gives the following formula which is suitable when $X(1-\theta)$ is large:

(23) $$\begin{aligned}\sum_{j=X+1}^{\infty} (\theta^j/j) \doteqdot \theta^{X+1}(1-\theta)^{-1}[&X^{-1} - 1!\{X(X-1)\}^{-1}(1-\theta)^{-1} \\ &+ 2!\{X(X-1)(X-2)\}^{-1}(1-\theta)^{-2} - \cdots \\ &+ (-1)^r r!\{X(X-1)\cdots(X-r)\}^{-1}(1-\theta)^{-r}].\end{aligned}$$

The error is of size $(-1)^{r+1}$ and smaller in magnitude than

$$\frac{r+1}{(X-r+1)\theta} \times \text{(last term in the summation in brackets)}.$$

*Note that the formula $\sum_{j=1}^{X} j^{-1} = \gamma + \frac{1}{2}X^{-1} - \frac{1}{12}X^{-2} + \log X$ has an error less than 10^{-6} for $X \geq 10$; less than 10^{-10} for $X \geq 100$.

5. Tables

Table 4 (from Patil [20]) gives the expected value (5.1) of the logarithmic series distribution (1), as a function of θ.

TABLE 4

Expected Value of Logarithmic Series Distribution

$\theta = 0.01(0.01)0.59$

θ	00	01	02	03	04	05	06	07	08	09
00	—	1.0050	1.0102	1.0154	1.0207	1.0261	1.0316	1.0372	1.0429	1.0487
10	1.0546	1.0606	1.0667	1.0730	1.0794	1.0858	1.0925	1.0992	1.1061	1.1132
20	1.1204	1.1277	1.1352	1.1429	1.1507	1.1587	1.1669	1.1752	1.1838	1.1926
30	1.2016	1.2108	1.2202	1.2299	1.2398	1.2500	1.2604	1.2711	1.2821	1.2934
40	1.3051	1.3170	1.3294	1.3421	1.3551	1.3687	1.3825	1.3968	1.4116	1.4269
50	1.4427	1.4591	1.4760	1.4935	1.5117	1.5306	1.5503	1.5706	1.5919	1.6140

Fisher [9] gives a table from which the index of diversity, δ, can be calculated given M and S, from equation (14). This equation may be written

$$e^{(M/\delta)(S/M)} = 1 + M/\delta. \tag{24}$$

Fisher's tables give values of $\log_{10}(M/\delta)$ given $\log_{10}(M/S) = 0.4(0.1)3.5$. Alternatively the tables of Barton et al. [3] may be entered with $p = S/M$, reading off $b = S/\delta$.

Tables of the individual probabilities (1) and cumulative probabilities have been provided by Williamson and Bretherton [34]. The argument of the tables is the expected value ($\mu = -\theta(1-\theta)^{-1}[\log(1-\theta)]^{-1}$) and not θ. However, for $\mu = 1.0(0.1)10.0(1)50$ values of θ corresponding to given μ (to 5 decimal places) are provided. Values of second and fourth differences are also given. Individual probabilities and cumulative probabilities are given (also to 5 decimal places) for

$$\mu = 1.1(0.1)2.0(0.5)5.0(1)10$$

the tabulation in each case being continued until the cumulative probability exceeds 0.999.

Extensive tables of these probabilities are available in Patil et al. [22]. These give values to six decimal places for $\theta = 0.01(0.01)0.70(0.005)0.900(0.001)0.999$.

6. Estimation

We will consider the problem of estimating θ given values of n independent random variables $\mathbf{x}_1, \mathbf{x}_2, \ldots \mathbf{x}_n$, each having distribution (1). The maximum likelihood estimator $\hat{\theta}$ satisfies the equation

$$\bar{x} = n^{-1} \sum_{j=1}^{n} x_j = \frac{\hat{\theta}}{-(1 - \hat{\theta}) \log (1 - \hat{\theta})}. \tag{25}$$

This equation can also be obtained by equating the sample and population means. They should be equal since (1) is a generalized power series distribution.

The tables of Williamson and Bretherton [34] may be used to obtain $\hat{\theta}$, given $\bar{x}$ (entering the table with $\mu = \bar{x}$). Patil and Wani [23] give $\hat{\theta}$ to four decimal places for $\bar{x} = 1.02(0.02)2.00(0.05)4.00(0.1)8.0(0.2)16.0(0.5)30.0(2)40$ $(5)60(10)140(20)200$. They also tabulate the bias and standard error of $\hat{\theta}$ in a few cases. (See also Patil et al. [22]). In the tables of Barton et al. [3] solutions of the equation (in b)

$$e^b = 1 + b\bar{x}$$

are given to seven decimal places for $p = 1 - \bar{x}^{-1} = 0(0.001)0.999$. Then $\hat{\theta}$ is calculated as

$$\hat{\theta} = b\bar{x}/(1 + b\bar{x}). \tag{26}$$

If only an approximate value of $\hat{\theta}$ is required, this can be obtained by inverse interpolation in Table 4. For $\bar{x} < 25$, the approximation

$$\hat{\theta} \doteqdot 1 - \{1 + [(\tfrac{5}{3} - \tfrac{1}{16} \log \bar{x})(\bar{x} - 1) + 2] \log \bar{x}\}^{-1}$$

gives good results (Birch [4]).

The variance of $\hat{\theta}$ is approximately

$$(\theta^2/\mu_2)n^{-1}$$

where μ_2 is the variance of the distribution (1), and is given explicitly in formula (6.1). To estimate this variance it is the usual practice to replace θ by $\hat{\theta}$.

Other estimators of θ (based on relationships which are exactly true in the population) are

(*a*) $1 -$ (proportion of x's equal to 1)$/\bar{x}$

(*b*) $1 - \left(n^{-1} \sum_{j=1}^{n} x_j^2\right)/\bar{x}$

(*c*) $\sum_{j=2}^{\infty} \{j/(j - 1)\} \cdot$ (proportion of x's equal to j).

Estimators (*a*) and (*b*) are asymptotically unbiased. Estimator (*c*) is unbiased for all sizes of sample, and has variance

$$n^{-1}\left[\sum_{j=2}^{\infty} \{j/(j - 1)\}^2 \Pr[x = j] - \theta^2\right]. \tag{27}$$

The asymptotic efficiencies of these estimators (relative to $\hat{\boldsymbol{\theta}}$) for certain values of θ are shown in Table 5.

TABLE 5

Asymptotic Efficiencies (in %) of estimators of θ.

	Value of θ		
Estimator	0.1	0.5	0.9
(*a*)	98.3	89.7	73.9
(*b*)	22.8	44.9	48.8
(*c*)	89.5	44.7	5.7

It appears that (*a*) is preferable to (*b*) or (*c*) on grounds both of accuracy and simplicity.

The minimum variance unbiased estimator, $\boldsymbol{\theta}^*$, of θ is given by the following equation

$$\boldsymbol{\theta}^* = b\left(\sum_1^n \mathbf{x}_j - 1\right) \Big/ b\left(\sum_1^n \mathbf{x}_j\right)$$

if $\sum_{j=1}^n \mathbf{x}_j > n$, zero if $\sum_{j=1}^n \mathbf{x}_j = n$, where $b(m)$ is the coefficient of θ^m in

$$[-\log(1 - \theta)]^n.$$

(See (26), Chapter 2). Tables of $\boldsymbol{\theta}^*$, given $\sum_1^n \mathbf{x}_j$, for small n, to four decimal places, are given in Patil et al. [22].

7. Characterization

The following characterization of the logarithmic series distribution has been given by Patil and Wani [24]:

"Let $\mathbf{x}$ and $\mathbf{y}$ be two independent discrete random variables each taking the value 1 with non-zero probability; then if

$$\Pr[\mathbf{x} = X \mid \mathbf{x} + \mathbf{y} = Z] = \frac{[X^{-1} + (Z - X)^{-1}]\beta^X}{\sum_{j=1}^{X-1} [X^{-1} + (Z - j)^{-1}]\beta^j} \tag{28}$$

for $0 < \beta < \infty$,

$\mathbf{x}$ and $\mathbf{y}$ each has the logarithmic series distribution (1) with parameters in the ratio β."

8. Truncated Logarithmic Series Distribution

The most common form of truncation is by exclusion of values greater than a specified value, r. This kind of truncation is likely to be encountered with distributions of logarithmic series type, as they have long positive tails and it is not always practicable to evaluate each large observed value. Paloheimo [17] describes circumstances in which a truncated logarithmic series distribution might be appropriate.

If $\mathbf{x}_1, \mathbf{x}_2, \ldots, \mathbf{x}_n$ are independent random variables each with the distribution

$$\Pr[\mathbf{x} = k] = (\theta^k/k)\left[\sum_{j=1}^{r} (\theta^j/j)\right]^{-1} \qquad (k = 1,2,\ldots r) \tag{29}$$

then the maximum likelihood estimator, $\hat{\boldsymbol{\theta}}$, of θ satisfies the equation

$$\bar{\mathbf{x}} = n^{-1}\sum_{j=1}^{n} \mathbf{x}_j = \frac{\hat{\boldsymbol{\theta}}(1 - \hat{\boldsymbol{\theta}}^r)}{(1 - \hat{\boldsymbol{\theta}})\left[\sum_{j=1}^{r} (\hat{\boldsymbol{\theta}}^j/j)\right]}. \tag{30}$$

Patil and Wani [23] provide tables giving the value of $\boldsymbol{\theta}$ corresponding to selected values of $\bar{\mathbf{x}}$ for $r = 4(1)8(2)12,15(5)40(10)60(20)100,200,500,1000$. They also give some numerical values of the bias and standard deviation of $\hat{\boldsymbol{\theta}}$.

Alternatively a method based on equating sample and population moments may be used. The sth population moment about zero is

$$K\sum_{j=1}^{r} j^{s-1}\theta^j \qquad \left(K = \left[\sum_{j=1}^{r} (\theta^j/j)\right]^{-1}\right).$$

Now since

$$j^2 - (r+2)j + (r+1) = (j - r - 1)(j - 1)$$

it follows that

$$\begin{aligned}\sum_{j=1}^{r} [j^2 - (r+2)j + (r+1)]\theta^j &= \sum_{j=1}^{r+1} (j - r - 1)(j - 1)\theta^j \\ &= \theta\sum_{j=0}^{r} (j - r)j\theta^j \\ &= \theta\sum_{j=1}^{r} (j^2 - rj)\theta^j.\end{aligned}$$

This leads to using

$$\frac{\mathbf{m}_3' - (r+2)\mathbf{m}_2' + (r+1)\mathbf{m}_1'}{\mathbf{m}_3' - r\mathbf{m}_2'} \tag{31}$$

as an estimator of θ $\left(\text{where } \mathbf{m}_s' = n^{-1}\sum_{j=1}^{n} \mathbf{x}_j^s\right)$. Further discussion of properties of these estimators will be found in a paper by Ahmad [1].

9. Related Distributions

The logarithmic series distribution can be derived as a limiting form of the truncated negative binomial distribution (Section 2). Another relationship among the logarithmic series, Poisson and negative binomial distributions was pointed out by Quenouille [26]. If the number of 'groups' of individuals has a Poisson distribution with expected value ϕ and the number of individuals per group has the logarithmic series distribution (1) then the probability that there is a total of k individuals is the coefficient of t^k in

$$(32)\qquad \sum_{j=0}^{\infty} [e^{-\phi}\phi^j/j!][-\alpha \log(1-\theta t)]^j = e^{-\phi-\alpha\phi\log(1-\theta t)}$$
$$= e^{-\phi}(1-\theta t)^{-\alpha\phi}$$
$$= [(1-\theta t)/(1-\theta)]^{-\alpha\phi}$$
$$(\text{since } e = (1-\theta)^{-\alpha}).$$

This probability is therefore

$$\frac{\Gamma(\alpha\phi + k)}{k!\Gamma(\alpha\phi)}(1-\theta)^{\alpha\phi}\theta^k$$

and the distribution of the number of individuals is negative binomial with parameters $\alpha\phi$, $\theta/(1-\theta)$.

The distribution of the sum ($\mathbf{s}_n$) of n independent random variables, each having the logarithmic series distribution (1) has the probability generating function $[-\alpha \log(1-\theta t)]^n$ (as used in deriving (32)). From this it follows that

$$(33)\quad \Pr[\mathbf{s}_n = k] = \frac{\text{coefficient of } t^k \text{ in } [-\log(1-\theta t)]^n}{[-\log(1-\theta)]^n} \qquad (k = n, n+1, \ldots).$$

The coefficient of t^k in the expansion of $[-\log(1-\theta t)]^n$ is

$$\theta^k(n!/k!)|S_k^{(n)}|$$

where $S_k^{(n)}$ is the Stirling number of the first kind with arguments n and k defined by

$$S_k^{(n)} = \frac{1}{n!}\left[\frac{d^n}{dx^n}\left(\prod_{t=1}^{k}(x-t+1)\right)\right]_{x=0}$$

(see Jordan [11] and Chapter 1, Section 2).

For this reason Patil and Wani [24] have called the distribution (33) the *first type Stirling distribution*. They state the following properties of the distribution:

(a) If $\theta < 2n^{-1}$, the distribution has a unique mode at $k = n$ (the smallest possible value), and the values of $\Pr[\mathbf{x} = k]$ decrease as k increases;

(b) If $\theta = 2n^{-1}$ there are two equal modal values at $k = n$ and $k = n + 1$;

(c) If $\theta > 2n^{-1}$, the value of $\Pr[\mathbf{x} = k]$ increases with k to a maximum (or pair of equal maxima) and thereafter decreases as k increases.

In view of the definition of this distribution, it is clear that as n tends to infinity the standardized distribution corresponding to (33) tends to the unit normal distribution.

A distribution obtained by assigning a lognormal distribution to the parameter of a Poisson distribution is termed a *discrete lognormal distribution.* It is a compound Poisson distribution and, as such, is discussed in Chapter 8. It is mentioned here because, as noted in Section 2, a truncated form (obtained by omission of the zero class) is an important competitor of the logarithmic series distribution as a model for the distribution of observed abundancies of species, and similar phenomena. Bliss [6] made comparisons of the fidelity with which the two distributions represented five* sets of data from moth trap experiments, and found that in each case the truncated discrete lognormal distribution gave the better fit (as judged by χ^2 probabilities) though both distributions gave acceptable representations (again judged by χ^2).

*Actually four *distinct* sets, and the combined data.

REFERENCES

[1] Ahmad, M. (1965). Estimation of parameters when a few observations are missing, II — The truncated lognormal distribution, *35th Session of the International Statistical Institute, Belgrade.* Preprint No. 52.

[2] Anscombe, F. J. (1950). Sampling theory of the negative binomial and logarithmic series distributions, *Biometrika*, **37,** 358–382.

[3] Barton, D. E., David, F. N. and Merrington, M. (1963). Tables for the solution of the exponential equation $\exp(b) - b/(1 - p) = 1$, *Biometrika*, **50,** 169–172.

[4] Birch, M. W. (1963). An algorithm for the logarithmic series distribution, *Biometrics*, **19,** 651–652.

[5] Blackman, G. E. (1935). A study of statistical methods of species in a grassland association, *Annals of Botany* (*New Series*), **49,** 749–777.

[6] Bliss, C. I. (1963). An analysis of some insect trap records, *Proceeding of the International Symposium on Discrete Distributions, Montreal*, 385–397. (See [16])

[7] Buxton, P. A. (1940). Studies on populations of head lice, III, Material from South India, *Parasitology*, **32,** 296–302.

[8] Chatfield, C., Ehrenberg, A. S. C. and Goodhardt, G. J. (1966). Progress on a simplified model of stationary purchasing behavior (with discussion), *Journal of the Royal Statistical Society, Series A*, **129,** 317–367.

[9] Fisher, R. A. Corbet, A. S., and Williams, C. B. (1943). The relation between the number of species and the number of individuals in a random sample of an animal population, *Journal of Animal Ecology*, **12,** 42–57.

[10] Gower, J. C. (1961). A note on some asymptotic properties of the logarithmic series distribution, *Biometrika*, **48,** 212–215.

[11] Jordan, C. (1950). *Calculus of Finite Differences* (2nd edition), New York: Chelsea.

[12] Kendall, D. G. (1948). On some modes of population growth leading to R. A. Fisher's logarithmic series distribution, *Biometrika*, **35,** 6–15.

[13] Khatri, C. G. (1959). On certain properties of power series distributions, *Biometrika*, **46,** 486–488.

[14] Nelson, W. C. and David, H. A. (1967). The logarithmic distribution: A review, *Virginia Journal of Science*, **18,** 95–102.

[15] Noack, A. (1950). A class of random variables with discrete distributions, *Annals of Mathematical Statistics*, **21,** 127–132.

[16] Owen, A. R. G. (1963). The summation of class frequencies *Proceedings of the International Symposium on Discrete Distributions, Montreal*, 395–397 (Appendix to [6]).

[17] Paloheimo, J. E. (1963). On statistics of search, *Bulletin of the International Statistical Institute*, **40(2)**, 1060–1061.

[18] Patil, G. P. (1962). Estimation by two-moments method for generalized power series distribution and certain applications, *Sankhyā, Series B*, **24,** 201–214.

[19] Patil, G. P. (1962). Certain properties of the generalized power series distribution, *Annals of the Institute of Statistical Mathematics, Tokyo*, **14,** 179–182.

[20] Patil, G. P. (1962). Some methods of estimation for the logarithmic series distribution, *Biometrics*, **18**, 68–75.

[21] Patil, G. P. (1962). On homogeneity and combined estimation for the generalized power series distribution and certain applications, *Biometrics*, **18**, 365–374.

[22] Patil, G. P., Kamat, A. R. and Wani, J. K. (1964). *Certain Studies on the Structure and Statistics of the Logarithmic Series Distribution and Related Tables*, ARL 64–197, Aerospace Research Laboratories, Wright-Patterson Air Force Base, Ohio.

[23] Patil, G. P. and Wani, J. K. (1963). Maximum likelihood estimation for the complete and truncated logarithmic series distributions, *Proceedings of the International Symposium on Discrete Distributions, Montreal*, 398–409. (Also *Sankhyā, Series A*, **27**, 281–292.)

[24] Patil, G. P. and Wani, J. K. (1965). On certain structural properties of the logarithmic series distribution and the first type Stirling distribution, *Sankhyā, Series A.*, **27**, 271–280.

[25] Preston, F. W. (1948). The commonness, and rarity, of species, *Ecology*, **29**, 254–283.

[26] Quenouille, M. H. (1949). A relation between the logarithmic, Poisson, and negative binomial series, *Biometrics*, **5**, 162–164.

[27] Rowe, J. A. (1942). Mosquito light trap catches from ten American cities, 1940, *Iowa State College Journal of Science*, **16**, 487–518.

[28] Siromoney, G. (1962). Entropy of logarithmic series distributions, *Sankhyā, Series A*, **24**, 419–420.

[29] *Tables of Sine, Cosine and Exponential Integrals*, W.P.A., New York, 1940.

[30] Williams, C. B. (1944). Some applications of the logarithmic series and the index of diversity to ecological problems, *Journal of Ecology*, **32**, 1–44.

[31] Williams, C. B. (1947). The logarithmic series and its applications to biological problems, *Journal of Ecology*, **34**, 253–272.

[32] Williams, C. B. (1953). The relative abundance of different species in a wild animal population, *Journal of Animal Ecology*, **22**, 14–31.

[33] Williams, C. B. (1960). The range and pattern of insect abundance, *American Naturalist*, **94**, 137–151.

[34] Williamson, E. and Bretherton, M. H. (1964). Tables of the logarithmic series distribution, *Annals of Mathematical Statistics*, **35**, 284–297.

[35] Witherby, H. F., Jourdain, F. C. R., Tichihurst, N. F. and Tucker, B. W. (1941). *The Handbook of British Birds*, **5**, London: H. F. and G. Witherby, Ltd.

8

Some Compound, Generalized and Modified Discrete Distributions

1. Introduction

In most chapters of this book there are sections on modified forms of the distributions to which the major part of the chapter is devoted, particularly in Chapters 3–7, which deal with the more important discrete distributions. These sections, however, discuss only a few distributions of special interest. The variety of modified forms of these distributions is such that it was decided that a separate chapter should be assigned to them.

Some of the distributions described here are used infrequently. We do not enter into much detail in the discussion of such distributions. Certain distributions (the Poisson-binomial, and Polya-Aeppli ("Poisson-Pascal") distributions, in particular) are given much fuller treatment. Much of this treatment can be applied to the less well-known distributions; especially to the calculation of moments. Frequent use will be made of probability generating functions and (by differentiating and setting the argument, t, equal to 1) the factorial moments derived from them. A brief discussion of general aspects of this technique is contained in Chapter 2.

Although particular examples of these distributions have been known for a considerable time, systematic development dates from about 1943 (Feller [18]).

2. Compound Distributions

The most important classes of distributions to be discussed in this chapter are formed by 'mixtures' of discrete distributions. The notion of 'mixing' often has a simple and direct interpretation in terms of the physical situation under investigation. But sometimes 'mixing' is just a mechanism for constructing new distributions for which empirical justification must be sought later. For discrete distributions "compounding" is commonly used in place of "mixing," and the resultant distributions are called *compound* distributions.

The meaning of 'mixing' (or 'compounding') has already been described in Chapter 1 (Section 7.3). If the cumulative distribution function of a random variable is $F(\mathbf{x} \mid \theta_1,\theta_2,\ldots,\theta_k)$, depending on the k parameters $\theta_1, \theta_2, \ldots, \theta_k$, then a 'compound' (or 'mixture') distribution is constructed by ascribing to some, or all, of the θ's, a probability distribution. The new distribution has the cumulative distribution function $E[F(\mathbf{x} \mid \boldsymbol{\theta}_1,\boldsymbol{\theta}_2,\ldots,\boldsymbol{\theta}_k)]$, the expectation being taken with respect to the joint distribution of the $\boldsymbol{\theta}$'s. It can still depend on the values of those θ's (if any) which do not vary, and also on parameters of the joint distribution of the other $\boldsymbol{\theta}$'s.

It will be convenient to denote compound distributions in the symbolic form

$$F_1 \underset{\boldsymbol{\theta}}{\wedge} F_2$$

where F_1 represents the original distribution, $\boldsymbol{\theta}$ the varying parameters, and F_2 the 'compounding' (mixing) distribution.

Thus, for example,

$$\text{(i)} \qquad \text{Poisson }(\boldsymbol{\theta}) \underset{\boldsymbol{\theta}}{\wedge} \text{Gamma }(\alpha,\beta)$$

means a compound Poisson distribution formed by ascribing the gamma distribution with probability density function

$$p_{\boldsymbol{\theta}}(\theta) = [\beta^\alpha\Gamma(\alpha)]^{-1}\theta^{\alpha-1} \exp[-\theta/\beta] \qquad (0 \leq \theta;\, \alpha,\beta > 0)$$

to the expected value $\boldsymbol{\theta}$, of a Poisson distribution. This compound Poisson distribution is, in fact, a negative binomial distribution (see Chapter 5, Section 4).

We now give a brief treatment of a number of compound distributions. It would be easy to extend this catalogue by taking further combinations of distributions. We have been guided by the amount of attention given to the various distributions by other workers, as well as by our own predilections as to what is interesting and useful.

$$\text{(ii)} \qquad \text{Poisson }(\boldsymbol{\theta}) \underset{\boldsymbol{\theta}}{\wedge} \text{Rectangular }(a,b).$$

This distribution was proposed by Bhattacharya and Holla [5] as a possible alternative to (i) above in the theory of accident proneness. It is supposed that the probability density function of $\boldsymbol{\theta}$ is:

$$p_{\boldsymbol{\theta}}(\theta) = (b-a)^{-1} \qquad (a \leq \theta \leq b).$$

This leads to a distribution defined by

$$\text{(1)} \qquad \Pr[\mathbf{x} = k] = [k!(b-a)]^{-1}\int_a^b \theta^k e^{-\theta}\, d\theta \qquad (k = 0,1,2,\ldots).$$

Using the relationship between the Poisson and gamma distributions (Chapter 4,

Section 7), the right-hand side of (1) can be written

$$(2) \qquad (b-a)^{-1}\left[e^{-a}\left(1+\frac{a}{1!}+\frac{a^2}{2!}+\cdots+\frac{a^k}{k!}\right) - e^{-b}\left(1+\frac{b}{1!}+\frac{b^2}{2!}+\cdots+\frac{b^k}{k!}\right)\right].$$

It will be noted that

$$(3) \qquad \Pr[\mathrm{x}=k+1] = \Pr[\mathrm{x}=k] + (e^{-a}a^{k+1} - e^{-b}b^{k+1})/(k+1)!.$$

The distribution is unimodal, with mode at the greatest integer k (including zero) for which

$$k \leq (b-a)[\log (b/a)]^{-1}.$$

(If $(b-a)[\log (b/a)]^{-1}$ is an integer, there are two equal modal values.)

The expected values and first three moments of the distribution are

$$(4.1) \qquad \mu_1' = \tfrac{1}{2}(a+b)$$

$$(4.2) \qquad \mu_2 = \tfrac{1}{2}(a+b) + \tfrac{1}{12}(b-a)^2$$

$$(4.3) \qquad \mu_3 = \tfrac{1}{2}(a+b) + \tfrac{1}{4}(b-a)^2$$

(4.4)

$$\mu_4 = \tfrac{1}{2}(a+b) + \tfrac{1}{2}(4b^2+ab+4a^2) + \tfrac{1}{4}(a+b)(b-a)^2 + \tfrac{1}{80}(b-a)^4.$$

(iii) Poisson $(\boldsymbol{\theta}) \underset{\boldsymbol{\theta}}{\wedge}$ Truncated normal (ξ,σ).

This is the *Poisson-normal* distribution (Patil [39]). It has also been called the *Gauss-Poisson* distribution (Berlyand et al. [4]).

The point of truncation (from the left) of the normal distribution is usually taken at zero, ensuring that θ has zero probability of being negative. The parameters ξ and σ represent the expected value and standard deviation, respectively, of the normal distribution before truncation.

For this compound distribution:

$$(5) \qquad \Pr[\mathrm{x}=k] = \frac{\exp[\frac{1}{2}\sigma^2-\xi]}{k!}\,\frac{\displaystyle\int_0^\infty \theta^k \exp\left[-\frac{1}{2}\left(\frac{\theta-\xi+\sigma^2}{\sigma}\right)^2\right]d\theta}{\displaystyle\int_0^\infty \exp\left[-\frac{1}{2}\left(\frac{\theta-\xi}{\sigma}\right)^2\right]d\theta}.$$

The partial moment in the numerator can be expressed as a finite sum [$(k+1)$ terms] of multiples of incomplete gamma functions.

(iv) Poisson $(\boldsymbol{\theta}) \underset{\boldsymbol{\theta}}{\wedge}$ Lognormal (ξ,σ,a).

This distribution is termed by Anscombe [2] the *discrete lognormal* distribution. (Bliss [7] calls the distribution obtained by omitting the zero class a *truncated discrete lognormal* (see Chapter 7, Sections 2 and 8).)

The lognormal distribution (see Chapter 14) employed is usually supposed to have a range of variation from 0 to ∞. This is ensured by taking $a = 0$. However, it is possible to take any positive value for a.

If we suppose that $\log \boldsymbol{\theta}$ is distributed normally with expected value ξ and standard deviation σ, then the expected value of the compound distribution is

(6.1) $$E(\boldsymbol{\theta}) = \exp(\xi + \tfrac{1}{2}\sigma^2).$$

The variance is

(6.2)
$$E(\boldsymbol{\theta}^2 + \boldsymbol{\theta}) - [E(\boldsymbol{\theta})]^2 = \exp(2\xi + 2\sigma^2) + \exp(\xi + \tfrac{1}{2}\sigma^2) - \exp(2\xi + \sigma^2).$$

(v) $$\text{Poisson}(\boldsymbol{\theta}) \underset{\theta/\phi}{\wedge} \text{Poisson}(\lambda).$$

This is known as *Neyman's Type A* distribution. It is conventionally, and for historical reasons, regarded as a 'contagious' distribution, and we will discuss it in the next chapter.

(vi) $$\text{Poisson}(\boldsymbol{\theta}) \underset{\theta/\phi}{\wedge} \text{Binomial}(N,p).$$

For this distribution

(7) $$\Pr[\mathbf{x} = k] = \phi^k(k!)^{-1} \sum_{j=0}^{N} e^{-j\phi} j^k \binom{N}{j} p^j q^{N-j} \qquad (k = 0,1,2,\ldots)$$

where $q = 1 - p$.

Since
$$\sum_{j=0}^{N} e^{-j\phi} j^{(k)} \binom{N}{j} = N^{(k)}(pe^{-\phi})^k(q + pe^{-\phi})^{N-k}$$

it follows that

(8) $$\Pr[\mathbf{x} = k] = (\phi^k/k!) \sum_{j=0}^{k} (\Delta^j 0^k/j!) N^{(j)}(pe^{-\phi})^j(q + pe^{-\phi})^{N-j}.$$

The probability generating function is

(9) $$E[\exp\{\mathbf{j}\phi(t-1)\}] = [q + p\exp\{\phi(t-1)\}]^N$$

(where $\mathbf{j}$ has a binomial distribution with parameters N,p). The moments may be found from (9). However, it is more convenient to use the formula

$$E(\mathbf{x}^{(r)}) = \phi^r E(\mathbf{j}^r).$$

The resulting formulas are

(10.1) $$\mu_1' = Np\phi$$

(10.2) $$\mu_2 = Np\phi + Npq\phi^2$$

(10.3) $$\mu_3 = Np\phi + 3Npq\phi^2 + Npq(q-p)\phi^3$$

(10.4) $$\mu_4 = 3(Np\phi + Npq\phi^2)^2 + Np\phi + 7Npq\phi^2 + 6Npq(q-p)\phi^3 + Npq(1-6pq)\phi^4.$$

This distribution should not be confused with the more commonly used Poisson-binomial distribution (see (xii) below).

(vii) Poisson $(\boldsymbol{\theta}) \underset{\theta/\phi}{\wedge}$ Negative binomial (N,P).

For this distribution

(11) $$\Pr[\mathbf{x} = k] = (\phi^k/k!) \sum_{j=0}^{\infty} e^{-j\phi} j^k \binom{N+j-1}{N-1} Q^{-N}(P/Q)^j \qquad (k = 0,1,2,\ldots)$$

where $Q = 1 + P$.

The probability generating function is

(12) $$E[\exp\{\mathbf{j}\phi(t-1)\}] = [Q - P\exp\{\phi(t-1)\}]^{-N}$$

where $\mathbf{j}$ has a negative binomial distribution with parameters N, P.

Moments can be obtained from formulas (10) by replacing N by $-N$, p by $-P$ and q by Q.

This distribution should not be confused with the Poisson-Pascal (generalized Pólya-Aeppli) distribution (see (xviii) below).

(viii) Poisson $(\boldsymbol{\theta}) \underset{\theta/\phi}{\wedge}$ Logseries (λ).

("Logseries" is used as an abbreviation for "Logarithmic series.")

For this distribution:

(13) $$P_k = \Pr[\mathbf{x} = k] = [-\log(1-\lambda)]^{-1} \sum_{j=1}^{\infty} (\lambda^j/j) e^{-j\phi}[(j\phi)^k/k!]$$
$$= [-\log(1-\lambda)]^{-1}(\phi^k/k!) \sum_{j=1}^{\infty} j^{k-1}(\lambda e^{-\phi})^j \qquad (k = 0,1,2,\ldots;\ \lambda < 1).$$

Noting that $$\sum_{j=1}^{\infty} j^{(r)}(\lambda e^{-\phi})^j = r!(\lambda e^{-\phi})^r(1-\lambda e^{-\phi})^{-(r+1)}$$

the right-hand side of (13) can be written (for $k > 0$)

(14) $$[-\log(1-\lambda)]^{-1}(\phi^k/k!) \sum_{j=0}^{k-1} (\Delta^j 0^{k-1})(\lambda e^{-\phi})^j(1-\lambda e^{-\phi})^{-(j+1)}$$

while, for $k = 0$,

(15.1) $$P_0 = [\log(1-\lambda e^{-\phi})]/[\log(1-\lambda)].$$

We note, in particular, from (14):

$$P_1 = [-\log(1-\lambda)]^{-1}\lambda\phi e^{-\phi}(1-\lambda e^{-\phi})^{-1}. \tag{15.2}$$

Another interesting representation is

$$P_k = (\phi^k/k!)[\log(1-\lambda e^{-\phi})/\log(1-\lambda)]\mu'_k \tag{16}$$

where μ'_k is the kth moment about zero of a logseries distribution with parameter $\lambda e^{-\phi}$. The probability generating function is

$$[\log\{1-\lambda e^{\phi(t-1)}\}]/[\log(1-\lambda)]. \tag{17}$$

The expected value is

$$[-\log(1-\lambda)]^{-1}(1-\lambda)^{-1}\phi\lambda \tag{18.1}$$

and the variance is

$$\begin{aligned}&[-\log(1-\lambda)^{-1}](1-\lambda)^{-1}\phi\lambda\\ &\quad+[-\log(1-\lambda)]^{-1}(1-\lambda)^{-2}\phi^2\lambda[1-\lambda\{-\log(1-\lambda)\}^{-1}].\end{aligned} \tag{18.2}$$

Katti and Rao [32] have prepared tables of $P_k/(1-P_0)$ — (i.e. truncated distribution, omitting zeroes).

(ix) Poisson $(\boldsymbol{\theta}) \underset{\boldsymbol{\theta}/\phi}{\wedge}$ Hypergeometric (n, X, N).

For this distribution

$$\Pr[\mathbf{x}=k] = \binom{N}{n}^{-1}(\phi^k/k!)\sum_j e^{-j\phi}(j\phi)^k\binom{X}{j}\binom{N-X}{n-j}; \tag{19}$$

the summation is taken over the range for which $0 \le j \le X$ and $0 \le n-j \le N-X$. The case $n = 1$ is, of course, identical with the Poisson $\wedge$ Binomial [(vi) above] with $N = 1$, $p = X/N$.

The probability generating function is

$$\frac{(N-n)!(N-X)!}{N!(N-n-X)!}F(-n,-X;N-X-n+1;e^{\phi(t-1)}). \tag{20}$$

Chistyakov [8] considered the following model:

Given n balls which are independently placed into N boxes with the probability of "hitting" the kth box being equal to a_k (with $\sum a_k = 1$).

Denote by $\boldsymbol{\mu}_0$ the numbers of empty boxes. If $a_1 = a_2 = \cdots = a_N = N^{-1}$, then the distribution of $\boldsymbol{\mu}_0$ approaches a Poisson distribution as $n, N \to \infty$ with n/N constant.

By generalizing, Chistyakov [8] obtained the following:

Let $n, N \to \infty$ and $\dfrac{n}{N^2} \to 0$.

If $E(\boldsymbol{\mu}_0) \to m =$ constant and $(1 - a_k)^n \to \gamma_k$, $k = 1, 2, \ldots$, then

$$E(t^{\boldsymbol{\mu}_0}) \to e^{\lambda(t-1)} \prod_{k=1}^{\infty} (1 - \gamma_k - \gamma_k t) \tag{21}$$

uniformly for $t \in [0,1]$ where $\lambda = m = \sum_{k=1}^{\infty} \gamma_k \geq 0$. The distribution corresponding to the right-hand side of (21) is a composition of the Poisson distribution (with parameter λ) with two independent random variables each taking values 0 and 1 with probabilities $1 - \gamma_k$ and γ_k respectively.

We now consider compound binomial distributions. The binomial distribution has two parameters, N and p, and either or both of these may be supposed to have probability distributions. We will not discuss cases in which both N and p vary, although it is easy to construct such examples.

In most cases discussed in statistical literature, N has discrete distributions, while p is continuous. This former restriction is natural, but the latter is not essential. The reader will recall that for the Poisson expected value (θ) both continuous ((i)–(iv)) and discrete ((v)–(ix)) distributions have been used. However, discrete distributions for p have not been found of much use, nor have they attracted much attention from a theoretical point of view (but see (xii) below).

(x) Binomial $(N,\mathbf{p}) \underset{\mathbf{p}}{\wedge}$ Beta (See Chapter 24)

This is the *Pólya-Eggenberger* distribution which will be discussed in Chapter 9, Section 4. It has already been described in Chapter 3, Section 11 and Chapter 6, Section 8.5. It is termed the *binomial-beta* distribution by Ishii and Hayakawa [27].

This distribution has been used with models of variation of the number of defective items found in routine sampling inspection. Horsnell [26] has also proposed the use of a rectangular compounding distribution with range not extending over the entire interval 0 to 1. Horsnell has also used

(xi) Binomial $(N,\mathbf{p}) \underset{\mathbf{p}}{\wedge}$ Triangular (See Chapter 25)

in the same connection.

Hald [25] has obtained a general approximation for compound binomial probabilities with continuous compounding distributions. For a compounding distribution of $\mathbf{p}$ with density function $w(p)$, Hald obtains the formula

$$\binom{n}{r} \int_0^1 p^r(1 - p)^{n-r} w(p)\, dp$$
$$= n^{-1} w(r/n)[1 + b_1(r/n)n^{-1} + b_2(r/n)n^{-2} + 0(n^{-3})]$$

where

$$b_1(h) = [w(h)]^{-1}[-w(h) + (1 - 2h)w'(h) + \tfrac{1}{2}h(1 - h)w^{(2)}(h)]$$

and

$$b_2(h) = [w(h)]^{-1}[w(h) - 3(1 - 2h)w'(h) + \{1 - 6h(1 - h)\}w^{(2)}(h) + \tfrac{5}{6}h(1 - h)(1 - 2h)w^{(3)}(h) + \tfrac{1}{8}h^2(1 - h)^2w^{(4)}(h)].$$

We now come to one of the more important of the distributions considered in this chapter:

(xii) $$\text{Binomial } (\mathbf{N},p) \underset{\mathbf{N}/n}{\wedge} \text{Poisson } (\lambda)$$

This distribution is known as the *Poisson-binomial* distribution, and was discussed by Skellam [51] in 1952. Since 1957, when McGuire et al. [37] used the distribution to represent variation in numbers of corn-borer larvae in a randomly chosen area, it has found an increasing number of applications. Computational difficulties have impeded its wider adoption, but these are becoming less important with improvements in computational facilities, and the development of formulas better adapted to computational requirements.*

The distribution is defined by

(22) $$P_k = \Pr[\mathbf{x} = k] = e^{-\lambda} \sum_{j \geq k/n} \binom{nj}{k} p^k q^{nj-k}(\lambda^j/j!)$$
$$= e^{-\lambda}(p/q)^k(k!)^{-1} \sum_{j \geq k/n} (nj)^{(k)}[(\lambda q^n)^j/j!].$$

The probability generating function is

(23) $$E[(q + pt)^{n\mathbf{j}}] = \exp[\lambda\{(q + pt)^n - 1\}]$$

(where $\mathbf{j}$ has a Poisson distribution with expected value λ).

Putting $t = 0$, we see that $P_0 = \exp[-\lambda(1 - q^n)]$. Differentiating with respect to t and setting $t = 0$ leads to $P_1 = n\lambda pq^{n-1} \exp[-\lambda(1 - q^n)]$. Further analysis gives $P_2 = \tfrac{1}{2}n\lambda p^2[n - 1 + n\lambda q^n]q^{n-2} \exp[-\lambda(1 - q^n)]$. Note that if $n\lambda pq^{n-1} < 1$ but $n - 1 + n\lambda q^n > 2q/p$ then $P_0 > P_1 < P_2$ so there is a local minimum at p_1 ($k = 1$). (Multimodality is a rather typical property of discrete mixtures of unimodal distributions.)

The recurrence relation

(24) $$P_k = \frac{n\lambda p}{k + 1} \sum_{j=0}^{k} \binom{n - 1}{j} p^j q^{n-j-1} P_{k-j}$$

may be used to calculate successive P_k's. However, to avoid possible accumulations of errors, Shumway and Gurland [46] propose using formula (22) written

*Some confusion can arise with the distribution described in Chapter 3, Section 11 (see footnote on page 78).

in the form

$$P_k = \exp[-\lambda(1 - q^n)] \cdot (p/q)^k (k!)^{-1} \mu'_{(k)} \tag{25}$$

where $\mu'_{(k)}$ is the k-th factorial moment of $n\mathbf{j}$ ($\mathbf{j}$ having a Poisson distribution with expected value λq^n). From (25) it follows that

$$\frac{P_{k+1}}{P_k} = \frac{p}{(k+1)q} R_{[k]} \qquad \left(R_{[k]} = \frac{\mu'_{(k+1)}}{\mu'_{(k)}}\right). \tag{26}$$

Shumway and Gurland [46] give tables of $R_{[k]}$, to 5 decimal places, for $k = 0(1)9$ and $\lambda q^n = 0.10(0.02)1.10$.

The same authors [47], have suggested yet another way of calculating the P_k's. This is based on direct calculation of $\mu'_{(k)}$ in (25). Firstly, $(nj)^{(k)}$ is expanded in powers of nj, using the formula

$$(nj)^{(k)} = \sum_{i=1}^{k} S_i^{(k)} (nj)^i$$

where $S_i^{(k)}$ are Stirling numbers of the first kind (as defined in Chapter 1, Section 2). Thus

$$\mu'_{(k)} = \sum_{i=1}^{k} S_i^{(k)} n^i E(\mathbf{j}^i)$$

whence

$$\begin{aligned} \mu'_{(k)} &= \sum_{i=1}^{k} S_i^{(k)} n^i \sum_{t=1}^{i} (\Delta^t 0^i / i!) E(\mathbf{j}^{(t)}) \\ &= \sum_{i=1}^{k} S_i^{(k)} n^i \sum_{t=1}^{i} (\Delta^t 0^i / i!) (\lambda q^n)^t \\ &= \sum_{t=1}^{k} A_{kt} (\lambda q^n)^t \end{aligned} \tag{27}$$

where $A_{kt} = \sum_{i=t}^{k} S_i^{(k)} (\Delta^t 0^i / t!) n^i$.

Schumway and Gurland [46] have provided tables of A_{kt} for $n = 3, 4$ with $k, t = 1(1)10$ ($k > t$).

The expected value and first three central moments of the Poisson-binomial distribution are

$$\mu'_1 = n\lambda p \tag{28.1}$$

$$\mu_2 = n^2 p^2 \lambda + npq\lambda \tag{28.2}$$

$$\mu_3 = n^3 p^3 \lambda + 3n^2 p^2 q\lambda + npq(q - p)\lambda \tag{28.3}$$

$$\begin{aligned} \mu_4 = 3(n^2p^2\lambda + npq\lambda)^2 + 6n^3p^3q\lambda + n^2p^2q(7 - 11p)\lambda \\ + npq(1 - 6pq)\lambda. \end{aligned} \tag{28.4}$$

Given a set of observed values of m independent random variables $\mathbf{x}_1, \mathbf{x}_2, \ldots \mathbf{x}_m$, each having distribution (22) the equations satisfied by the maximum likelihood estimators $\hat{\boldsymbol{\lambda}}, \hat{\mathbf{p}}$ of λ, p (n being supposed known) were obtained in the following form by Sprott [52]:

$$\bar{\mathbf{x}} = m^{-1} \sum_{j=1}^{m} \mathbf{x}_j = n\hat{\boldsymbol{\lambda}}\hat{\mathbf{p}} \tag{29.1}$$

$$\sum_{j=1}^{m} (\mathbf{x}_j + 1)(\hat{\mathbf{P}}_{\mathbf{x}_j+1}/\hat{\mathbf{P}}_{\mathbf{x}_j}) = mn\hat{\boldsymbol{\lambda}}\hat{\mathbf{p}} \tag{29.2}$$

(where $\hat{\mathbf{P}}_k = P_k$ with λ, p replaced by $\hat{\boldsymbol{\lambda}}$, $\hat{\mathbf{p}}$ respectively).

These equations have to be solved iteratively. An initial pair $\hat{\boldsymbol{\lambda}}_1, \hat{\mathbf{p}}_1$ are obtained by equating first and second sample and population moments. This gives

$$\mathbf{p}_1 = (\mathbf{s}^2 - \bar{\mathbf{x}})/[(n-1)\bar{\mathbf{x}}] \tag{30.1}$$

$$\left(\mathbf{s}^2 = (m-1)^{-1} \sum_{i=1}^{m} (\mathbf{x}_i - \bar{\mathbf{x}})^2\right)$$

$$\hat{\boldsymbol{\lambda}}_1 = \bar{\mathbf{x}}/(n\hat{\mathbf{p}}_1). \tag{30.2}$$

The variances and covariance of $\hat{\boldsymbol{\lambda}}$ and $\hat{\mathbf{p}}$ are, approximately

$$\operatorname{var}(\hat{\mathbf{p}}) \doteqdot \frac{q^2A + [np + (n-1)pq](n\lambda)^{-1}}{[n\lambda(n + q/p)A - (n-1)^2]m} \tag{31.1}$$

$$\operatorname{var}(\hat{\boldsymbol{\lambda}}) \doteqdot \frac{n^2\lambda^2A - n\lambda(n - 1 + p^{-1})}{[n\lambda(n + q/p)A - (n-1)^2]m} \tag{31.2}$$

$$\operatorname{cov}(\hat{\boldsymbol{\lambda}},\hat{\mathbf{p}}) = -\frac{-n\lambda qA + nq + p}{[n\lambda(n + q/p)A - (n-1)^2]m} \tag{31.3}$$

where

$$A = -1 + \left[\sum_{j=0}^{\infty} (j+1)^2(P_{j+1}^2/P_j)\right](n\lambda p)^{-2}.$$

An alternative method of estimation uses the equations

$$n\tilde{\boldsymbol{\lambda}}\tilde{\mathbf{p}} = \bar{\mathbf{x}} \tag{32.1}$$

$$n\tilde{\boldsymbol{\lambda}}\tilde{\mathbf{p}}\tilde{\mathbf{q}}^{n-1} = \mathbf{f}_1/\mathbf{f}_0 \tag{32.2}$$

where $\mathbf{f}_j$ denotes the proportion of j's ($j = 0,1$) among the m observations. From (32.1) and (32.2)

$$\tilde{\mathbf{p}} = 1 - \tilde{\mathbf{q}} = 1 - [\mathbf{f}_1/(\mathbf{f}_0\bar{\mathbf{x}})]^{1/(n-1)} \tag{33.1}$$

$$\tilde{\boldsymbol{\lambda}} = \bar{\mathbf{x}}/(n\tilde{\mathbf{p}}). \tag{33.2}$$

Instead of the ratio $\mathbf{f}_1/\mathbf{f}_0$ the proportion $\mathbf{f}_0$ alone might be used. Distinguishing this case by primes, we have

$$n\tilde{\boldsymbol{\lambda}}'\tilde{\mathbf{p}}' = \bar{\mathbf{x}} \tag{34.1}$$

$$\exp\left[-\tilde{\lambda}'(1 - \tilde{\mathbf{q}}'^n)\right] = \mathbf{f}_0 \tag{34.2}$$

whence

$$\bar{\mathbf{x}}/[-\log \mathbf{f}_0] = n\tilde{\mathbf{p}}'(1 - \tilde{\mathbf{q}}'^n)^{-1}. \tag{35}$$

Equation (35) can be solved by iteration.

Katti and Gurland [31] find that this last method is markedly superior to that using first and second moments (see equations (30)). Tables 1 and 2 show approximate efficiencies (ratios of generalized variances) relative to maximum likelihood estimators.

TABLE 1

Efficiency of the Method of Moments for the Poisson Binomial Distribution

		λ				
n	p	.1	.3	.5	1.0	2.0
2	.1	.928	.865	.843	.840	.870
2	.3	.732	.569	.525	.533	.635
2	.5	.494	.307	.264	.267	.392
3	.1	.896	.823	.793	.779	.810
3	.3	.658	.501	.452	.446	.542
3	.5	.426	.268	.231	.231	.333
5	.1	.816	.726	.688	.671	.715
5	.3	.527	.379	.337	.332	.435
5	.5	.345	.210	.178	.176	.277

TABLE 2

Efficiency of the Method of the First Moment and First Frequency for the Poisson Binomial Distribution

		λ				
n	p	.1	.3	.5	1.0	2.0
2	.1	.984	.974	.977	.991	.947
2	.3	.937	.888	.883	.923	.981
2	.5	.862	.740	.700	.717	.851
3	.1	.994	.986	.984	.944	.994
3	.3	.968	.930	.918	.935	.974
3	.5	.896	.798	.763	.765	.850
5	.1	.995	.987	.985	.993	.989
5	.3	.969	.924	.905	.911	.950
5	.5	.889	.769	.716	.690	.793

In the special case $n = 2$, we have a class of distributions termed *Hermite* distributions by Kemp and Kemp [33]. They consider a general class of distributions with probability generating functions of form $\exp\left[\sum_{i=1}^{\infty} a_i(t^i - 1)\right]$. This same class of distributions were termed *composed Poisson distributions* by Janossy et al. [28]. They show that for every distribution in this class (*except* Poisson) the variance exceeds the mean. This is a very wide class, including the negative binomial, Neyman Types A, B and C, as well as Poisson-binomial and other distributions. Poisson distributions, themselves, correspond to the case $a_i = 0$ for $i > 1$. In the case when $a_i = 0$ for $i > 2$, Kemp and Kemp [33] show that, necessarily, $a_1 \geq 0$, $a_2 \geq 0$. Putting $a_1 = 2\lambda pq$ and $a_2 = \lambda p^2$ gives the Poisson-binomial probability generating function $\exp[\lambda\{(q + pt)^2 - 1\}]$. By appropriate choice of λ and p, any pair of positive values for a_1 and a_2 can be attained.

In the even more special case $n = 1$, the probability generating function (23) becomes $\exp[-\lambda p(1 - t)]$. In this case the Poisson-binomial distribution is just a simple Poisson distribution with expected value λp.

If n is not known, but must be estimated, then the problems of fitting become very complicated. Indeed even when n is known, fitting is quite troublesome if $n \geq 3$.

Fitting a truncated Poisson-binomial distribution, with the zero class missing, has been described by Shumway and Gurland [46].

(xiii) $\qquad$ Binomial $(\mathbf{N},p) \underset{\mathbf{N}/n}{\wedge}$ Binomial (N',p').

The probability generating function of this distribution is $[q' + p'(q + pt)^n]^{N'}$ and the distribution is defined by

$$\Pr[\mathbf{x} = k] = \sum_{j \geq k/n} \binom{N'}{j} p'^j q'^{N'-j} \binom{nj}{k} p^k q^{nj-k} \tag{36}$$

$$(k = 0,1,2,\ldots N'n;\ q = 1 - p;\ q' = 1 - p').$$

If $n = 1$, the probability generating function is $(q' + p'q + pp't)^{N'}$ and the distribution is a binomial with parameters N', pp'. This result is also evident on realizing that the model corresponds to N' repetitions of a two-stage experiment with probabilities of success p, p' at the two stages independently.

The distribution (36) has not been much used in statistical work. The fact that there are four parameters p, p', n, N to be estimated (three, even if n is known) makes fitting the distribution a discouraging task.

(xiv) $\qquad$ Binomial $(\mathbf{N},p) \underset{\mathbf{N}/n}{\wedge}$ Negative binomial (N',P').

This distribution, like (xiii) has found little application in statistical work. The probability generating function is

$$[Q' - P'(q + pt)^n]^{-N'}$$

where $Q' = 1 + P'$; $q = 1 - p$. If $n = 1$, this becomes $[Q' - P'q - P'pt]^{-N'}$, corresponding to a negative binomial distribution with parameters N', $P'p$. Khatri [34] has studied the distributions obtained if $\mathbf{N}$ has a binomial or negative binomial distribution truncated by omission of the zero class. These are just binomial or negative binomial, respectively, "with zeroes" in the sense described in Section 4.

(xv) $$\text{Binomial }(\mathbf{N},p) \underset{\mathbf{N}/n}{\wedge} \text{Logseries }(\theta).$$

For this distribution

$$\Pr[\mathbf{x} = k] = [-\log(1-\theta)]^{-1} \sum_{j \geq k/n} (\theta^j/j) \binom{nj}{k} p^k q^{nj-k} \quad (k = 0,1,2,\ldots; 0 < \theta < 1). \tag{37}$$

The probability generating function is

$$[\log\{1 - \theta(q+pt)^n\}]/[\log(1-\theta)]. \tag{38}$$

If $n = 1$ we have an interesting special case:

$$\begin{cases} \Pr[\mathbf{x} = 0] = [\log(1 - q\theta)]/[\log(1-\theta)] \\ \Pr[\mathbf{x} = k] = [-\log(1-\theta)]^{-1}[p\theta/(1-q\theta)]^k/k \quad (k = 1,2,\ldots). \end{cases} \tag{39}$$

This is a logarithmic series distribution with parameter $p\theta(1 - q\theta)^{-1}$, 'modified' (as described in Section 4) by the addition of a proportion of zeroes (see Patil [39]).

(xvi) $$\text{Binomial }(N,\mathbf{p}) \underset{n\mathbf{p}}{\wedge} \text{Hypergeometric }(n,X,N').$$

This model corresponds to sampling without replacement from a population of size N' containing X "defective" items, followed by sampling (with sample size N) with replacement from the resulting set of n individuals.

The distribution is defined by the equation

$$\Pr[\mathbf{x} = k] = \left[\binom{N}{k} \Big/ \binom{N'}{n}\right] \sum_y \binom{X}{y} \binom{N'-X}{n-y} \left(\frac{y}{n}\right)^k \left(1 - \frac{y}{n}\right)^{N-k}.$$

The range of summation for y is

$$\max(0, n - N' + X) \quad \text{to} \quad \min(X, n).$$

(xvii) $$\text{Negative binomial }(N,\mathbf{P}) \underset{\mathbf{Q}^{-1}}{\wedge} \text{Beta}.$$

Using the definition of the negative binomial distribution as it is given in equation (1) of Chapter 5, it is supposed that $\mathbf{p} = \mathbf{Q}^{-1}$ has the probability density function

$$[B(\alpha,\beta)]p^{\alpha-1}(1-p)^{\beta-1}.$$

The resulting distribution is a generalized hypergeometric distribution of Type IV according to Kemp and Kemp's system of classification (see Chapter 6, Section 8).

(xviii) Negative binomial $(\mathbf{N},P) \bigwedge_{\mathbf{N}}$ Gamma.

(See also Section 3(iii).)

The probability generating function is

$$(40) \quad (\beta^{\alpha}\Gamma(\alpha))^{-1} \int_0^{\infty} N^{\alpha-1}e^{-N/\beta}(Q - Pt)^{-N}\, dN = [1 + \beta^{-1}\log(Q - Pt)]^{-\alpha}$$

(xix) Negative binomial $(\mathbf{N},P) \bigwedge_{\mathbf{N}/K}$ Poisson (θ).

This distribution is probably the most commonly used after the Poisson-binomial, of the new distributions described in this chapter. It is sometimes called the *Poisson-Pascal* distribution.

The probability generating function is

$$(41) \quad E[(Q - Pt)^{-\mathbf{j}K}] = \exp[\theta\{(Q - Pt)^{-K} - 1\}] \qquad (Q = P + 1)$$

where $\mathbf{j}$ has a Poisson distribution with expected value θ. Individual probabilities are

$$(42.1) \quad P_0 = \Pr[\mathbf{x} = 0] = \exp[-\theta(1 - Q^{-K})]$$

$$(42.2) \quad P_k = \Pr[\mathbf{x} = k] = e^{-\theta}(P/Q)^k(k!)^{-1}\sum_{j=1}^{\infty}(Kj)^{[k]}\{(\theta Q^{-K})^j/j!\} \qquad (k = 1,2,\ldots).$$

Shumway and Gurland [46] have provided tables to assist in the direct calculation of P_k from (42.2). Their analysis is very similar to that for the Poisson-binomial distribution (xii), except that instead of (descending) factorial moments we have *ascending* factorial moments.

Equation (42.2) can be written

$$P_k = [(P/Q)^k/k!]\mu_{[k]}\exp[-\theta(1 - Q^{-K})]$$

where

$$\mu_{[k]} = E[(K\mathbf{j})^{[k]}]$$

with $\mathbf{j}$ distributed as a Poisson variable with expected value θQ^{-K}. By an analysis similar to that leading to (27) it can be shown that

$$(43) \quad \mu_{[k]} = \sum_{t=1}^{k} A_{kt}^{*}(\theta Q^{-K})^t$$

where

$$A_{kt}^{*} = \sum_{i=1}^{k} S_i^{*k}(\Delta^t 0^i/t!)K^i$$

with S_i^{*k} defined by

$$(Kj)^{[k]} = \sum_{i=1}^{k} S_i^{*k}(Kj)^i.$$

Shumway and Gurland [47] give the coefficients $S_i^{*k}(\Delta^t 0^i/t!)$ for $k, t = 1(1)10$ $(k \geq t)$.

If the negative binomial be defined in the form

$$(44) \qquad \Pr[\mathbf{x} = k] = \binom{k-1}{N-1} Q^{-k} P^{k-N} \qquad (k = N, N+1, \ldots)$$

(i.e. $\mathbf{x}$ is increased by N as compared with the standard form) then the resulting distribution is defined by

$$(45.1) \qquad \Pr[\mathbf{x} = 0] = e^{-\theta}$$

$$(45.2) \qquad \Pr[\mathbf{x} = k] = e^{-\theta}(P/Q)^k \sum_{j \leq k/K} \binom{k-1}{Kj-1} [(\theta P^{-K})^j/j!] \qquad (k \geq K).$$

This form is known as the *generalized Pólya-Aeppli* distribution. The special case $K = 1$ corresponds to the *Pólya-Aeppli* distribution, and we will devote special attention to this case. It arises in a model formed by supposing that objects (which are to be counted) occur in clusters, the number of clusters having a Poisson distribution while the number of objects per cluster has the geometric distribution:

$$(46) \quad \Pr[\mathbf{y} = k] = qp^{k-1} \qquad (k = 1,2,\ldots; q = 1 - p; p = P/Q, q = Q^{-1}).$$

The Pólya-Aeppli distribution was described by Pólya [41] in 1931. He ascribed the derivation of the distribution to Aeppli [1] in a thesis completed in 1924.

The distribution is defined by

$$(47.1) \qquad P_0 = \Pr[\mathbf{x} = 0] = e^{-\theta}$$

$$(47.2) \qquad P_k = \Pr[\mathbf{x} = k] = e^{-\theta} p^k \sum_{j=1}^{k} \binom{k-1}{j-1} [(\theta q/p)^j/j!] \qquad (k \geq 1).$$

Evans [17] gives an alternative form, in which θ is replaced by $2m(2 + a)^{-1}$ with $m = \theta/p$, so that $a = 2q/p$ and $2 + a = 2p^{-1}$. He gives the recurrence relationship

$$(48) \qquad kP_k - [2(k-1)q + \lambda p]P_{k-1} - (k-2)q^2 P_{k-2} = 0.$$

Values of P_k may be calculated from (48), or from the direct formula,

$$(49) \qquad P_k = P_0 q^k (\lambda p/q) e^{-\lambda p/q} M(k+1; 2; \lambda p/q).$$

where $M(\cdot)$ is a confluent hypergeometric function, defined by equation (43) of Chapter 1. (See also Philipson [40].)

Evans also gave the following inequalities which may be used in estimating the cumulative probability $\Pr[\mathbf{x} \leq R - 1] = \sum_{k=0}^{R-1} P_k$:

$$(50) \qquad p^{-2}[P_{R+1} - (q - p)P_R] < 1 - \sum_{k=0}^{R-1} P_k$$
$$< [p^2 - p(\lambda - 2q)R^{-1}]^{-1}$$
$$\times [P_{R+1} - \{2q + (\lambda p - 2q)R^{-1} - 1\}P_R].$$

Douglas [15] obtained the approximate formula

$$P_k \doteqdot (z_k^{k+1} p\sqrt{2\pi})^{-1}[2\theta q(1 - pz_k)^{-3} + k(pz_k)^{-2}]^{-\frac{1}{2}} \exp[\theta\{q(1 - pz_k)^{-1} - 1\}]$$

where z_k is the smaller root of

$$\theta pqz_k = k(1 - pz_k)^2.$$

The expected value and first three central moments of the distribution are most easily found by using the factorial cumulant generating function. The factorial moment generating function of the geometric distribution (46) is

$$\sum_{k=1}^{\infty} (1 + t)^k qp^{k-1} = q(1 + t)\{1 - p(1 + t)\}^{-1}$$
$$= q(1 + t)(q - pt)^{-1}$$

and so the factorial moment generating function of the Pólya-Aeppli distribution is

$$(51) \qquad E[q^{\mathbf{j}}(1 + t)^{\mathbf{j}}(q - pt)^{-\mathbf{j}}] = \exp[\theta t(q - pt)^{-1}]$$

(where $\mathbf{j}$ has a Poisson distribution with expected value θ). Hence the factorial cumulant generating function is

$$\theta t(q - pt)^{-1}$$

and the rth factorial cumulant is

$$\kappa_{(r)} = \theta q^{-1}(p/q)^{r-1} \qquad (r = 1,2,\ldots).$$

Hence the ordinary cumulants are

$$(52.1) \qquad \kappa_1 = \theta q^{-1} = \mu_1'$$

$$(52.2) \qquad \kappa_2 = \theta q^{-1}(1 + p/q) = \theta q^{-2} = \mu_2$$

$$(52.3) \qquad \kappa_3 = \theta q^{-1}(1 + 3p/q + p^2/q^2) = \theta(1 + 2pq)q^{-3} = \mu_3$$

$$(52.4) \qquad \kappa_4 = \theta[1 + (4 - p)pq]q^{-4}.$$

From these equations we find

$$\alpha_3 = \sqrt{\beta_1} = (1 + 2pq)\theta^{-\frac{1}{2}} \tag{53.1}$$

$$\alpha_4 = \beta_2 = 3 + [1 + (4 - p)pq]\theta^{-1}. \tag{53.2}$$

Note that

$$\frac{\beta_2 - 3}{\beta_1} = \frac{1 + (4 - p)pq}{(1 + 2pq)^2} = 1 - \frac{(4q + 1)p^2q}{(1 + 2pq)^2}$$

so that for given p, the (β_1,β_2) points lie on a line through the normal point (0,3) with slope less than 1. As θ tends to infinity the (β_1,β_2) point tends to the normal point.

Note also that

$$P_1 = e^{-\theta}p(\theta q/p) = \theta q e^{-\theta} \tag{54.1}$$

and

$$P_2 = e^{-\theta}p^2[(\theta q/p) + \tfrac{1}{2}(\theta q/p)^2] = \theta q(p + \tfrac{1}{2}\theta q)e^{-\theta}. \tag{54.2}$$

The distribution has a local minimum at $k = 1$ if $\theta q < 1$, and $p + \frac{1}{2}\theta q > 1$, that is $\theta > 2$ and $q > \theta^{-1}$.

This distribution is called the *geometric Poisson* distribution by Sherbrooke [45], who gives tables of individual and cumulative probabilities to 4 decimal places for

$$\mu_1' = \theta q^{-1} = 0.10, 0.25(0.25)1.00(0.5)3.0(1)10;$$
$$q^{-1} = 1.0(0.5)5.0(1)7.$$

We now consider the problem of estimating p and θ, given n independent random variables $\mathbf{x}_1, \mathbf{x}_2, \ldots, \mathbf{x}_n$, each having distribution (46).

A simple method of estimation may be based on (47.1) and (54.1). From these two formulas we have $\theta = -\log P_0$; $q = \log(P_1/P_0)/[-\log P_0]$. This suggests using as estimators

$$\tilde{\boldsymbol{\theta}} = -\log \mathbf{f}_0; \; \tilde{\mathbf{p}} = \log(\mathbf{f}_1/\mathbf{f}_0)/[-\log \mathbf{f}_0] \tag{55}$$

where $\mathbf{f}_0, \mathbf{f}_1$ are the observed proportions of 0, 1 respectively.

Equations for the maximum likelihood estimators are

$$\hat{\boldsymbol{\theta}}(1 - \hat{\mathbf{p}})^{-1} = \bar{\mathbf{x}}\left(= n^{-1}\sum_{j=1}^{n}\mathbf{x}_j\right) \tag{56.1}$$

$$\hat{\boldsymbol{\theta}} = \bar{\mathbf{x}} - n^{-1}\hat{\mathbf{p}}\sum_{j=1}^{n}(\mathbf{x}_j - 1)(\hat{\mathbf{P}}_{\mathbf{x}_j-1}/\hat{\mathbf{P}}_{\mathbf{x}_j}) \tag{56.2}$$

(where $\hat{\mathbf{P}}_k$ denotes P_k according to (47) with θ replaced by $\hat{\boldsymbol{\theta}}$ and p by $\hat{\mathbf{p}}$).

Formulas (56) are obtained by using the relations

$$\begin{aligned} \frac{\partial P_k}{\partial \theta} &= (k\theta^{-1} - 1)P_k - (k - 1)\theta^{-1}pP_{k-1} \\ \frac{\partial P_k}{\partial p} &= q^{-1}[(k - 1)P_{k-1} - kP_k]. \end{aligned} \tag{57}$$

Returning now to the more general distribution (42) where θ, P and K have to be estimated, Katti and Gurland [30] describe three methods of fitting (other than maximum likelihood).

Method (a): Estimation from first three sample moments

$$\begin{cases} \hat{\mathbf{K}} = \hat{\kappa}_{(3)}\hat{\kappa}_{(1)}[\hat{\kappa}_{(3)}\hat{\kappa}_{(1)} - \hat{\kappa}_{(2)}^2]^{-1} - 2 \\ \hat{\mathbf{P}} = \hat{\kappa}_{(2)}\hat{\kappa}_{(1)}^{-1}(\hat{\mathbf{K}} + 1)^{-1} \\ \hat{\boldsymbol{\theta}} = \hat{\kappa}_{(1)}(\hat{\mathbf{K}}\hat{\mathbf{P}})^{-1} \end{cases} \tag{58}$$

where $\hat{\kappa}_{(j)}$ is the jth sample factorial cumulant.

Method (b): Estimation from first two sample moments and proportion of zeros ($\mathbf{f}_0$) *in sample*

$$\begin{cases} \hat{\mathbf{P}} \log [1 + (\hat{\kappa}_{(2)}\hat{\kappa}_{(1)}^{-1} - \hat{\mathbf{P}})\hat{\kappa}_{(1)}^{-1} \log \mathbf{f}_0] = (\hat{\kappa}_{(2)}\hat{\kappa}_{(1)}^{-1} - \hat{\mathbf{P}}) \log (1 + \hat{\mathbf{P}}) \\ \hat{\mathbf{K}} = \hat{\kappa}_{(2)}(\hat{\kappa}_{(1)}\hat{\mathbf{P}})^{-1} - 1 \\ \hat{\boldsymbol{\theta}} = \hat{\kappa}_{(1)}(\hat{\mathbf{K}}\hat{\mathbf{P}})^{-1}. \end{cases} \tag{59}$$

Method (c): Estimation from first two sample moments and ratio of proportion of ones ($\mathbf{f}_1$) *to proportion of zeroes in the sample*

$\hat{\mathbf{P}}$ is calculated from

$$\hat{\mathbf{P}}^{-1} \log (1 + \hat{\mathbf{P}}) = \hat{\kappa}_{(1)}\hat{\kappa}_{(2)}^{-1} \log \{\hat{\kappa}_{(1)}\mathbf{f}_0/\mathbf{f}_1\}. \tag{60}$$

Then $\hat{\mathbf{K}}$ and $\hat{\boldsymbol{\theta}}$ are calculated from the last two equations of (58).

Katti and Gurland [30] calculated the asymptotic efficiency (ratio of generalized variances) of each method relative to the method of maximum likelihood for $\theta = 0.1, 0.5, 1.0, 5.0$ and $P = 0.1, 0.3, 0.5, 1.0, 2.0$. For small θ (≤ 1) they find method (c) to be generally the best of the three, with an efficiency greater than 90% for $K \leq 1, P \leq 0.5$. For larger θ, (b) is better. In general (a) is worst. More detailed and precise comparisons of moment and maximum likelihood estimators have been carried out by Shenton and Bowman [44]. Their calculations indicate that asymptotic comparisons may not be reliable when applied to sample sizes less than 100.

(xx) Negative binomial $(\mathbf{N},P) \underset{\mathbf{N}/n}{\wedge}$ Binomial (N',p).

The probability generating function is

$$E[(Q - Pt)^{-n\mathbf{j}}] = [q + p(Q - Pt)^{-n}]^{N'}$$

(where **j** has a binomial distribution with parameters N', p).

This distribution is not often used. The same is true of:

(xxi) Negative binomial $(\mathbf{N},P) \underset{\mathbf{N}/n}{\wedge}$ Negative binomial (N',P')

and

(xxii) Negative binomial $(\mathbf{N},P) \underset{\mathbf{N}/n}{\wedge}$ Hypergeometric (n',X,N').

It is possible to envision a model giving rise to (xx), but the number of parameters (four) is likely to make fitting a rather difficult matter.

(xxiii) Negative binomial $(\mathbf{N},P) \underset{\mathbf{N}/n}{\wedge}$ Logseries (θ).

The probability generating function is

$$E[(Q - Pt)^{-n\mathbf{j}}] = [\log\{1 - \theta(Q - Pt)^{-n}\}]/[\log(1 - \theta)]$$

(where **j** has a logarithmic series distribution with parameter θ).

If $n = 1$, the probability generating function can be written

$$\frac{\log(Q - Pt - \theta) - \log(Q - Pt)}{\log(1 - \theta)}$$

and the rth factorial moment is

$$\mu_{(r)} = [-\log(1 - \theta)]^{-1} r![(Q - \theta)^{-r} - Q^{-r}]. \tag{61}$$

We will not discuss compound logarithmic series distributions. They do not lend themselves conveniently to analysis, and are not at present used in statistical analysis. We conclude this Section with brief notices of some compound hypergeometric distributions. More details are given by Hald [24]. In all cases it is the parameter **X** which is supposed to have a 'compounding' distribution Hald points out that the rth factorial moment of a compound hypergeometric distribution is

$$\frac{n^{(r)}}{N^{(r)}} \times (r\text{th factorial moment of } \mathbf{X}) \tag{62}$$

(xxiv) Hypergeometric $(n,\mathbf{X},N) \underset{\mathbf{X}}{\wedge}$ Binomial (N,p).

This distribution is a binomial with parameters n, p. This is evident on realizing that it represents the results of choosing a random subset of a random sample of fixed size.

(xxv) $\quad$ Hypergeometric $(n,\mathbf{X},N) \underset{\mathbf{X}}{\wedge}$ Hypergeometric (N,X',N').

By an argument similar to that in (xxiv), it can be seen that this is a hypergeometric distribution with parameters n, X', N'.

(xxvi) $\quad$ Hypergeometric $(n,\mathbf{X},N) \underset{\mathbf{X}}{\wedge}$ Discrete rectangular $(0,1;\ (N+1)^{-1})$.

The distribution $\Pr[\mathbf{X} = x] = (N+1)^{-1}(x = 0,N^{-1},2N^{-1},\ldots 1)$ is assumed for $\mathbf{X}$ (see Chapter 10, Section 2).

(xxvii) $\quad$ Hypergeometric $(n,\mathbf{X},N) \underset{\mathbf{X}}{\wedge}$ Pólya-Eggenberger (See Chapter 5).

(xxviii) $\quad$ Hypergeometric $(n,\mathbf{X},N) \underset{\mathbf{X}}{\wedge}$ Mixed binomial.

It is supposed that $\mathbf{X}$ has a compound binomial distribution of form

$$\Pr[\mathbf{X} = x] = \sum_{i=1}^{m} \omega_i \binom{N}{x} p_i^x q_i^{N-x}$$

with $0 < p_i < 1$, $0 < \omega_i < 1$, $q_i = 1 - p_i$, $\sum_{i=1}^{m} \omega_i = 1$. In view of (xxiv) above, it can be seen that the compounded distribution is of similar form.

3. 'Generalized' Distributions

Since 1943, (Feller [18]) the adjective 'generalized' has very often been used in a restricted sense for discrete distributions.* We shall describe this special meaning of the word.

If, in the probability generating function, $g_1(t)$ of a distribution F_1, the argument, t, is replaced by the probability generating function, $g_2(t)$, of another (or the same) distribution F_2, then the resulting function, $g_1(g_2(t))$, is also a probability generating function. It is easy to see that $g_1(g_2(t))$ is a polynomial function of t with non-negative coefficients. Further, if $t = 1$, then $g_2(1) = 1$ and $g_1(g_2(1)) = g_1(1) = 1$. So there is a probability distribution corresponding to $g_1(g_2(t))$. This distribution is called a *'generalized' F_1 distribution*. More precisely it is called an F_1 distribution generalized by the 'generalizer' (or generalizing distribution) F_2. It is written in the symbolic form

$$F_1 \vee F_2.$$

We need not discuss all the pairs of distributions described in Section 2 with $\wedge$ replaced by $\vee$, because a number of the latter pairs are, in fact, included

*It should be realized, however, that 'generalized' is still used with other meanings; as, for example, in 'generalized Poisson' (see Section 10.2 of Chapter 4, and Gerstenkorn [19]).

in the former class of distributions (see Gurland [20]). This is so if F_2 is a distribution which depends on a parameter ϕ such that $[g_2(t \mid \phi)]^j$ is the probability generating function of F_2 with parameter $j\phi$ (that is $[g_2(t) \mid \phi]^j = g_2(t \mid j\phi)$). For, in this case, denoting $\Pr[\mathbf{x} = k]$ for distribution F_1 by P_k, we have the following equation

$$g_1(t) = \sum_k P_k t^k$$

and

$$g_1(g_2(t \mid \phi)) = \sum_k P_k \cdot [g_2(t \mid \phi)]^k = \sum_k P_k \cdot g_2(t \mid k\phi) \tag{63}$$

and this is the probability generating function of

$$F_2 \underset{\theta/\phi}{\wedge} F_1$$

(θ denoting the value of the parameter of F_2).

Since the Poisson, binomial and negative binomial distributions are all of the type just described, it follows that when one of these is the 'generalizing' distribution, the resultant distribution is a compound distribution, and need not be discussed separately in this section. We therefore restrict our attention to cases in which the generalizing distribution is a hypergeometric, or a logarithmic series distribution.

(i) Poisson (λ) $\vee$ Logseries (θ).

The probability generating function is

$$\exp\left[\lambda\left\{\frac{\log(1 - t\theta)}{\log(1 - \theta)} - 1\right\}\right] = e^{-\lambda}(1 - t\theta)^{\lambda/\log(1-\theta)} \tag{64}$$
$$= [(1 - \theta)^{-1} - \theta(1 - \theta)^{-1}t]^{\lambda/\log(1-\theta)}.$$

This is recognized as the probability generating function of the negative binomial distribution with parameters $\lambda[-\log(1 - \theta)]^{-1}$, $\theta(1 - \theta)^{-1}$. (See Quenouille [42].)

(ii) Binomial (N,p) $\vee$ Logseries (θ).

The probability generating function is

$$\left(q + p\frac{\log(1 - t\theta)}{\log(1 - \theta)}\right)^N. \tag{65}$$

The expected value is $Np\theta(1 - \theta)^{-1}[-\log(1 - \theta)]^{-1}$ and the variance is

$$Np\theta(1 - \theta)^{-2}[-\log(1 - \theta)]^{-1}\{1 - p\theta[-\log(1 - \theta)]^{-1}\}.$$

(iii) Negative binomial (N,P) $\vee$ Logseries (θ).

The probability generating function is

$$\left(Q - P\frac{\log(1-\theta t)}{\log(1-\theta)}\right)^{-N} \tag{66}$$

$$= \left[1 + \{-\log(1-\theta)\}^{-1}\left\{1 + \log\left(\frac{1}{1-\theta} - \frac{\theta t}{1-\theta}\right)\right\}\right]^{-N}.$$

On comparison with (40) we see that this is a compound negative binomial distribution with a gamma compounding distribution.

The expected value of the distribution is $NP\theta(1-\theta)^{-1}[-\log(1-\theta)]^{-1}$, and the variance is

$$NP\theta(1-\theta)^{-2}[-\log(1-\theta)]^{-1}[1 + P\theta\{-\log(1-\theta)\}^{-1}]. \tag{67}$$

(iv) Logseries (λ) $\vee$ Logseries (θ).

The probability generating function is

$$\frac{\log[1 - \lambda\{\log(1-\theta t)\}/\{\log(1-\theta)\}]}{\log(1-\lambda)}. \tag{68}$$

The expected value is $\lambda\theta(1-\theta)^{-1}(1-\lambda)^{-1}[\log(1-\theta)\log(1-\lambda)]^{-1}$ and the variance is

$$\frac{\lambda\theta}{(1-\theta)^2(1-\lambda)\log(1-\theta)\log(1-\lambda)} \times\left[1 - \frac{\lambda\theta}{(1-\lambda)\log(1-\theta)}\left\{1 + \frac{1}{\log(1-\lambda)}\right\}\right].$$

4. 'Modified' Distributions

Many of the distributions described in Sections 2 and 3 have been developed in recent years (most since 1940). A major motivating force was the empirical observation that many distributions obtained in the course of experimental investigations often had an excess of zeroes as compared with a Poisson distribution with the same mean. This phenomenon is to be expected when some kind of clustering is present, and, indeed, many of the distributions described in Sections 2 and 3 do possess the property that the proportion in the zero class is greater than exp [$-$(expected value)], which is the value which would be predicted on the basis of a Poisson distribution.

The simplest way of increasing the proportion of zeroes is just to add an arbitrary proportion of zeroes, decreasing the remaining proportions in an appropriate constant ratio. Thus, if, for the original distribution

$$\Pr[\mathbf{x} = k] = P_k \qquad (k = 0,1,2,\ldots)$$

then for the *modified* distribution (denoted by primes)

$$(69) \qquad \begin{cases} P_0' = \omega + (1 - \omega)P_0 \\ P_k' = (1 - \omega)P_k \qquad (k \geq 1) \end{cases}$$

with $0 < \omega < 1$.

It is possible to take ω *less* than zero (*decreasing* the proportion of zeroes), provided

$$\omega + (1 - \omega)P_0 \geq 0$$

that is

$$\omega \geq -P_0/(1 - P_0).$$

If $\omega = -P_0/(1 - P_0)$, we simply have truncation on the left (by omission of the zero class). However, usually ω is greater than zero.

These distributions are sometimes called *modified*, but it seems unreasonable to restrict the term to such a special kind of modification. A more explicit description is obtained by adding the words "with (added) zeroes" to the name of the original distribution. (The word 'added' may often be omitted.) Thus a *Poisson with zeroes* distribution is defined by

$$(70) \qquad \begin{cases} \Pr[x = 0] = \omega + (1 - \omega)e^{-\lambda} \\ \Pr[x = k] = (1 - \omega)e^{-\lambda}\lambda^k/k! \qquad (k \geq 1). \end{cases}$$

A negative binomial "with zeroes" distribution has been already mentioned in Chapter 6, Section 3.3. Other names for these distributions are *inflated* (Singh [49], Pandey [38]) and '*pseudo-contagious*' (Cohen [13]).

The probability generating function, moments, etc. of a "with zeroes" distribution are easily derived from those of the original distribution. For example, if $g(t)$ be the original probability generating function, then that of the modified distribution is

$$\omega + (1 - \omega)g(t).$$

Similarly, the rth moment about zero of the modified distribution is equal to

$$(1 - \omega) \cdot (r\text{th moment about zero of original distribution})$$

We note that

$$(F_1 \text{ with zeroes}) \wedge F_2$$

is the same as

$$(F_1 \wedge F_2) \text{ with zeroes}$$

and

$$(F_1 \text{ with zeroes}) \vee F_2$$

is the same as

$$(F_1 \vee F_2) \text{ with zeroes}$$

respectively (with the same value of ω in each case).

In fitting a 'with zeroes' distribution, the estimation of parameters other than ω can be carried out, ignoring the observed frequency in the zero class, and using the technique appropriate to the original distribution truncated by omission of the zero class. The value of ω is then estimated so as to equate the expected and observed frequencies in the zero class. (Note that in the 'modified' distribution an arbitrary probability is, in effect, assigned to the zero class.)

Thus for the Poisson with zeroes (as in (70)), the estimator $\hat{\lambda}$ of λ is obtained from the (maximum likelihood) equation

(71.1) $$\text{(sample mean, excluding all zeroes)} = \hat{\lambda}[1 - \exp(-\hat{\lambda})]^{-1}$$

and then $\hat{\omega}$ is chosen to satisfy the equation

(71.2) $$\hat{\omega} + (1 - \hat{\omega})e^{-\hat{\lambda}} = \text{observed proportion of zeroes}$$

that is $$\hat{\omega} = (1 - e^{-\hat{\lambda}})^{-1}\,[(\text{observed proportion of zeroes}) - e^{-\hat{\lambda}}].$$

Singh [49] obtained the approximate formulas

(72.1) $$\operatorname{var}(\hat{\lambda}) \doteqdot (1 - \omega)^{-1}\lambda(1 - e^{-\lambda})(1 - e^{-\lambda} - \lambda e^{-\lambda})^{-1}$$

(72.2) $$\operatorname{var}(\hat{\omega}) \doteqdot (1 - \omega)[\omega(1 - \lambda e^{-\lambda}) + (1 - \omega)e^{-\lambda}](1 - e^{-\lambda} - \lambda e^{-\lambda})^{-1}.$$

Cohen [10] gives some examples of fitting this distribution to empirical data.

Cohen [9], in another paper, considers another distribution which arises in a model representing a situation in which (with probability P) a value $(c + 1)$ is classified as c. The distribution is

(73)
$$\Pr[\mathbf{x} = k] = \begin{cases} e^{-\lambda}(\lambda^k/k!) & (k = 0,1,2,\ldots,c - 1,c + 2,\ldots) \\ e^{-\lambda}(\lambda^c/c!)\{1 + P\lambda/(c + 1)\} & (k = c) \\ e^{-\lambda}(\lambda^{c+1}/(c + 1)!)(1 - P) & (k = c + 1). \end{cases}$$

Maximum likelihood estimators $\hat{\lambda}$, $\hat{\mathbf{P}}$ of λ, P respectively are

(74.1) $$\hat{\lambda} = \tfrac{1}{2}[\bar{\mathbf{x}} - (c + 1) + \mathbf{f}_c + \{[\bar{\mathbf{x}} - c + \mathbf{f}_c]^2 + 4(c + 1)(\bar{\mathbf{x}} - \mathbf{f}_{c+1})\}^{\frac{1}{2}}]$$

(74.2) $$\hat{\mathbf{P}} = [\mathbf{f}_c - (c + 1)\mathbf{f}_c\hat{\lambda}^{-1}](\mathbf{f}_c + \mathbf{f}_{c+1})^{-1}$$

where $\bar{\mathbf{x}}$ is the sample mean, and $\mathbf{f}_c$, $\mathbf{f}_{c+1}$ the proportions of observed values equal to c, $c + 1$ respectively (in a random sample of size N). Asymptotic formulas for the variances and covariance of $\hat{\lambda}$ and $\hat{\mathbf{P}}$ (for large N) are:

(75)
$$\begin{aligned} \operatorname{var}(\hat{\lambda}) &\doteqdot Q_{22}(Q_{11}Q_{22} - Q_{12}^2)^{-1}N^{-1} \\ \operatorname{var}(\hat{\mathbf{P}}) &= Q_{22}(Q_{11}Q_{22} - Q_{12}^2)^{-1}N^{-1} \\ \operatorname{cov}(\hat{\lambda},\hat{\mathbf{P}}) &\doteqdot -Q_{12}(Q_{11}Q_{22} - Q_{12}^2)^{-1}N^{-1} \end{aligned}$$

where

$$\begin{aligned} Q_{11} &= (c + 1)^{-1}[\lambda^{-1}(c + 1 - P\psi) - P^2\psi(c + 1 + \lambda P)^{-1}]; \\ Q_{12} &= -\psi(c + 1 + \lambda P)^{-1} \\ Q_{22} &= (c + 1)^{-1}\lambda\psi[\lambda(c + 1 + \lambda P)^{-1} + (1 - \lambda)^{-1}] \\ &\qquad (\psi = e^{-\lambda}(\lambda^c/c!)). \end{aligned}$$

In further papers [11] [12], Cohen has considered other similarly modified Poisson distributions.

Yoneda [54] has considered a general extended modification of the Poisson distribution. This allows the frequencies of the values 0, 1, 2, . . . , K to take arbitrary values, the remaining frequencies (for $K+1, K+2, \ldots$) being proportional to Poisson probabilities. The distributions depend on $(K+1)$ parameters $\omega_0, \omega_1, \ldots \omega_k$ as well as on the parameter, λ, of the Poisson 'upper tail', and are defined by

$$
(76)\quad \begin{cases} \Pr[\mathbf{x} = k] = (1 - \omega_k)e^{-\lambda}(\lambda^k/k!)\left[1 - e^{-\lambda}\sum_{j=1}^{K}\omega_j(\lambda^j/j!)\right]^{-1} & (k = 0,1,2,\ldots,K) \\ \Pr[\mathbf{x} = k] = e^{-\lambda}(\lambda^k/k!)\left[1 - e^{-\lambda}\sum_{j=1}^{K}\omega_j(\lambda^j/j!)\right]^{-1} & (k = K+1, K+2,\ldots). \end{cases}
$$

The equations satisfied by the maximum likelihood estimators $\hat{\lambda}, \hat{\omega}_0, \hat{\omega}_1, \ldots \hat{\omega}_K$ of λ and the ω's depend on the arithmetic mean of all observed $\mathbf{x}$'s greater than K: $\bar{\mathbf{x}}(K)$, say. The equations are

$$
(77.1)\quad \bar{\mathbf{x}}(K) = \hat{\lambda}\left[1 + (\hat{\lambda}^K/K!)\left\{\sum_{j=K+1}^{\infty}(\hat{\lambda}^j/j!)\right\}^{-1}\right]
$$

$$
(77.2)\quad \hat{\omega}_i = 1 - \frac{\text{Proportion of } \mathbf{x}\text{'s equal to } i}{\text{Proportion of } \mathbf{x}\text{'s greater than } K}\;\frac{\sum_{j=K+1}^{\infty}(\hat{\lambda}^j/j!)}{\hat{\lambda}^i/i!} \qquad (i = 0,1,2,\ldots K).
$$

(Note that if the observed proportions follow a Poisson distribution exactly then $\hat{\omega}_i = 0$.)

Yoneda [54] gives tables (reproduced in Table 3 below) of

$$
\lambda\left[1 + (\lambda^K/K!)\left\{\sum_{j=K+1}^{\infty}(\lambda^j/j!)\right\}^{-1}\right]
$$

from which equation (77.1) can be solved for $\hat{\lambda}$, by inverse interpolation, with accuracy sufficient for most purposes.

Khatri [34] has shown that for the "log series with zeroes" distribution

$$
\Pr[\mathbf{x} = 0] = \omega
$$
$$
\Pr[\mathbf{x} = k] = (1 - \omega)\theta^k[-k \log(1 - \theta)]^{-k} \qquad (k \geq 1)
$$

the maximum likelihood estimators of ω and θ are given by

$$
\hat{\omega} = 1 - \mathbf{f}_0
$$
$$
(\text{sample mean, excluding zeroes}) = \hat{\theta}(1 - \hat{\theta})^{-1}[-\log(1 - \hat{\theta})]^{-1}.
$$

TABLE 3

Numerical Values of $\dfrac{\lambda P[\mathbf{y} \geq K]}{P[\mathbf{y} > K]} = \lambda\left[1 + \left(\dfrac{\lambda^K}{K!}\right)\left\{\sum_{j=K+1}^{\infty} (\lambda^j/j!)\right\}^{-1}\right]$

	λ								
K	0.1	0.2	0.3	0.4	0.5	0.6	0.7	0.8	0.9
0	1.0508	1.1033	1.1575	1.2133	1.2708	1.3298	1.3906	1.4528	1.5166
1	2.0339	2.0689	2.1051	2.1424	2.1810	2.2208	2.2618	2.3039	2.3475
2	3.0254	3.0514	3.0784	3.1062	3.1347	3.1642	3.1943	3.2256	3.2576
3	4.0202	4.0412	4.0625	4.0844	4.1071	4.1300	5.1539	4.1783	4.2033
4		5.0341	5.0518	5.0700	5.0883	5.1075	5.1269	5.1470	5.1674
5				6.0596	6.0756	6.0916	6.1080	6.1248	6.1420
6					7.0656	7.0795	7.0938	7.1081	7.1229

(values of λ continued)

	λ									
K	1	2	3	4	5	6	7	8	9	10
0	1.5820	2.3130	3.1572	4.0748	5.0340	6.0149	7.0064	8.0027	9.0011	10.0005
1	2.3922	2.9114	3.5595	4.3224	5.1756	6.0908	7.0451	8.0215	9.0100	10.0045
2	3.2906	3.6744	4.1652	4.7692	5.4811	6.2854	7.1611	8.0870	9.0453	10.0228
3	4.2290	4.5258	4.9053	5.3796	5.9550	6.6807	7.3974	8.2392	9.1379	10.0765
4	5.1883	5.4272	5.7285	6.1056	6.5680	7.1234	7.7721	8.5086	9.3212	10.1948
5	6.1594	6.3576	6.6042	6.9096	7.2846	7.7386	8.2785	8.9062	9.6181	10.4056
6	7.1380	7.3068	7.5129	7.7660	8.0742	8.4479	8.8954	9.4230	10.0337	10.7248
7	8.1219	8.2680	8.4441	8.6572	8.9158	9.2266	9.5991	10.0414	10.5598	11.1552
8	9.1086	9.2374	9.3912	9.5744	9.7931	10.0558	10.3689	10.7407	11.1783	11.6877
9		10.2134	10.3482	10.5080	10.6972	10.9213	11.1880	11.5026	11.8742	12.3080
10		11.1932	11.3136	11.4548	11.6203	11.8145	12.0432	12.3135	12.6299	13.0005
11		12.1768	12.2853	12.4112	12.5571	12.7277	12.9268	13.1601	13.4326	13.7511
12			13.2615	13.3736	13.5063	13.6564	13.8315	14.0345	14.2712	14.5471
13			14.2413	14.3444	14.4623	14.5969	14.7529	14.9316	15.1395	15.3785
14				15.3176	15.4252	15.5463	15.6855	15.8456	16.0290	16.2405
15				16.2956	16.3931	16.5044	16.6294	16.7727	16.9357	17.1231
16				17.2761	17.3658	17.4668	17.5815	17.7106	17.8574	18.0238
17					18.3419	18.4352	18.5396	18.6565	18.7898	18.9396
18					19.3205	19.4068	19.5029	19.6110	19.7314	19.8678
19						20.3818	20.4709	20.5699	20.6796	20.8045
20						21.3602	21.4428	21.5342	21.6369	21.7482
21							22.4167	22.5026	22.5965	22.7001
22							23.3952	23.4744	23.5612	23.6576
23								24.4478	24.5300	24.6204
24								25.4257	25.5020	25.5852
25									26.4765	26.5548
26										27.5265
27										28.5019

Evidently $$\text{var}(\hat{\omega}) = \omega(1 - \omega)n^{-1}.$$

Also $$n\,\text{var}(\hat{\boldsymbol{\theta}}) \doteqdot \theta[(1 - \theta)\log(1 - \theta)]^2(1 - \omega)^{-1}(-\log(1 - \theta) - \theta]^{-1}$$

and $$\lim_{n\to\infty} n\,\text{cov}(\omega,\hat{\boldsymbol{\theta}}) = 0.$$

5. Special Distributions

In this section we will describe two distributions which are not of the simple form $F_1 \wedge F_2$ or $F_1 \vee F_2$, with F_1, F_2 standard discrete distributions but which are related to these kinds of distributions. As already mentioned, it is possible to construct unlimited numbers of distributions by combinations of the operations described in Sections 2, 3 and 4. The two distributions in this Section were selected because they seemed relatively useful.

5.1 *Thomas Distribution*

This is similar to the Pólya-Aeppli distribution (Section 2 (xix)) except that the geometric distribution is replaced by the distribution of a Poisson random variable *increased by 1.* Symbolically the distribution may be represented by

$$\boldsymbol{\theta}\phi^{-1} + \text{Poisson}\,(\boldsymbol{\theta}) \underset{\boldsymbol{\theta}/\phi}{\wedge} \text{Poisson}\,(\lambda).$$

In this form a similarity to the Neyman Type A distribution (Section 2(v)) is also apparent.

This distribution was used by Thomas [53] in constructing a model for the distribution of numbers of plants of a given species in randomly placed quadrats. It is well suited to situations in which one 'parent' as well as 'offspring' are included in the count for each cluster. If there are no clusters then there are no plants so

$$\Pr[\mathbf{x} = 0] = e^{-\lambda}.$$

Given that there are j clusters, the conditional distribution of the number of plants ($\mathbf{x}$) is that of

$$j + (\text{Poisson variable with expected value } j\phi).$$

The conditional expected values of $\mathbf{x}$ and $\mathbf{x}^2$ are

$$E(\mathbf{x} \mid j) = j(1 + \phi);\; E(\mathbf{x}^2 \mid j) = j^2(1 + \phi)^2 + j\phi.$$

Hence

$$E(\mathbf{x}) = \lambda(1 + \phi) \tag{78.1}$$

$$\text{var}(\mathbf{x}) = \lambda(1 + 3\phi + \phi^2). \tag{78.2}$$

By similar analysis we find

$$\mu_3(\mathbf{x}) = \lambda(1 + 7\phi + 6\phi^2 + \phi^3) \tag{78.3}$$

$$\mu_4(\mathbf{x}) = 3\lambda^2(1 + 3\phi + \phi^2)^2 + \lambda(1 + 15\phi + 25\phi^2 + 10\phi^3 + \phi^4). \tag{78.4}$$

Hence the distribution is positively skew, and

$$\alpha_3^2 = \beta_1 = \lambda^{-1}(1 + 7\phi + 6\phi^2 + \phi^3)^2(1 + 3\phi + \phi^2)^{-3}$$

$$\alpha_4 = \beta_2 = 3 + \lambda^{-1}(1 + 15\phi + 25\phi^2 + 10\phi^3 + \phi^4)(1 + 3\phi + \phi^2)^{-2}.$$

We note that

$$\frac{\beta_2 - 3}{\beta_1} = \frac{(1 + 15\phi + 25\phi^2 + 10\phi^3 + \phi^4)(1 + 3\phi + \phi^2)}{(1 + 7\phi + 6\phi^2 + \phi^3)^2} \tag{79}$$

so that for constant ϕ, the (β_1,β_2) point lies on a line passing through the normal (Gaussian) point (0,3). Values of the ratio (79) for various values of ϕ are shown in the following table. The values are remarkably stable. A maximum of about 1.19 is attained near $\phi = 0.25$.

ϕ	0.1	0.5	1	2	3	4	5	10
$(\beta_2 - 3)/\beta_1$	1.166	1.177	1.156	1.130	1.114	1.101	1.092	1.063

Thomas [53] applied the distribution to fit observed distributions of plants per quadrat (*Armeria maritima* and *Plantago maritima*) and obtained marked improvement over Poisson distributions. In particular, Thomas distributions can be bimodal, and bimodality was a feature in both observed distributions. (We note that $P_0 = e^{-\lambda}$; $P_1 = \lambda e^{-\phi} \cdot e^{-\lambda}$; $P_2 = \lambda e^{-\phi}(\phi + \frac{1}{2}\lambda e^{-\phi})e^{-\lambda}$.)

If only the proportions $\mathbf{f}_0, \mathbf{f}_1$ of zeroes and unit values, respectively, among N independent observed values are used, the maximum likelihood estimators $\hat{\boldsymbol{\phi}}, \hat{\boldsymbol{\lambda}}$ of ϕ, λ are

$$\hat{\boldsymbol{\lambda}} = -\log \mathbf{f}_0$$

$$\hat{\boldsymbol{\phi}} = -\log (\mathbf{f}_1/\hat{\boldsymbol{\lambda}}).$$

If observed and theoretical first and second moments are equated, the resulting equations are easy to solve. This method requires an exhaustive count of actual numbers of plants in each quadrat, but in some circumstances this might not be a severe objection. Thomas [53] gave the following (approximate) values of the ratios of standard errors of estimation of the expected value $\lambda(1 + \phi)$, using these two methods.

The moment generating function of the distribution is

$$\exp [\lambda\{\exp (t - \phi + \phi e^t) - 1\}].$$

The moments (78.1)–(78.4) could be calculated directly from this formula, but the process is little, if any, simpler than the direct method outlined above.

TABLE 4

Ratio of Standard Errors of Estimators of $\lambda(1 + \phi)$
(S.E. of moment estimator)/(S.E. of "maximum likelihood" estimator)*

	$E(\mathbf{f}_0)$									
$E(\mathbf{f}_1)$	0.05	0.10	0.20	0.30	0.40	0.50	0.60	0.70	0.80	0.90
0.05	0.311	0.410	0.510	0.580	0.640	0.695	0.749	0.805	0.867	0.947
0.10	0.280	0.450	0.581	0.657	0.718	0.772	0.824	0.877	0.935	—
0.20	—	0.387	0.635	0.736	0.801	0.855	0.906	0.956	—	—
0.30	—	—	0.601	0.766	0.850	0.911	0.963	—	—	—

*It should be borne in mind that "maximum likelihood" does not refer to the original counts, only to the reduced data $\mathbf{f}_0, \mathbf{f}_1$.

5.2 *Log-Zero-Poisson Distribution*

This is the name given to a distribution constructed by Katti and Rao [32]. It can be obtained by 'modifying' a Poisson $(\boldsymbol{\theta}) \underset{\theta/\phi}{\wedge}$ Logseries (λ) distribution (see Section 2(viii)) by "adding zeroes". From (13)

$$\Pr[\mathbf{x} = k] = [-\log(1 - \lambda)]^{-1}(\phi^k/k!) \sum_{j=1}^{\infty} j^{k-1}(\lambda e^{-\phi})^j \qquad (k = 0,1, \ldots).$$

This is now modified by adding extra probability to the event ($\mathbf{x} = 0$). The resultant *log-zero-Poisson* distribution is

$$(80.1) \qquad P_0 = \Pr[\mathbf{x} = 0] = \omega + (1 - \omega)[\log(1 - \lambda e^{-\phi})][\log(1 - \lambda)]^{-1}$$

$$(80.2) \qquad P_k = \Pr[\mathbf{x} = k] = (1 - \omega)[-\log(1 - \lambda)]^{-1}(\phi^k/k!) \sum_{j=1}^{\infty} j^{k-1}(\lambda e^{-\phi})^j$$

$$(k \geq 1; 0 < \omega < 1).$$

The probability generating function is

$$\omega + (1 - \omega)[\log(1 - \lambda e^{\phi(t-1)})][\log(1 - \lambda)]^{-1}.$$

Katti and Rao [32] have prepared tables of the truncated distribution,

$$\frac{P_k}{1 - P_0} = \frac{\Pr[\mathbf{x} = k]}{1 - \Pr[\mathbf{x} = 0]} = \frac{(\phi^k/k!) \sum_{j=1}^{\infty} j^{k-1}(\lambda e^{-\phi})^j}{-\log(1 - \lambda e^{-\phi})}, \qquad (k \geq 1)$$

from which the values of individual probabilities can easily be calculated, given ω. These authors claim that by appropriate choice of the parameters ω, λ and ϕ a wide variety of distributions can be modelled. They describe a method of fitting the distribution by maximum likelihood (using original counts) and use it to fit 35 different sets of data. They also fit Poisson, Poisson with zeroes,

Neyman Type A, negative binomial, and Poisson-binomial distributions with $n = 2$ and $n = 3$. The log-zero-Poisson emerges quite well from these comparisons.

The maximum likelihood equations are

$$\hat{\phi}\hat{\mathbf{p}}_1\hat{\mathbf{p}}_2 = \bar{\mathbf{x}} \tag{81.1}$$

$$\hat{\mathbf{p}}_1 \log [\hat{\mathbf{q}}_2 - \hat{\mathbf{p}}_2 e^{-\hat{\phi}}] = 1 - \mathbf{f}_0 \tag{81.2}$$

$$\sum_{j=0}^{\infty} (j + 1)\mathbf{f}_j(\hat{\mathbf{P}}_{j+1}/\hat{\mathbf{P}}_j) = \bar{\mathbf{x}} \tag{81.3}$$

where

$$\bar{\mathbf{x}} = \text{arithmetic mean of } N \text{ (independent) observed values}$$

$$\mathbf{f}_j = \text{proportion of } j\text{'s among these } N \text{ values}$$

$$\hat{\mathbf{p}}_1 = (1 - \hat{\omega})/[-\log (1 - \hat{\lambda})]$$

$$\hat{\mathbf{p}}_2 = \hat{\lambda}/(1 - \hat{\lambda}); \ \hat{\mathbf{q}}_2 = 1/(1 - \hat{\lambda})$$

while $\hat{\mathbf{P}}_j$ is given by (80.1) and (80.2) with ω, λ, ϕ replaced by $\hat{\omega}, \hat{\lambda}, \hat{\phi}$ respectively.

REFERENCES

[1] Aeppli, A. (1924). Ph.D. Thesis, University of Zürich.

[2] Anscombe, F. J. (1950). Sampling theory of the negative binomial and logarithmic series distribution, *Biometrika*, **37,** 358–382.

[3] Beckmann, M. J. and Bobkoski, F. (1957). Airline demand: An analysis of some frequency distributions, *Naval Research Logistics Quarterly*, **5,** 43–51.

[4] Berlyand, O. S., Nazarov, I. M. and Pressman, A. Ya. (1962). i^n erfc distribution — or mixed Gauss-Poisson distribution, *Doklady Akademii Nauk SSSR*, **147,** 1005–1007 (In Russian).

[5] Bhattacharya, S. K. and Holla, M. S. (1965). On a discrete distribution with special reference to the theory of accident proneness, *Journal of the American Statistical Association*, **60,** 1060–1066.

[6] Blischke, W. R. (1963). Mixtures of discrete distributions, *Proceedings of the International Symposium on Discrete Distributions, Montreal*, 351–372.

[7] Bliss, C. I. (1963). An analysis of scme insect trap records, *Proceedings of the International Symposium on Discrete Distributions, Montreal*, 385–397.

[8] Chistyakov V. P. (1967). Limiting discrete distributions in the problem with the balls with arbitrary probability of deposition, *Matematicheskie Zametki*, **1,** 9–16.

[9] Cohen, A. C. (1959). Estimation in the Poisson distribution when sample values of $(c + 1)$ are sometimes erroneously reported as c, *Annals of the Institute of Statistical Mathematics, Tokyo*, **11,** 189–193.

[10] Cohen, A. C. (1960). Estimating the parameters of a modified Poisson distribution, *Journal of the American Statistical Association*, **55,** 139–143.

[11] Cohen, A. C. (1960). Estimation in the truncated Poisson distribution when zeros and ones are missing, *Journal of the American Statistical Association*, **55,** 342–348.

[12] Cohen, A. C. (1960). An extension of a truncated Poisson distribution, *Biometrics*, **16,** 446–450.

[13] Cohen, A. C. (1960). *On a class of pseudo-contagious distributions*, University of Georgia, Institute of Statistics, Technical Report No. 11.

[14] Cohen, A. C. (1963). Estimation in mixtures of discrete distributions, *Proceedings of the International Symposium on Discrete Distributions, Montreal*, 373–378.

[15] Douglas, J. B. (1963). Asymptotic expansions for some contagious distributions, *Proceedings of the International Symposium on Discrete Distributions, Montreal*, 291–302.

[16] Dubey, S. D. (1965). *The compound Pascal distribution.* Unpublished manuscript, Procter and Gamble Co.

[17] Evans, D. A. (1953). Experimental evidence concerning contagious distributions in ecology, *Biometrika*, **40,** 186–211.

[18] Feller, W. (1943). On a general class of 'contagious' distributions, *Annals of Mathematical Statistics*, **14,** 389–400.

[19] Gerstenkorn, T. (1962). On the generalized Poisson distribution, *Prace Lodzkie Towarzystwo Nauk*, **3,** No. 85.

[20] Gurland, J. (1957). Some interrelations among compound and generalized distributions, *Biometrika*, **44**, 265–268.

[21] Gurland, J. (1958). A generalized class of contagious distributions, *Biometrics*, **14**, 229–249.

[22] Gurland, J. (1963). A method of estimation for some generalized Poisson distributions, *Proceedings of the International Symposium on Discrete Distributions, Montreal*, 141–158.

[23] Haight, F. A. (1959). The generalized Poisson distribution, *Annals of the Institute of Statistical Mathematics, Tokyo*, **11**, 101–105.

[24] Hald, A. (1960). The compound hypergeometric distribution and a system of single sampling inspection plans based on prior distributions and costs, *Technometrics*, **2**, 275–340.

[25] Hald, A. (1968). The mixed binomial distribution and the posterior distribution of p for a continuous prior distribution, *Journal of the Royal Statistical Society, Series B*, **30**, 359–367.

[26] Horsnell, G. (1957). Economic acceptance sampling schemes, *Journal of the Royal Statistical Society, Series A*, **120**, 148–191.

[27] Ishii, G. and Hayakawa, R. (1960). On the compound binomial distribution, *Annals of the Institute of Statistical Mathematics, Tokyo*, **12**, 69–80.

[28] Janossy, L., Rényi, A. and Aczél, J. (1950). On composed Poisson distributions, I. *Acta Mathematica, Hungarian Academy of Science*, **1**, 209–224.

[29] Katti, S. K. and Gurland, J. (1958). *Some families of contagious distributions*, Air Force Technical Note TN59–4AD208184.

[30] Katti, S. K. and Gurland, J. (1961). The Poisson Pascal distribution, *Biometrics*, **17**, 527–538.

[31] Katti, S. K. and Gurland, J. (1962). Some methods of estimation for the Poisson binomial distribution, *Biometrics*, **18**, 42–51.

[32] Katti, S. K. and Rao, A. V. (1965). *The log-zero-Poisson distribution*, Florida State University, Statistics Report M106.

[33] Kemp, C. D. and Kemp, A. W. (1965). Some properties of the 'Hermite' distribution, *Biometrika*, **52**, 381–394.

[34] Khatri, C. G. (1961). On the distributions obtained by varying the number of trials in a binomial distribution, *Annals of the Institute of Statistical Mathematics, Tokyo*, **13**, 47–51.

[35] Khatri, C. G. (1962). A fitting procedure for a generalized binomial distribution, *Annals of the Institute of Statistical Mathematics, Tokyo*, **14**, 133–141.

[36] Khatri, C. G. and Patel, I. R. (1961). Three classes of univariate discrete distributions, *Biometrics*, **17**, 567–575.

[37] McGuire, J. U., Brindley, T. A. and Bancroft, T. A. (1947). The distribution of the European corn-borer larvae, Pyrausta Neubilalis (H*bn*), in field corn, *Biometrics*, **13**, 65–78.

[38] Pandey, K. N. (1965). On generalized inflated Poisson distribution, *Journal of Scientific Research, Benares Hindu University*, **15**, 157–162.

[39] Patil, G. P. (1964). On certain compound Poisson and compound binomial distributions, *Sankhyā, Series A*, **26**, 293–294.

[40] Philipson, C. (1960). The theory of confluent hypergeometric functions and its application to compound Poisson processes, *Skandinavisk Aktuarietidskrift*, **43,** 136–162.

[41] Pólya, G. (1931). Sur quelques points de la théorie des probabilitiés, *Annales de l'Institut Henri Poincaré*, **1,** 117–161.

[42] Quenouille, M. H. (1949). A relation between the logarithmic, Poisson and negative binomial series, *Biometrics*, **5,** 162–164.

[43] Satterthwaite, F. E. (1942). Generalized Poisson distribution, *Annals of Mathematical Statistics*, **13,** 410–417.

[44] Shenton, L. R., Bowman, K. O., and Reinfields, J. (1969). Sampling moments of moment and maximum likelihood estimators for discrete distributions, *Bulletin of the International Statistical Institute*, **37,** 359–361.

[45] Sherbrooke, C. C. (1966). *Discrete compound Poisson processes and tables of the geometric Poisson distribution.* RAND Memorandum RM–4831–PR.

[46] Shumway, R. and Gurland, J. (1960). A fitting procedure for some generalized Poisson distributions, *Skandinavisk Aktuarietidskrift*, **43,** 87–108.

[47] Shumway, R. and Gurland, J. (1960). Fitting the Poisson binomial distribution, *Biometrics*, **16,** 522–533.

[48] Singh, M. P. (1966). A note on generalized inflated binomial distribution *Sankhyā, Series A*, **28,** 99.

[49] Singh, S. N. (1963). A note on inflated Poisson distribution, *Journal of the Indian Statistical Association*, **1,** 140–144.

[50] Skellam, J. G. (1948). A probability distribution derived from the binomial distribution by regarding the probability of success as variable between the sets of trials, *Journal of the Royal Statistical Society, Series B*, **10,** 257–261.

[51] Skellam, J. G. (1952). Studies in statistical ecology, I. Spatial pattern, *Biometrika*, **39,** 346–362.

[52] Sprott, D. A. (1958). The method of maximum likelihood applied to the Poisson binomial distribution, *Biometrics*, **14,** 97–106.

[53] Thomas, Marjorie (1949). A generalization of Poisson's binomial limit for use in ecology, *Biometrika*, **36,** 18–25.

[54] Yoneda, K. (1962). Estimations in some modified Poisson distributions, *Yokohama Mathematical Journal*, **10,** 73–96.

9

Contagious Distributions

1. Introduction and Historical Remarks

The use of the term 'contagious' applied to discrete distributions dates from 1939 when Neyman [29] used it to describe a model (see Section 2(v), Chapter 8) of the distribution of larvae in a randomly chosen area on a field. The model was constructed by assuming that the variation (in number of groups of eggs per unit area) could be represented by a Poisson distribution, while the number of larvae developing from a group could be represented by independent random variables, each having a common Poisson distribution.

This is, in fact, a model representing *heterogeneity*. More recently, a substantial body of opinion has arisen that heterogeneity should be distinguished from 'true contagion,' that is situations in which the events under observation depend on the pattern of previous occurrences of the events. Whether this corresponds more to the sense of the word 'contagion' is arguable. We have compromised, and have included in the present chapter Neyman's contagious distributions (of Types A, B, & C), but otherwise, only distributions arising from 'contagion' of the second kind. Thomas' distribution, which is allied to Neyman's Type A distribution, is not included here, but it has been discussed in Chapter 8, Section 5.1.

2. Neyman's Type A Distribution

2.1 *Definition*

This distribution is obtained by compounding a Poisson distribution by a Poisson distribution. It may be represented symbolically as

$$\text{Poisson}\,(\boldsymbol{\theta}) \underset{\theta/\phi}{\wedge} \text{Poisson}\,(\lambda) \tag{1}$$

and is defined by

$$(2.1)\qquad P_k(\lambda,\phi) = \Pr[\mathbf{x} = k] = \sum_{j=1}^{\infty} e^{-\lambda}(\lambda^j/j!)e^{-j\phi}[(j\phi)^k/k!] \qquad (k > 0)$$

and

$$(2.2)\qquad P_0(\lambda,\phi) = \Pr[\mathbf{x} = 0] = e^{-\lambda} + \sum_{j=1}^{\infty} e^{-\lambda}(\lambda^j/j!)e^{-j\phi}$$

$$= \exp[-\lambda(1 - e^{-\phi})].$$

(David and Moore [10] call ϕ the "*Index of Clumping*".)

Equation (2.1) can be written

$$(3)\qquad \Pr[\mathbf{x} = k] = (\phi^k/k!)\sum_{j=1}^{\infty} e^{-\lambda}(\lambda^j/j!)e^{-j\phi}\sum_{r=1}^{k} j^{(r)}[\Delta^r 0^k/r!]$$

$$= (\phi^k/k!)e^{-\lambda}\sum_{r=1}^{k}\frac{\Delta^r 0^k}{r!}\sum_{j=r}^{\infty}\frac{(\lambda e^{-\phi})^j}{(j-r)!}$$

$$= (\phi^k/k!)e^{-\lambda}\sum_{r=1}^{k}(\lambda e^{-\phi})^r[\Delta^r 0^k/r!]\exp(\lambda e^{-\phi}).$$

In particular,

$$\Pr[\mathbf{x} = 1] = \lambda\phi e^{-\phi}\exp[-\lambda(1 - e^{-\phi})]$$

and

$$\Pr[\mathbf{x} = 2] = \tfrac{1}{2}\lambda\phi^2 e^{-\phi}(1 + \lambda e^{-\phi})\exp[-\tfrac{1}{2}(1 - e^{-\phi})].$$

It will be noticed that $\Pr[\mathbf{x} = 1] < \Pr[\mathbf{x} = 0]$ if $\lambda\phi e^{-\phi} < 1$ and $\Pr[\mathbf{x} = 1] < \Pr[\mathbf{x} = 2]$ if $\frac{1}{2}\phi(1 + \lambda e^{-\phi}) > 1$. For any given value of $\lambda\phi e^{-\phi}$ it is possible to choose ϕ and λ so that the second inequality is satisfied (i.e. so that the distribution has a local minimum at $\mathbf{x} = 1$). Further remarks on the modality of the distribution are included in Section 2.2.

The probability generating function is

$$(4)\qquad \exp[-\lambda\{1 - e^{-\phi(1-t)}\}].$$

From (4), or from (2.1) and (2.2), it is possible to derive the recurrence relation (Beall [4]):

$$(5.1)\qquad \Pr[\mathbf{x} = k] = (\lambda\phi/k)e^{-\lambda}\sum_{j=0}^{k-1}(\lambda^j/j!)\Pr[\mathbf{x} = k - j - 1].$$

Douglas [11] has pointed out that from equation (3) and equation (7) of Chapter 4,

$$(5.2)\qquad \Pr[\mathbf{x} = k] = (\phi^k/k!)\mu_k'^{*}\Pr[\mathbf{x} = 0]$$

where $\mu_k'^{*}$ is the kth moment about zero of a Poisson distribution with expected value $\lambda e^{-\phi}$.

2.2 *Moments and Other Properties*

The rth factorial moment is

$$\mu_{(r)} = E[\phi^r \mathbf{j}^{[r]}] = \phi^r \sum_{j=1}^{r} \frac{\Delta^j 0^r}{j!} \lambda^j \tag{6}$$

(where $\mathbf{j}$ has a Poisson distribution with expected value λ).

From (6)

$$E(\mathbf{x}) = \lambda\phi \tag{7.1}$$

$$\operatorname{var}(\mathbf{x}) = \lambda\phi(1 + \phi) \tag{7.2}$$

$$\mu_3(\mathbf{x}) = \lambda\phi(1 + 3\phi + \phi^2) \tag{7.3}$$

$$\mu_4(\mathbf{x}) = 3\lambda^2\phi^2(1 + \phi)^2 + \lambda\phi(1 + 7\phi + 6\phi^2 + \phi^3). \tag{7.4}$$

Comparison with formulas (78.1)–(78.4) of Chapter 8 is interesting.

The moment ratios are

$$\sqrt{\beta_1} = \alpha_3 = \mu_3/\mu_2^{\frac{3}{2}} = (1 + 3\phi + \phi^2)(1 + \phi)^{-\frac{3}{2}}(\lambda\phi)^{-\frac{1}{2}} \tag{8.1}$$

$$\beta_2 = \alpha_4 = \mu_4/\mu_2^2 = 3 + (1 + 7\phi + 6\phi^2 + \phi^3)(1 + \phi)^{-2}(\lambda\phi)^{-1}. \tag{8.2}$$

Shenton [36] has established the recurrence relations among the cumulants:

$$\kappa_{r+1} = \phi\left\{\sum_{j=0}^{r-1}\binom{r}{j}\kappa_{r-j} + \lambda\right\} \tag{9.1}$$

$$\kappa_{r+1} = \phi\left(\kappa_r + \frac{\partial\kappa_r}{\partial\lambda}\right). \tag{9.2}$$

For a fixed value of ϕ the point (β_1, β_2) lies on the line with equation

$$\frac{\beta_2 - 3}{\beta_1} = \frac{(1 + 7\phi + 6\phi^2 + \phi^3)(1 + \phi)}{(1 + 3\phi + \phi^2)^2}.$$

As λ tends to infinity the point approaches the 'normal point' (0,3).

The value of the ratio (9) remains remarkably stable. For $\phi = 0.1$ it is equal to 1.135 and rises to a maximum value of 1.215 near $\phi = \frac{1}{2}$, falling slowly thereafter to 1 as ϕ tends to infinity. (See below.)

ϕ	$(\beta_2 - 3)/\beta_1$
1	1.200
2	1.164
3	1.141
4	1.124
5	1.110
6	1.099
7	1.090
10	1.071
20	1.042

These results are similar to those for the Thomas distribution (Chapter 8, Section 5.1) which has a similar genesis.

The narrow limits of the ratio restricts the field of applicability of the distribution. The Neyman Type B and Type C, and the generalized contagious distributions described later in this chapter extend the range covered by this class of distributions.

Barton [3] has made a systematic study of the modality of the Type A distribution. His results are summarized in Figure 1. It will be noted that it is possible for the distribution to have *three or more* modes (including one at zero).

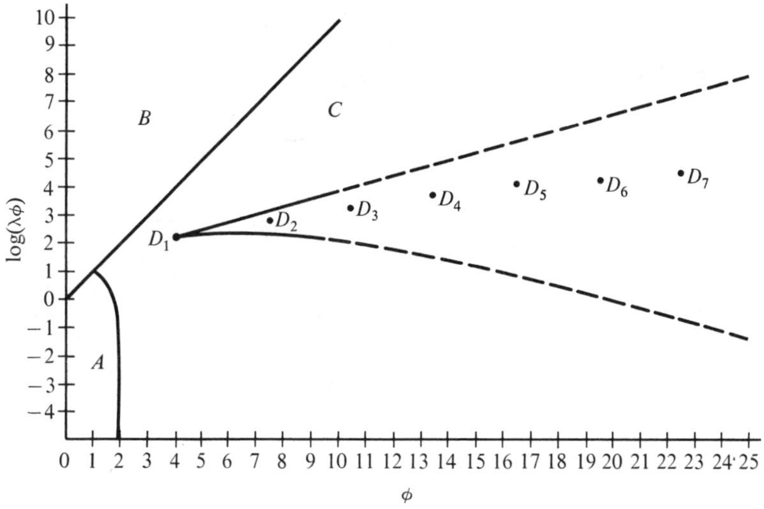

FIGURE 1

Notes: $A \equiv$ 1 mode (at zero)
$B \equiv$ 1 mode (not at zero)
$C \equiv$ 2 modes (one at zero)
$D_r \equiv$ double point (cusp) of boundary between regions with $(r + 1)$ and with $(r + 2)$ modes (one at zero in each case)
(D_r (for $r \geq 2$) and boundary (associated with D_r) were calculated approximately.)

Figures 2a–d show four Type A contagious distributions with common expected value 2. It will be noticed that bimodality appears as ϕ increases and λ decreases. This is to be expected since, for larger ϕ, the accretion of each additional 'group' makes a larger contribution to the overall distribution.

The modal values of $\mathbf{x}$ are *approximately* multiples of ϕ. Shenton and Bowman [37] have given the modal values of $\mathbf{x}$ for a few Type A distributions, which illustrate this point. For example with $\lambda = 7$ and $\phi = 25$, modal values are 0, 25, 50, 76, 103, 129, 153, 173; with $\lambda = 20$ and $\phi = 25$ they are 0, 25, 52, 486.

FIGURE 2a

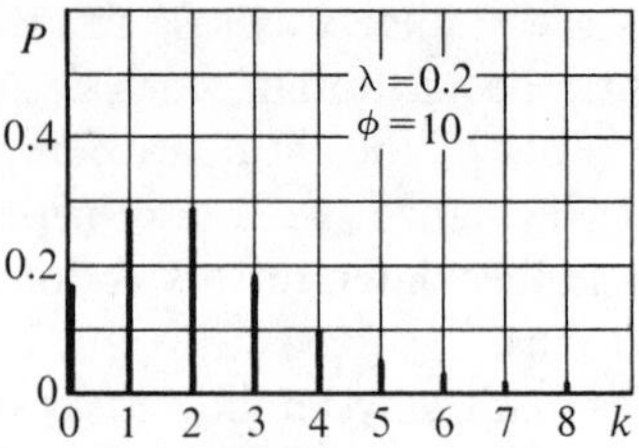

FIGURE 2b

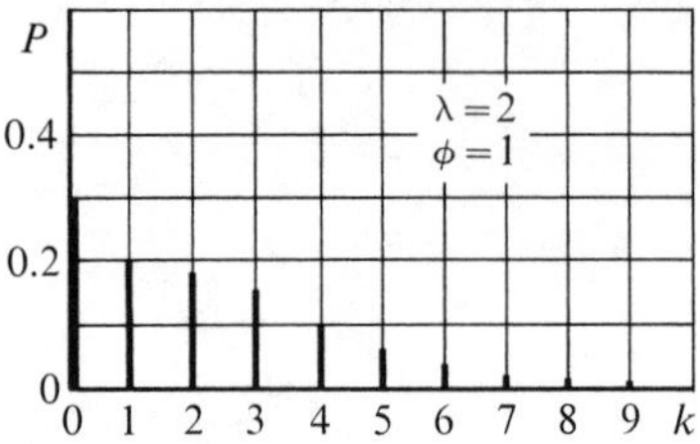

FIGURE 2c

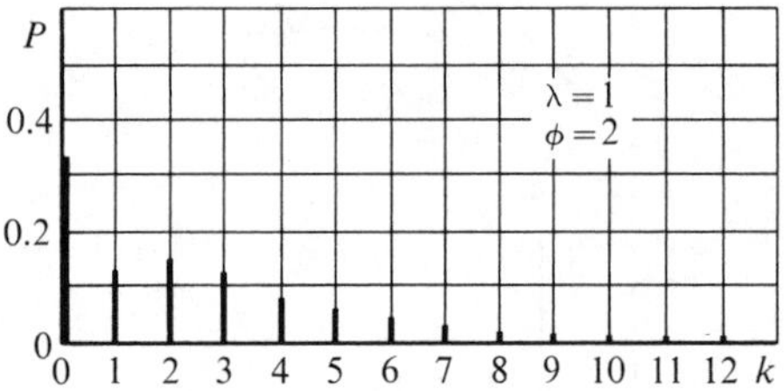

FIGURE 2d

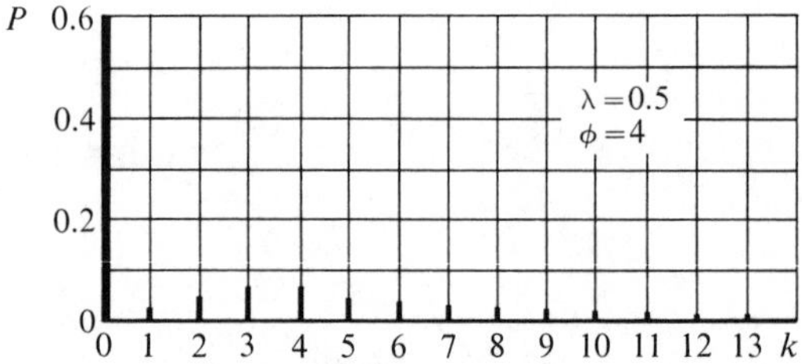

2.3 *Tables and Approximations*

Grimm [18] gives values of $\Pr[\mathbf{x} = k]$ for

$$E(\mathbf{x}) = \lambda\phi = 0.1(0.1)1.0(0.2(4.0, 6.0, 10.0$$
$$\phi = 0.2, 0.5, 1.0, 2, 3, 4, 5.$$

to 5 decimal places, up to $\Pr[\mathbf{x}] = 0.99900$.

Douglas [11] has given tables of $\mu'^{*}_{k+1}/\mu^{*}_{k}$, to assist in using equation (5.2), to 2–4 decimal places for $k = 0(1)19$; $\lambda e^{-\phi} = 0.000(0.001)0.030$, 0.01(0.01) 0.30(0.1)3.0.

Martin and Katti [28] have considered three approximations to the Neyman Type A distribution, applicable when the parameters λ and ϕ take 'extreme' values. (These are, in fact, limiting forms of the distribution.) It is hoped the error resulting from their use will be less than that from the application of approximations directly to the exact distribution.

Limiting Form I

The distribution of the standardized variable

$$\mathbf{y} = (\mathbf{x} - \lambda\phi)[\lambda\phi(1 + \phi)]^{-\frac{1}{2}}$$

is approximately unit normal.

This approximation is useful if λ is large and ϕ not too small.

Limiting Form II

If λ is *small* then $\mathbf{x}$ is approximately distributed as a *modified* Poisson variable ("Poisson with zeroes" (Chapter 8, Section 4)) with

$$\Pr[\mathbf{x} = 0] \doteqdot (1 - \lambda) + \lambda e^{-\phi}$$
$$\Pr[\mathbf{x} = k] \doteqdot \lambda(\phi^k/k!)e^{-\phi} \qquad (k > 0).$$

Limiting Form III

If ϕ is small then $\mathbf{x}$ is approximately distributed as a Poisson variable with expected value $\lambda\phi$.

The following diagrams (taken from Martin and Katti [28]) give a general picture of the regions in which these three limiting forms give useful practical approximations.

If we define $\xi^2 = \sum_i (P_i - P_i^*)^2$ where P_i and P_i^* are the probabilities in the Neyman Type A and the approximating distribution respectively, then ξ^2 can be considered as a measure of the degree of exactness of the approximation. Another index used to measure the degree of fit is $\zeta^2 = \sum_i \frac{(P_i - P_i^*)^2}{P_i}$. Some contours of ξ^2 and ζ^2 are given in Figures 3 and 4 for the first three approximations mentioned. For more details of these approximation methods and a few examples of their fit, see Martin and Katti [28].

Douglas [12] has given the approximate formula

$$(10.1) \qquad P_k(\lambda,\phi) \doteqdot \frac{e^{-\lambda}}{\sqrt{(2\pi)}} \cdot \frac{\phi^k \exp[k/g(k)]}{[g(k)]^k[k(1 + g(k))]^{\frac{1}{2}}}$$

where

$$g(k) \exp[g(k)] = k(\lambda e^{-\phi})^{-1}.$$

Bowman and Shenton [7] quote the following formula due to Philpot [31]:

$$(10.2) \qquad P_k(\lambda,\phi) \doteqdot P_0(\lambda,\phi) \cdot \frac{\phi^k}{k!} \cdot \frac{x_0}{(x_0 + k - \frac{1}{2})^{\frac{1}{2}}} \{\exp[f(x_0) - \lambda e^{-\phi}]\}$$

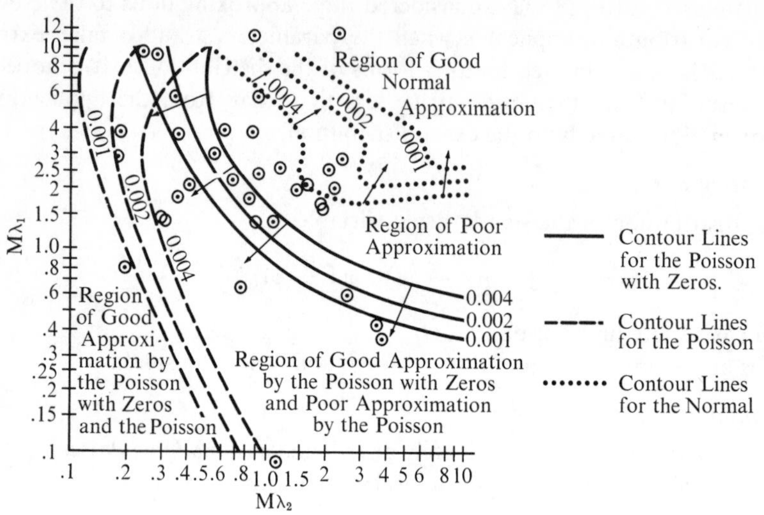

FIGURE 3

Contour Lines of ξ^2 for the

(1) *Poisson approximation*
(2) *Modified Poisson approximation*
(3) *Normal approximation*

where
$$x_0 \log [x_0 e^{\phi}/\lambda] = x - \tfrac{1}{2}$$
and
$$f(x_0) = x_0 + (k - \tfrac{1}{2})\{\log (\lambda e^{-\phi}) + x_0^{-1}(k - x_0 - \tfrac{1}{2})\}.$$

These authors also consider approximations of form

$$P_k(\lambda,\phi) \doteqdot \sum_{j=1}^{s} A_{j,s} e^{-\theta_j}(\theta_j^k/k!) \tag{10.3}$$

with the A's and θ's chosen to give correct values for the first $(2s - 1)$ moments.

2.4 *Estimation*

The two parameters of the Type A distribution are λ and ϕ. Given observations on n independent random variables $\mathbf{x}_1, \mathbf{x}_2, \ldots, \mathbf{x}_n$, each having distribution (2.1)–(2.2), the maximum likelihood estimators $\hat{\lambda}$, $\hat{\phi}$ of λ, ϕ respectively, satisfy the equations

$$n^{-1} \sum_{i=1}^{n} \mathbf{x}_i = \bar{\mathbf{x}} = \hat{\lambda}\hat{\phi} \tag{11.1}$$

$$\sum_{i=1}^{n} (\mathbf{x}_i + 1) \frac{P_{\mathbf{x}_i+1}(\hat{\lambda},\hat{\phi})}{P_{\mathbf{x}_i}(\hat{\lambda},\hat{\phi})} = n\bar{\mathbf{x}}. \tag{11.2}$$

Shenton [36] solved these equations using a Newton-Raphson iterative procedure. Tables given by Douglas [11] facilitate the calculations considerably. Equation (11.2) is usually rather complicated in form, and estimators based on equation of first and second sample and population moments are frequently used in practice. These estimators are:

$$\boldsymbol{\lambda}^* = \bar{\mathbf{x}}/\boldsymbol{\phi}^* \tag{12.1}$$

$$\boldsymbol{\phi}^* = (\mathbf{s}^2 - \bar{\mathbf{x}})/\bar{\mathbf{x}} \tag{12.2}$$

where

$$\mathbf{s}^2 = (n-1)^{-1} \sum_{i=1}^{n} (\mathbf{x}_i - \bar{\mathbf{x}})^2.$$

(Note that equations (11.1) and (12.1) are identical.)

If the sample size (n) is large the variances and covariances of these estimators are approximately

$$\operatorname{var}(\boldsymbol{\lambda}^*) \doteqdot \lambda[2 + \phi^2 + 2\lambda(1+\phi)^2]/(\phi^2 n) \tag{13.1}$$

$$\operatorname{var}(\boldsymbol{\phi}^*) \doteqdot [2 + \phi + 2\lambda(1+\phi)^2]/(\lambda n) \tag{13.2}$$

$$\operatorname{cov}(\boldsymbol{\lambda}^*, \boldsymbol{\phi}^*) \doteqdot -2[1 + \lambda(1+\phi)^2]/(\phi n). \tag{13.3}$$

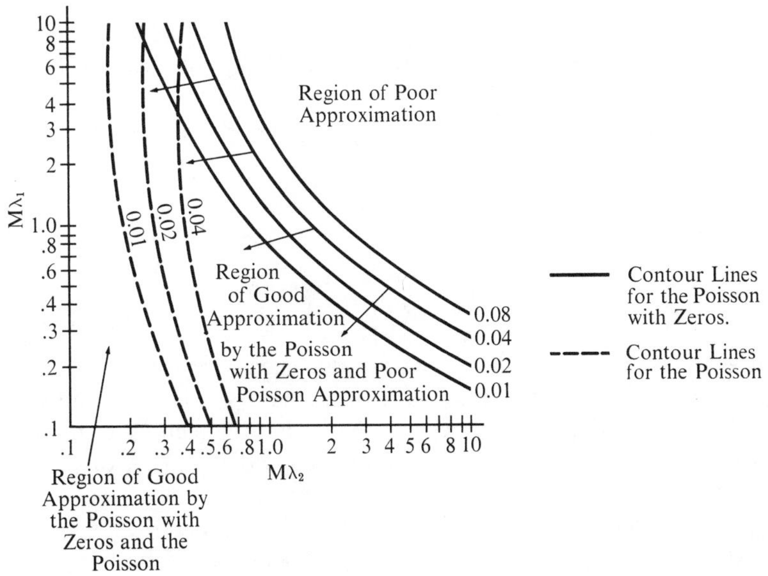

FIGURE 4

Contour Lines of ζ^2 for the

(1) *Poisson approximation*

(2) *Modified Poisson approximation*

For $\phi < 0.2$ the asymptotic efficiency of the moment estimators is at least 85% (whatever be the value of λ), while for $0.2 < \phi < 1.0$ the efficiency is generally between 75% and 90%. A table of efficiencies is given by Shenton [36].

Shenton and Bowman [37] [7] have shown that even in samples of size 100, there is substantial bias in both the maximum likelihood and moment estimators. For λ the bias is positive; for ϕ it is negative. Generally the biases in the two kinds of estimators are of comparable magnitude, that is the moment estimators of usually being slightly larger. The bias in estimators of ϕ does not greatly depend on λ and is of order -1 to -2 per cent. The bias in estimators of λ decreases as ϕ increases; for $\phi = 1$ it is about 10%.

A third method of estimation uses the sample mean ($\bar{\mathbf{x}}$) and the observed proportion of zeroes ($\mathbf{f}_0$). The equations for the estimators $\tilde{\lambda},\tilde{\phi}$ are

$$\bar{\mathbf{x}} = \tilde{\lambda}\tilde{\phi} \qquad \text{(Cf. (11.1), (12.1))} \tag{14.1}$$

$$\mathbf{f}_0 = \exp[-\tilde{\lambda}(1 - e^{-\tilde{\phi}})] \tag{14.2}$$

$\tilde{\lambda}$ may be eliminated between these equation giving the equation (to be solved for $\tilde{\phi}$).

$$\bar{\mathbf{x}}/[-\log \mathbf{f}_0] = \tilde{\phi}(1 - e^{-\tilde{\phi}})^{-1}. \tag{14.3}$$

A fourth method uses the ratio of frequencies ($\mathbf{f}_1/\mathbf{f}_0$) of ones and zeroes, in place of $\mathbf{f}_0$, leading to the formulas

$$\bar{\mathbf{x}} = \tilde{\lambda}'\tilde{\phi}' \tag{15.1}$$

$$\mathbf{f}_1/\mathbf{f}_0 = \tilde{\lambda}'\tilde{\phi}'e^{-\tilde{\phi}'} \tag{15.2}$$

whence

$$\tilde{\phi}' = \log(\bar{\mathbf{x}}\mathbf{f}_0/\mathbf{f}_1). \tag{15.3}$$

Katti and Gurland [26] have tabled the efficiencies (relative to the moment estimators (λ^*, ϕ^*) of the two latter pairs of estimators. Figures 5 and 6 (taken from their paper) summarize the results. Note that the method of moments is better (for most values of ϕ) if $\lambda > 5$ (approx.) For $\lambda < 4.5$, the estimators $\tilde{\lambda}$, $\tilde{\phi}$ seem to be preferable.

Katti [25] has suggested the different methods of estimation might be combined, with some gain in average accuracy. In particular he suggests using statistics which are functions of the sample mean and variance, *and* of the observed proportion ($\mathbf{f}_0$) of zeroes. His proposed method requires solution of the simultaneous equations (in matrix notation).

$$\frac{(\hat{\phi},\hat{\phi}^2,e^{-\hat{\phi}} - 1)\hat{\Omega}^{-1}(\kappa_{(1)},\kappa_{(2)}, \log \mathbf{f}_0)'}{(\hat{\phi},\hat{\phi}^2 e^{-\hat{\phi}} - 1)\hat{\Omega}^{-1}(\hat{\phi},\hat{\phi}^2,e^{-\hat{\phi}})'} = \tilde{\lambda} = \frac{(1,2\hat{\phi},-e^{-\hat{\phi}})\hat{\Omega}^{-1}(\kappa_{(1)},\kappa_{(2)}, \log \mathbf{f}_0)'}{(1,2\hat{\phi},-e^{-\hat{\phi}})\hat{\Omega}^{-1}(\hat{\phi},\hat{\phi}^2,e^{-\hat{\phi}})'}$$

where $\hat{\lambda}$, $\hat{\phi}$ are the estimators of λ, ϕ respectively and $\kappa_{(1)}$, $\kappa_{(2)}$ are the first and

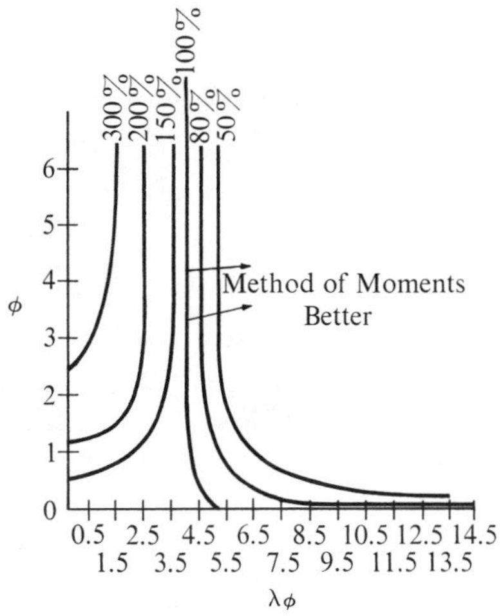

FIGURE 5

First moment and frequency vs Moments

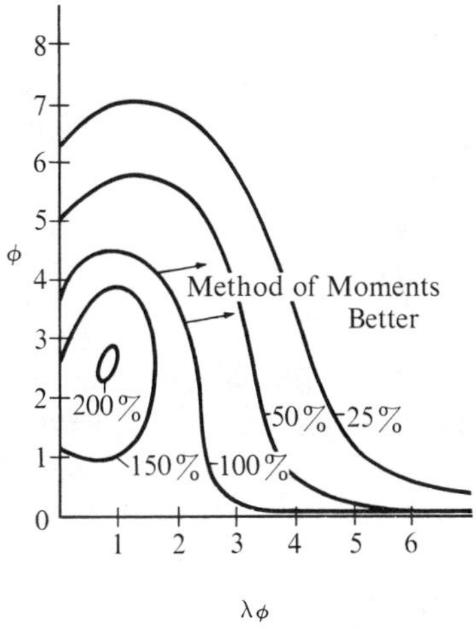

FIGURE 6

First moment and ratio of first two frequencies vs Moments

second sample factorial cumulants (sample mean, and [(sample variance) − (sample mean)]) respectively, and

$$\hat{\boldsymbol{\Omega}} = \begin{bmatrix} 1 & 0 & 0 \\ -1-2\hat{\mu}'_1 & 1 & 0 \\ 0 & 0 & 1 \end{bmatrix} \begin{bmatrix} \hat{\mu}'_2 - \hat{\mu}_1'^2 & & \\ \hat{\mu}'_3 - \hat{\mu}'_1\hat{\mu}'_2 - \hat{\mu}'_1 & \hat{\mu}'_4 - \hat{\mu}_2'^2 - \hat{\mu}'_2 & \\ & & \mathbf{f}_0^{-1}(1-\mathbf{f}_0) \end{bmatrix} \times \begin{bmatrix} 1 & -1-2\hat{\mu}'_1 & 0 \\ 0 & 1 & 0 \\ 0 & 0 & 1 \end{bmatrix}$$

(In $\hat{\boldsymbol{\Omega}}$, $\hat{\mu}'_r$ denotes the rth moment about zero of distribution (2) with λ, ϕ replaced by $\hat{\boldsymbol{\lambda}}$, $\hat{\boldsymbol{\phi}}$ respectively).

These equations must be solved by some form of iterative process (Katti suggests the use of graphical representation). The process could well be tedious, but Katti gives approximate values of efficiency (relative to maximum likelihood solution) which are very high (at least 99.5% for $\lambda \le 2$ and $\phi \le 1$; at least 97% for $\lambda \le 2$ and $1 \le \phi \le 2$). (See also Chapter 2, Section 5). A modified form of this estimator described by Hinz and Gurland [23] is easier to calculate and appears to have high efficiency also.

If a series of sets of sample values are available, and it is only desired to estimate ϕ (the "mean number per group") another method of fitting has found favor among practical workers. This is simply based on plotting log $\mathbf{f}_0$ against

$\bar{\mathbf{x}}$. Since

$$\frac{\log P_0(\lambda,\phi)}{E(\mathbf{x})} = -(1 - e^{-\phi})/\phi \tag{16}$$

a value of ϕ can be obtained from this ratio. A similar method can be used for the Thomas distribution. Pielou [32] has discussed these methods of estimation.

2.5 *Applications*

The Neyman Type A distribution is often used to describe plant distributions, especially when reproduction of the species produces 'clusters'. This frequently happens when the species is generated by seeds or by offshoots of parent plants, or by groups of seeds carried by living creatures, as is the case with many fruits. However, Archibald [1] found that there was not sufficient evidence to make induction from the type of fitted distribution to the type of reproduction. Evans [15] found that while Type A gives good results for plant distributions, negative binomial distributions (Chapter 5) were better for insect distributions. (See also Wadley [43]).

Pielou [32] investigated the use of Neyman Type A distributions in ecology; he compared them with the Thomas distributions discussed in the previous chapter. He found that neither family of distributions is likely to be applicable to describe plant distributions unless the 'clusters' of plants are so compact as not to lie across the edge of the quadrat used in selecting sample areas. The choice of quadrat size can greatly affect the results. Pielou found that the Type A distributions fitted a wider variety of distribution than did the Thomas distributions.

It has been suggested by David and Moore [10] that complete distributions need not be fitted to ecological data if one only needs to estimate indices of 'clustering' (or 'contagiousness') or the mean number of plants per quadrant. Among the methods they discuss is the regression method of estimating ϕ, described at the end of Section 2.4.

3. Generalizations of Neyman's Type A Distribution

The rather severe restriction on the shape of distribution (2) implied by the limits on the value of $(\beta_2 - 3)/\beta_1$, has caused some practical inconvenience, and various generalizations of this distribution have been considered. We now describe a generalization constructed by Gurland [20].

It is natural to seek a suitable generalization by modifying the assumptions described in Section 1. If we suppose, additionally, that while the number of larvae produced have a Neyman Type A distribution, the number surviving to be observed, given m are produced, has the binomial distribution

$$P(x \mid m) = \binom{m}{x} p^x(1 - p)^{m-x} \qquad (x = 0,1,\ldots m) \tag{17}$$

then (Feller [16]) the overall distribution is still a Type A distribution, with parameters λ, $p\phi$. So we do not get a new distribution in this way. However if we suppose that variation of the parameter p in (17) from group to group is represented by a beta distribution (so that the conditional distribution $P(\mathbf{x} \mid m)$ is of form Binomial $(m,\mathbf{p}) \bigwedge_{\mathbf{p}}$ Beta) then a new family of distributions is obtained.

In fact if the probability density function of $\mathbf{p}$ is

$$(B(\alpha,\beta))^{-1}p^{\alpha-1}(1-p)^{\beta-1} \qquad (0 \leq p \leq 1)$$

then the probability generating function of the number of larvae per group is

$$(18) \qquad (B(\alpha,\beta))^{-1}\int_0^1 e^{\phi p(t-1)}p^{\alpha-1}(1-p)^{\beta-1}\,dp = M(\alpha;\alpha+\beta;\phi(t-1))$$

where $M(\cdot)$ denotes the confluent hypergeometric function (equation (43) of Chapter 1).

It follows that the overall distribution of the total number ($\mathbf{x}$) of larvae has probability generating function

$$(19) \qquad \exp\{-\lambda[1 - M(\alpha;\alpha+\beta;\phi(t-1))]\}.$$

This corresponds to a family of distributions with four parameters. These are the original two parameters λ (expected number of groups) and ϕ (expected number of larvae per group) and α and β, defining the distribution of $\mathbf{p}$ ('probability of survival', or more generally, 'probability of being observed') from group to group.

If $\alpha = 1$, a sub-family is obtained which was studied by Beall and Rescia [5]. Members of this sub-family have probability generating functions of form

$$(20) \qquad e^{-\lambda}\exp\left[\lambda\Gamma(\beta+1)\sum_{j=0}^{\infty}\frac{\phi^j(t-1)^j}{\Gamma(\beta+j+1)}\right].$$

If β is put equal to zero, Neyman's Type A distribution is obtained. Corresponding to $\beta = 1$, $\beta = 2$ are two other distributions introduced by Neyman, called *Type B* and *Type C* respectively.

Distributions corresponding to (20) have three parameters. Beall and Rescia [5] suggested a method of fitting which essentially consists of first fixing β, then fitting λ and ϕ by equating first and second sample and population moments. (The equations for the estimators $\boldsymbol{\lambda}^* \equiv \boldsymbol{\lambda}^*(\beta)$ and $\boldsymbol{\phi}^* \equiv \boldsymbol{\phi}^*(\beta)$ are:

$$(21.1) \qquad \boldsymbol{\phi}^* = \tfrac{1}{2}(\beta+2)(\mathbf{s}^2 - \bar{\mathbf{x}})/\bar{\mathbf{x}}$$

$$(21.2) \qquad \boldsymbol{\lambda}^* = (\beta+1)\bar{\mathbf{x}}/\mathbf{s}^2$$

where $\mathbf{s}^2$, $\bar{\mathbf{x}}$ represent the sample variance and mean, as in (11) and (12).) The fitted and observed distributions are then compared by means of a χ^2 test, and $\Pr[\chi^2 >$ observed value$] = P_{\chi^2}(\beta)$ calculated. The process is repeated for a succession of integer values of β and then the value of β for which $P_{\chi^2}(\beta)$ is a maximum is selected as the estimated value of β (λ and ϕ having corresponding values, satisfying (21)). These calculations require evaluation of individual probabilities, $P_k(\lambda,\phi,\beta)$ say, of the distribution.

From (19) (with $\alpha = 1$) it follows that there is the following recurrence relation among the quantities $P_k \equiv P_k(\lambda,\phi,\beta)$:

(22) $$P_{k-1} = \lambda(k+1)^{-1} \sum_{j=0}^{k} f^{(j+1)}(0) P_{k-j}/j!$$

where $$f(t) = \Gamma(\beta+1) \sum_{j=0}^{\infty} \frac{\phi^j (t-1)^j}{\Gamma(\beta+j+1)}.$$

Furthermore

(23) $$P_0 = \exp[\lambda(f(0)-1)].$$

Beall and Rescia [5] suggest use of the recurrence formula

(24) $$f^{(j)}(0)/j! = j^{-1}(\phi+\beta+j)[f^{(j-1)}(0)/(j-1)!] - j^{-1}\phi f^{(j-2)}(0)/(j-2)! \qquad (j \geq 2)$$

to aid calculation.

Although β can be fractional, only integer values were used in the calculations of Beall and Rescia [5]. Fortunately, the optimal solution does not seem to be very sensitive to the exact value chosen for β, and taking the nearest integer value appears to give adequate results. Beall and Rescia [5] show that

(25) $$\beta = \frac{6(\mu_2^2 + \mu_1'\mu_2 - \mu_1'\mu_3 - \mu_1'^2)}{\mu_1'^2 + 2\mu_1'\mu_3 - 3\mu_2^2}.$$

Substituting sample moments for the population moments in (25), a formula from which β might be estimated is obtained. Beall and Rescia [5], however, reported 'unhappy results' using this formula, indicating that estimated values of β which are too low are obtained. They recommend rather, using the observed proportion of zeroes, $\mathbf{f}_0$, and the first and second sample moments.

As β tends to infinity and ϕ varies so that the first two moments remain fixed (see (28) below), the limiting distribution corresponding to (19) is a Pólya-Aeppli distribution (see Chapter 8, Section 2). This was pointed out by Gurland [20]. The same author also pointed out that for a fixed general value of α, the limiting distribution is a *generalized Pólya-Aeppli* distribution (Chapter 8, equation (45)).

Considering now the more general family of distributions with probability generating function of form (19), we see that the factorial cumulant generating function (obtained by replacing t by $(1+t)$ and taking natural logarithms) is

(26) $$\lambda[M(\alpha;\alpha+\beta;t)-1]$$

and so the rth factorial cumulant is

(27) $$\kappa_{(r)} = \lambda\phi^r\alpha^{[r]}/(\alpha+\beta)^{[r]}.$$

From (27) we find

(28.1) $$\mu_1' = \lambda\phi\alpha(\alpha+\beta)^{-1}$$

(28.2) $$\mu_2 = \lambda\phi\alpha(\alpha+\beta)^{-1}[1+\phi(\alpha+1)(\alpha+\beta+1)^{-1}]$$

(28.3) $$\mu_3 = \lambda\phi\alpha(\alpha+\beta)^{-1}[1+3\phi(\alpha+1)(\alpha+\beta+1)^{-1} + \phi^2(\alpha+1)(\alpha+2)(\alpha+\beta+1)^{-1}(\alpha+\beta+2)^{-1}].$$

For $\alpha = 1$

$$\mu_1' = \lambda\phi(\beta+1)^{-1};\ \mu_2 = \lambda\phi(\beta+1)^{-1}[1+2\phi(\beta+2)^{-1}]$$
$$\mu_3 = \lambda\phi(\beta+1)^{-1}[1+6\phi(\beta+2)^{-1}+6\phi^2(\beta+2)^{-1}(\beta+3)^{-1}].$$

There is a recurrence relation among the probabilities $P_k(\lambda,\phi,\alpha,\beta) \equiv P_k$.

(29) $$P_{k+1} = \lambda(k+1)^{-1}\sum_{j=0}^{k} F_jP_{k-j}$$

where $$F_j = \frac{\phi^{j+1}}{j!}\,\frac{\alpha^{[j+1]}}{(\alpha+\beta)^{[j+1]}}\,M(\alpha+j+1;\alpha+\beta+j+1;-\phi).$$

The F_j's are conveniently calculated from the recurrence formula

(30) $$F_j = j^{-1}(\phi+\alpha+\beta+j-1)F_{j-1} - \phi j^{-1}(j-1)^{-1}(\alpha+j-1)F_{j-2}.$$

(Formulas (29) and (30) reduce to (22) and (24) respectively on putting $\alpha = 1$.)

For estimation of λ and ϕ, *given* α and β, equation of first and second sample and population moments suffices. If all four parameters are to be estimated, Gurland [20] suggests using these two equations, together with those obtained by equating $\mathbf{f}_0$ and $\mathbf{f}_1$, to $P_0(\lambda,\phi,\alpha,\beta)$ and $P_1(\lambda,\phi,\alpha,\beta)$ respectively.

Darwin [9] extended Neyman's model to allow for movement of larvae from original positions, and also for non-Poisson distribution for number of developing larvae per group. A number of distributions based on this model are given by Thompson [42].

4. Pólya-Eggenberger Distribution

4.1 *Definition*

We now turn to 'truly contagious' distributions. Of the infinity of possibilities, only a few forms of dependence have been systematically studied and found acceptable for practical work. The distributions associated with these will be described here, and in succeeding Sections.

The earliest to be used was the *Pólya-Eggenberger distribution*, which was introduced in 1923 (Eggenberger and Pólya [13] — some further analysis was given by Pólya [33] in 1930). The genesis of this distribution is conveniently expressed in terms of random drawings of colored balls from an urn. Initially it is supposed that there are a white balls and b black balls in the urn. One ball is drawn at random, and then replaced, together with s balls of the same color. If this procedure is repeated n times, and $\mathbf{x}$ represents the total number of times a white ball is drawn, then the distribution of $\mathbf{x}$ is a Pólya-Eggenberger distribution with parameters n, a, b, s.

Evidently the distribution is defined by:

(31)

$$\Pr[\mathbf{x} = k] = \binom{n}{k} \frac{a(a+s)\cdots(a+\overline{k-1}\cdot s)b(b+s)\cdots(b+\overline{n-k-1}\cdot s)}{(a+b)(a+b+s)\cdots(a+b+\overline{n-1}\cdot s)}.$$

An alternative form of (31), in terms of parameters

$$n;\ P = a/(a+b);\ Q = 1 - P = b/(a+b);\ \text{and } \alpha = s/(a+b)$$

is

(31)′

$$\Pr[\mathbf{x} = k] = \binom{n}{k} \frac{P(P+\alpha)\dots(P+\overline{k-1}\cdot\alpha)Q(Q+\alpha)\dots \times (Q+\overline{n-k-1}\cdot\alpha)}{(1+\alpha)(1+2\alpha)\dots(1+\overline{n-1}\cdot\alpha)}$$

$$= \frac{\binom{n}{k}\prod_{j=0}^{k-1}(P+j\alpha)\prod_{j=0}^{n-k-1}(Q+j\alpha)}{\prod_{j=0}^{n-1}(1+j\alpha)} \qquad (k = 0, 1, \dots, n).$$

It is possible for s (and so α) to be negative. However s must satisfy the inequality

$$(a+b) + s(n-1) > 0.$$

Śródka [40] has obtained the following recurrence relation among the moments about zero of the distribution (31):

(32)

$$\mu'_{r+1} = (a+b+rs)^{-1}\sum_{j=0}^{r}\left[an\binom{r}{j} - (a-sn)\binom{r}{j+1} - s\binom{r}{j+2}\right]\mu'_{r-j}.$$

The rth factorial moment is

(33)

$$\mu_{(r)} = E[\mathbf{x}^{(r)}]$$

$$= n^{(r)}\prod_{j=0}^{r-1}\left[\frac{P+j\alpha}{1+j\alpha}\right]\sum_{k-r=0}^{n-r}\binom{n-r}{k-r}\frac{\prod_{j=0}^{k-r-1}(P'+j\alpha')\prod_{j=0}^{n-k-1}(Q'+j\alpha')}{\prod_{j=0}^{n-r-1}(1+j\alpha')}$$

$$(P' = (P + r\alpha)(1 + r\alpha)^{-1};\ Q' = Q(1 + r\alpha)^{-1};\ \alpha' = \alpha(1 + r\alpha)^{-1}).$$

In particular

(34.1) $$\mu_1' = nP$$

(34.2)
$$\mu_2 = n(n-1)P(P+\alpha)(1+\alpha)^{-1} + nP - (nP)^2 = nPQ(1+n\alpha)(1+\alpha)^{-1}$$

(34.3) $$\mu_3 = nPQ(Q-P)(1+n\alpha)(1+2n\alpha)(1+\alpha)^{-1}(1+2\alpha)^{-1}$$

(34.4)
$$\mu_4 = \frac{nPQ(1+n\alpha)[(1+2n\alpha)(1+3n\alpha)(1-3PQ) + (n-1)\{\alpha + 3PQ(1+n\alpha)\}]}{(1+\alpha)(1+2\alpha)(1+3\alpha)}.$$

From these formulas we obtain the moment ratios

(35) $$\alpha_3 = \sqrt{\beta_1} = \frac{(Q-P)(1+2n\alpha)^{\frac{1}{2}}(1+\alpha)^{\frac{1}{2}}}{[nPQ(1+n\alpha)]^{\frac{1}{2}}}$$

$$\alpha_4 = \beta_2 = \frac{3(n-1)(1+\alpha)}{n(1+2\alpha)(1+3\alpha)} + \frac{(1+\alpha)}{nPQ(1+2\alpha)(1+3\alpha)} \times [(1+2n\alpha)(1+3n\alpha)(1-3PQ) + (n-1)\alpha].$$

If $s = 0$, distribution (31) is a binomial distribution with parameters $(a + b)$, $a(a + b)^{-1}$; if $s = -1$, it is a hypergeometric distribution with parameters $n, a, (a + b)$. These relationships are apparent from the genesis of the distribution. Also, taking $a = b = s$, we obtain a discrete rectangular distribution.

The negative binomial distribution is obtained as a limiting distribution if $n \to \infty$, $a(a + b)^{-1} \to 0$ and $s(a + b)^{-1} \to 0$ in such a way that $na(a + b)^{-1}$ and $ns(a + b)^{-1}$ tend to finite non-zero values θ, ρ respectively. (The parameters of the limiting negative binomial distribution are $\theta\rho^{-1}$, ρ.) This limiting form of (31) is sometimes referred to as a "Pólya" distribution (e.g. Gnedenko [17], Eisenhart and Zelen [14] Arley and Buch [2], Hald [22]). On the other hand, Bosch [6] calls the general distribution (31) a "Pólya" distribution and reports further that it is sometimes called a *Skellam* distribution (see also Skellam [38]). Patil and Joshi [30] term the negative binomial a "Pólya-Eggenberger," and the distribution (31) simply a "Pólya" distribution. It is also the Type IIA generalized hypergeometric distribution of Kemp and Kemp [27], as already noted in Chapter 6, Section 8.5, where a method of estimating parameters by maximum likelihood has been described. It has been noted there, and also in Chapter 8, Section 2(x) that the distribution can be represented as

$$\text{Binomial}\ (n,\mathbf{p}) \underset{\mathbf{p}}{\wedge} \text{Beta}.$$

(See also Bosch [6].)

Sarkadi [35] has shown that the Pólya-Eggenberger distribution is also related to the distribution of the "number of exceedances," i.e. of the number of random variables, among a set of size n, the values of which are larger than the values of at least $(N - m + 1)$ out of N other random variables, all $(n + N)$ random variables being independent and having identical continuous distributions. Gumbel and von Schelling [19] showed that this distribution is defined by

(36)
$$\Pr[\mathbf{x} = k] = \binom{n}{k} m \binom{N}{m}\left[(n + N)\binom{n + N - 1}{k + m - 1}\right]^{-1} \qquad (k = 0,1,\ldots,n).$$

This is a Pólya-Eggenberger distribution with parameters n, $P = m/(N + 1)$ and $\alpha = (N + 1)^{-1}$ (see also Chapter 6, Section 1).

Karlin [24] used the term "Pólya type" for a wide class of distributions (including "exponential family, the noncentral t, the noncentral F and the noncentral chi-square", discussed in Chapters 18, 31, 30 and 28 respectively).

4.2 *Inverse Pólya-Eggenberger Distribution*

This distribution is related to the Pólya-Eggenberger distribution in the same way as the negative binomial is related to the binomial distribution. That is to say, instead of considering the probability of drawing k white balls in n drawings, as in (31), we consider how many drawings are needed to draw k white balls. The probability that exactly $(k + x)$ drawings are needed is

(37)
$$P(x) = \frac{a(a + s)\ldots(a + \overline{k - 1}\cdot s)b(b + s)\ldots(b + \overline{x - 1}\cdot s)}{(a + b)(a + b + s)\ldots(a + b + \overline{k + x - 1}\cdot s)}\binom{k + x - 1}{x}$$
$$(x = 0,1,2,\ldots).$$

Alternatively putting $P = a/(a + b)$; $Q = 1 - P = b/(a + b)$; $\alpha = s/(a + b)$:

$$(38)\qquad P(x) = \frac{P(P + \alpha)\ldots(P + \overline{k - 1}\cdot\alpha)Q(Q + \alpha)\ldots(Q + \overline{x - 1}\cdot\alpha)}{(1 + \alpha)(1 + 2\alpha)\ldots(1 + \overline{k + x - 1}\cdot\alpha)}$$
$$\times\binom{k + x - 1}{x}$$
$$= \frac{(P/\alpha)^{[k]}(Q/\alpha)^{[k]}}{(1/\alpha)^{[k+x]}}\binom{k + x - 1}{x}$$
$$= \frac{k}{k + x}\binom{-P/\alpha}{k}\binom{-Q/\alpha}{x}\binom{-1/\alpha}{k + x}.$$

This distribution belongs to the class of generalized hypergeometric distributions described by Kemp and Kemp [27] (see Chapter 6, Section 8.5).

5. Woodbury Distributions

In the genesis of the Pólya-Eggenberger distribution, the probability of drawing a white ball (a 'success') depended on both the number of previous successes and on the number of previous 'failures' (black balls). In fact, it was equal to

$$\frac{[a + s \cdot (\text{number of previous successes})]}{[a + b + s \cdot (\text{number of previous successes and failures, combined})]}.$$

Woodbury [44], on the other hand, considered a situation in which the probability of a success depends only on the number of previous successes and not on the number of previous failures (and so not on the total number of previous trials).

Letting p_r denote the probability of success, given that there have been exactly r previous successes, then the probability of exactly k successes in N trials is

(39)

$$P(k,N) = \left(\prod_{j=0}^{k-1} p_j\right)\left(\text{coefficient of } t^{N-k} \text{ in} \prod_{j=0}^{k} [1 - (1 - p_j)t]^{-1}\right).$$

Supposing no two of the p_j's are equal, we write

$$\prod_{j=0}^{k} [1 - (1 - p_j)t]^{-1} = \sum_{j=0}^{k} A_j[1 - (1 - p_j)t]^{-1}$$

where

$$\sum_{j=0}^{k} A_j \prod_{j' \neq j} [(1 - p_{j'})t] \equiv 1.$$

Putting $t = (1 - p_j)^{-1}$ we obtain

$$\begin{aligned} A_j &= \prod_{j' \neq j} [1 - (1 - p_{j'})/(1 - p_j)]^{-1} \\ &= (1 - p_j)^k \Big\{\prod_{j' \neq j} (p_{j'} - p_j)\Big\}^{-1}. \end{aligned}$$

Hence from (39)

$$(40) \qquad P(k,N) = \left(\prod_{j=0}^{k-1} p_j\right) \sum_{j=0}^{k} (1 - p_j)^N \left\{\prod_{j' \neq j} (p_{j'} - p_j)\right\}^{-1}.$$

The following recurrence relation is evident from the genesis of the distribution:

$$(41) \qquad P(k + 1,N + 1) = p_k P(k,N) + (1 - p_k)P(k + 1,N).$$

The general form of distribution (40) is rather complicated.

Rutherford [34] devoted particular attention to the special case in which the p_k's are determined by two parameters, P and c, through the equations

$$(42) \qquad p_k = P + ck \qquad (0 < P < 1 \text{ and if } c > 0,\ N < (1 - P)/c \text{ while if } c < 0,\ N < P/|c|).$$

In this case

(43)
$$P(k,N) = \frac{1}{k!}(Pc^{-1})(Pc^{-1}+1)\ldots(Pc^{-1}+k-1)\sum_{j=0}^{k}(-1)^j\binom{k}{j}(Q-cj)^N$$
$$(Q = 1 - P).$$

Equation (42) may be written in the form

$$P(k,N) = \frac{c^N}{k!}(Pc^{-1})^{[k]}\Delta^k(Q/c-k)^N \tag{43$'$}$$

where the difference operator applies to the variable Q/c. The probability generating function is

$$c^N\sum_{k=0}^{N}\frac{t^k}{k!}(Pc^{-1})^{[k]}\Delta^k(Q/c-k)^N = c^N\sum_{k=0}^{N}\frac{t^k}{k!}(Pc^{-1})^{[k]}\nabla^k(Q/c)^N \tag{44}$$

which may be written symbolically as

$$c^N(1-t\nabla)^{-P/c}(Q/c)^N. \tag{44$'$}$$

Chaddha [8] has considered two different special cases with (42) replaced by

$$p_k = (ck+P)/(ck+1) \tag{45}$$
$$p_k = P/(ck+1) \tag{46}$$

respectively.

(He related this choice of functions p_k to the construction of a model representing number of attendances at a sequence of committee meetings.)

Chaddha gave formulas for $P(0,N)$, $P(1,N)$, $P(2,N)$ and $P(N,N)$ for each case ((45) and (46)). He also tabulated values of $P(k,N)$, for each possible k to 4 significant figures, for $N = 1(1)15$; $c = 0.1(0.2)0.7$ and $1(0.5)3$; and $P = 0.1(0.1)0.9$ for each case. Some graphical representations of these probabilities are given in [8]. These distributions are unimodal unless P is small, when there is a local minimum at $x = 1$.

Chaddha described methods of estimation by maximum likelihood and by moments for the special case $N = 2$, but the general case appears to call for rather complicated calculations.

6. Some Other Related Distributions

It is interesting to note that the equations satisfied by maximum likelihood estimators $\hat{\lambda}$, $\hat{\phi}$ of the parameters λ. ϕ of a number of discrete distributions can be written in the form (11.1) and (11.2), by appropriate choice of parameters. Thus, the logarithmic binomial distribution (Section 2(xv) of Chapter 8) is such a distribution, if we take $\lambda = -\theta[(1-\theta)\log(1-\theta)]^{-1}$, $\phi = p$. Sprott [39] investigated general conditions under which this will be so. He found equation

(11.2) will be reached if the generating function $G(t;\lambda,\phi) = \sum_{j=0}^{\infty} P_j(\lambda,\phi)t^j$ is a function of λ and $(1 - t)$ times a function of ϕ (only). (In many cases the function of ϕ is simply ϕ.) For equation (11.1) also to be obtained, the distribution must be expressible in the form

$$\text{(47)} \qquad \text{Poisson } (\theta\mathbf{n}) \underset{\mathbf{n}}{\wedge} \text{Power series distribution}$$

(as defined in Chapter 2, Section 3). Since the probability generating function is of form

$$[\phi(\delta)]^{-1} \sum_j a_j \, \delta^j \exp[j\theta(t-1)] = \frac{\phi(\delta \exp[\theta(t-1)])}{\phi(\delta)}$$

and this can be written as

$$[\phi(\delta')]^{-1} \sum_j (A_j \, \delta'^j/j!)(q + pt)^j$$

$$\left(\delta' = e^{\delta};\ \theta = p\delta';\ A_j = \sum_{i=0}^{\infty} (-1)^i a_i i^j\right),$$

the distribution can also be expressed as

$$\text{(48)} \qquad \text{Binomial } (\mathbf{n},p) \underset{\mathbf{n}}{\wedge} \text{Power series distribution.}$$

The class of distributions introduced by Subrahmaniam [41] as an alternative to Gurland's [20] generalized contagious distributions (described in Section 3 of the present chapter) can also be considered as 'related' distributions.

In symbolic notation, Gurland's generalized contagious distributions may be written

$$\text{(49)} \qquad [\text{Binomial } (\mathbf{m},\mathbf{p}) \underset{\mathbf{m}}{\wedge} \{\text{Poisson } (\boldsymbol{\theta}) \underset{\theta/\phi}{\wedge} \text{Poisson } (\lambda)\}] \underset{\mathbf{p}}{\wedge} \text{Beta } (\alpha,\beta).$$

Subrahmaniam [41] has constructed another family of generalized contagious distributions which can be represented symbolically as

$$\text{(50)} \qquad [\text{Poisson } (\boldsymbol{\theta}) \underset{\theta/\phi}{\wedge} \{\text{Binomial } (\mathbf{n},\mathbf{p}) \underset{\mathbf{n}}{\wedge} \text{Poisson } (\lambda)\}] \underset{\mathbf{p}}{\wedge} \text{Beta } (\alpha,\beta).$$

The probability generating function of this distribution is

$$M(\alpha;\alpha + \beta;\ \lambda[\exp\{\phi(t-1)\} - 1])$$

which may be compared with (19).

For both Gurland's and Subrahmaniam's systems, the distribution inside the square brackets of (49) and (50) is a Neyman Type A distribution, but the parameter **p** which is compounded by the beta distribution appears in different ways in the two cases.

REFERENCES

[1] Archibald, E. E. A. (1948). Plant population, I: A new application of Neyman's contagious distributions, *Annals of Botany* (*New Series*), **47,** 221–235.

[2] Arley, N. and Buch, K. R. (1950). *Introduction to the Theory of Probability and Statistics*, New York: John Wiley & Sons, Inc.

[3] Barton, D. E. (1957). The modality of Neyman's contagious distribution of Type A, *Trabajos de Estadistica*, **8,** 13–22.

[4] Beall, G. (1940). The fit and significance of contagious distributions when applied to observations on larvae insects, *Ecology*, **21,** 460–474.

[5] Beall, G. and Rescia, R. (1953). A generalization of Neyman's contagious distribution, *Biometrics*, **9,** 354–386.

[6] Bosch, A. J. (1963). The Pólya distribution, *Statistica Neerlandica*, **17,** 201–213.

[7] Bowman, K. O. and Shenton, L. R. (1967). *Remarks on estimation problems for the parameters of the Neyman Type A distribution*, Oak Ridge National Laboratory, Report ORNL-4102.

[8] Chaddha, R. L. (1963). A case of contagion in binomial distribution, *Proceedings of the International Symposium on Discrete Distributions*, *Montreal*, 273–290.

[9] Darwin, J. H. (1951). Unpublished Ph.D. thesis, University of Manchester.

[10] David, F. N. and Moore, P. G. (1954). Notes on contagious distributions in plant populations, *Annals of Botany* (*New Series*), **53,** 47–53.

[11] Douglas, J. B. (1955). Fitting the Neyman Type A (two parameter) contagious distribution, *Biometrics*, **11,** 149–173.

[12] Douglas, J. B. (1963). Asymptotic expansions for some contagious distributions, *Proceedings of the International Symposium on Discrete Distributions*, *Montreal*, 291–302.

[13] Eggenberger, F. and Pólya, G. (1923). Über die Statistik verketteter Vorgänge, *Zeitschrift fur Angewandte Mathematik und Mechanik*, **1,** 179–289.

[14] Eisenhart, C. and Zelen, M. (1958). Chapter 12 in *Handbook of Physics* (1st Edition) New York: McGraw-Hill, Inc.

[15] Evans, D. A. (1953). Experimental evidence concerning contagious distributions, *Biometrika*, **40,** 186–211.

[16] Feller, W. (1943). On a general class of contagious distributions, *Annals of Mathematical Statistics*, **14,** 389–400.

[17] Gnedenko, B. V. (1961). *Probability Theory*, (3rd edition) Moscow: GITTL.

[18] Grimm, H. (1964). Tafeln der Neyman-Verteilung Typ A, *Biometrische Zeitschrift*, **6,** 10–23.

[19] Gumbel, E. J. and von Schelling, H. (1950). The distribution of the number of exceedances, *Annals of Mathematical Statistics*, **21,** 247–262.

[20] Gurland, J. (1958). A generalized class of contagious distributions, *Biometrics*, **14,** 229–249.

[21] Gurland, J. (1959). Some applications of the negative binomial and other contagious distributions, *American Journal of Public Health.*, **49,** 1388–1399.

[22] Hald, A. (1952). *Statistical Theory with Engineering Applications*, New York: John Wiley & Sons, Inc.

[23] Hinz, P. and Gurland, J. (1967). Simplified techniques for estimating parameters of some generalized Poisson distributions, *Biometrika*, **54,** 555–556.

[24] Karlin, S. (1957). Pólya type distributions II, *Annals of Mathematical Statistics*, **28,** 281–308.

[25] Katti, S. K. (1965). *Some estimation procedures for discrete distributions.* Statistical Report, The Florida State University, Tallahassee, Florida.

[26] Katti, S. K. and Gurland, J. (1962). Efficiency of certain methods of estimation for the negative binomial and the Neyman type A distributions, *Biometrika*, **49,** 215–226.

[27] Kemp, C. D. and Kemp, A. W. (1956). Generalized hypergeometric distributions, *Journal of the Royal Statistical Society, Series B*, **18,** 202–211.

[28] Martin, D. C. and Katti, S. K. (1962). Approximations to the Neyman Type A distribution for practical problems, *Biometrics*, **18,** 354–364.

[29] Neyman, J. (1939). On a new class of 'contagious' distributions, applicable in entomology and bacteriology, *Annals of Mathematical Statistics*, **10,** 35–57.

[30] Patil, G. P. and Joshi, S. W. (1968). *A Dictionary and Bibliography of Discrete Distributions*, Edinburgh and London: Oliver and Boyd.

[31] Philpot, J. W. (1964), Unpublished M.S. thesis, Virginia Polytechnic Institute.

[32] Pielou, E. C. (1957). The effect of quadrat size on the estimation of the parameters of Neyman's and Thomas' distribution, *Journal of Ecology*, **45,** 31–47.

[33] Pólya, G. (1930). Sur quelques points de la théorie des probabilitiés, *Annales de l'Institut Henri Poincaré*, **1,** 117–161.

[34] Rutherford, R. S. G. (1954). On a contagious distribution, *Annals of Mathematical Statistics*, **25,** 703–713.

[35] Sarkadi, K. S. (1957). On the distribution of the number of exceedances, *Annals of Mathematical Statistics*, **28,** 1021–1023.

[36] Shenton, L. R. (1949). On the efficiency of the method of moments and Neyman's type A distribution, *Biometrika*, **36,** 450–454.

[37] Shenton, L. R. and Bowman, K. O. (1967). Remarks on large sample estimators for some discrete distributions, *Technometrics*, **9,** 587–598.

[38] Skellam, J. G. (1948). A probability distribution derived from the binomial distribution by regarding the probability of success as variable between the sets of trials, *Journal of the Royal Statistical Society, Series B*, **10,** 257–261.

[39] Sprott, D. A. (1963). A class of contagious distributions and maximum likelihood estimation, *Proceedings of the International Symposium on Discrete Distributions, Montreal*, 337–350.

[40] Śródka, T. (1964). Wzór rekurencyjny na momenty zwykle w rozkladzie Pólyi, *Roczniki Polskiego Towarzystwa Matematycznego, Seria I*, **8,** 217–220.

[41] Subrahmaniam, K. (1966). On a general class of contagious distributions: The Pascal-Poisson distribution, *Trabajos de Estadistica*, **17,** 109–128.

[42] Thompson, H. R. (1954). A note on contagious distributions, *Biometrika*, **41,** 268–271.

[43] Wadley, F. M. (1950). Notes on the form of distribution of insect and plant populations, *Annals of the Entomological Society of America*, **43,** 581–586.

[44] Woodbury, M. A. (1949). On a probability distribution, *Annals of Mathematical Statistics*, **20,** 311–313.

10

Some Miscellaneous Discrete Distributions (Univariate)

1. Introduction

In this chapter are collected descriptions of various distributions which do not fall naturally into any of the major classes included in the foregoing chapters. Most of these distributions are, however, closer to some major classes than to others, and we have tried to point out such affinities.

Most of the distributions described in this chapter have arisen from consideration of specific physical situations, rather than from models of more general applicability. That is to say, they are of interest more from their connection with particular situations rather than on any general theoretical or methodological grounds.

In view of the diverse nature of these distributions, there will be no overriding arrangement of the sections in this chapter. But within each section we present distributions of similar form.

Haight [31] gives a summary of discrete distributions, including some which we do not discuss here.

2. Discrete Rectangular Distribution

The discrete form of the rectangular, or uniform, distribution (Chapter 25) is defined generally by

$$\Pr[\mathbf{x} = a + jh] = (M+1)^{-1} \qquad (j = 0,1,\ldots,M). \tag{1}$$

Various standard forms are in use. One is obtained by putting $a = 0, h = M^{-1}$ (so that the values taken by $\mathbf{x}$ are 0, M^{-1}, $2M^{-1}$,. . .,1). Another useful form, symmetrical about $\mathbf{x} = 0$, is obtained by making $(2a + Mh)$ equal to zero. This may be combined with taking a special value of h, such as 1. If this be done, then (1) becomes

(2) $$\Pr[\mathbf{x} = -\tfrac{1}{2}M + j] = (M+1)^{-1} \qquad (j = 0,1,\ldots,M).$$

The central moments of distribution (2) are easily obtained. The expected value, and all central moments of odd order are zero, while if r is even

$$\mu_r = (M+1)^{-1} \sum_{j=0}^{M} (j - \tfrac{1}{2}M)^r.$$

From Pierce [54] one can obtain the formulas

(3) $$\mu_2 = \tfrac{1}{12}\{(M+1)^2 - 1\}$$

(4) $$\mu_4 = \tfrac{1}{240}\{3(M+1)^2 - 7\} \cdot \{(M+1)^2 - 1\}$$

(5) $$\mu_6 = \tfrac{1}{112}\{3(M+1)^4 - 18(M+1)^2 + 31\} \cdot \{(M+1)^2 - 1\}.$$

The r-th central moment of the distribution (1) is obtained by multiplying μ_r for distribution (2) by h^r.

Alternatively, formulas for the moments of (1) may be derived from the moment generating function

(6) $$\begin{aligned} E[e^{t\mathbf{x}}] &= (M+1)^{-1} \sum_{j=0}^{M} \exp[t(a+jh)] \\ &= (M+1)^{-1} e^{ta}(e^{(M+1)th} - 1)(e^{th} - 1)^{-1} \\ &= (M+1)^{-1}[\exp\{t(a + \tfrac{1}{2}Mh)\}] \frac{\sinh\{\tfrac{1}{2}(M+1)th\}}{\sinh\{\tfrac{1}{2}th\}}. \end{aligned}$$

Note that for (2), $a + \tfrac{1}{2}Mh = 0$. By expanding the cumulant generating function it can be shown that

(7) $$\kappa_r = 0 \qquad \text{for } r \text{ odd}, r > 2$$

(8) $$\kappa_r = (-1)^{\frac{1}{2}r+1} r^{-1} B_{\frac{1}{2}r} h^r \{(M+1)^r - 1\} \qquad \text{for } r \text{ even}.$$

($B_{\frac{1}{2}r}$ is a Bernoulli number.)

The information generating function of distribution (1) is $(M+1)^{-u}$ and the entropy is $\log(M+1)$.

If $a = 0$ and $h = 1$, we have a distribution defined on non-negative integers:

(9) $$\Pr[\mathbf{x} = j] = (M+1)^{-1} \qquad (j = 0,1,\ldots,M).$$

The probability generating function of this distribution is

(10) $$(M+1)^{-1} \sum_{j=0}^{M} t^j = (M+1)^{-1}(1 - t^{M+1})(1 - t)^{-1}.$$

If $\mathbf{x}_1, \mathbf{x}_2 \ldots \mathbf{x}_n$ are independent random variables each having distribution

(9), then the probability that their range *equals* w is

(11)
$$P(w) = (M+1)\left[\left(\frac{w+1}{M+1}\right)^n - 2\left(\frac{w}{M+1}\right)^n + \left(\frac{w-1}{M+1}\right)^n\right]\left(1 - \frac{w}{M+1}\right) \qquad (w = 1,2,\ldots M)$$
$$P(0) = (M+1)^{-(n-1)}.$$

(This is evidently true if $h = 1$, whatever the value of a.)

Rider [57] has given tables containing values of $P(w)$ to four decimal places for $M = 9$, $n = 10$, $w = 0(1)9$. (He has also given corresponding values of the *probability density function* of range when the population distribution is continuous rectangular between 0 and 10.)

The discrete rectangular distribution can be obtained as a generalized hypergeometric distribution, as has been pointed out in Chapter 6 (Section 8).

3. Word Frequency Distributions

3.1 *Zeta Distribution* (*Zipf-Estoup Law*)

From observation and analysis of the frequency of different words in long sequences of text, the approximate formula

$$n_r \propto r^{-(\rho+1)} \qquad (r = 1,2,\ldots;\rho > 0)$$

for n_r the number of words appearing r times, was found to give a usefully accurate fit. In order to define a probability distribution appropriate to such a situation we define

(12)
$$\Pr[\mathbf{x} = r] = cr^{-(\rho+1)} \qquad (r = 1,2,\ldots)$$

with
$$c = \left[\sum_{r=1}^{\infty} r^{-(\rho+1)}\right] = [\zeta(\rho+1)]^{-1}$$

where $\zeta(\cdot)$ denotes the Riemann zeta function. This distribution is called the *zeta distribution.* When used in linguistic studies (as in the situation described at the beginning of this section), it is often called the *Zipf-Estoup law.* (Estoup [17], Zipf [75]). The use of the distribution in this connection has also been studied by Good [26]. It is found that ρ very often has values slightly in excess of 1. Seal [58] has applied the distribution to the number of insurance policies held by individuals.

The distribution (12) is analogous to the continuous Pareto (see Chapter 19) distribution, in the same way as the discrete rectangular is analogous to the continuous rectangular distribution. It is sometimes called the *discrete Pareto distribution.*

The sth moment of the distribution about zero is

(13)
$$\mu'_s = \zeta(\rho - s + 1)/\zeta(\rho + 1)$$

for $s < \rho$. If $s \geq \rho$ the moment is infinite. The zeta distribution, for ρ not large, has a long positive tail. Figures 1a-c compare zeta, logarithmic series and translated geometric distributions with the same expected value (equal to 2).

Shreider [60] has attempted to give a theoretical justification for the Zipf distribution, based on thermodynamical analogies. His arguments would

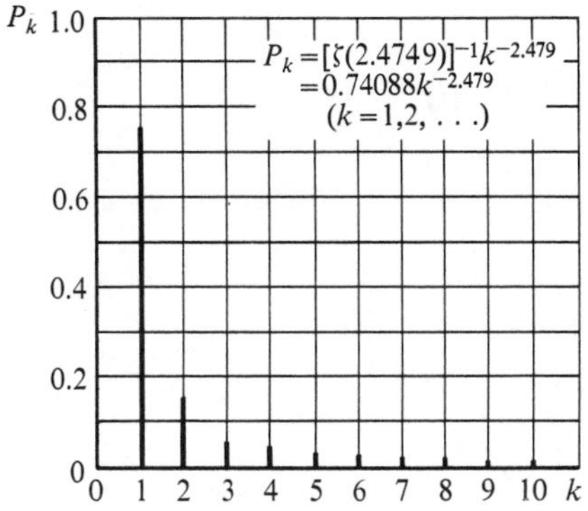

FIGURE 1a

Zeta Distribution

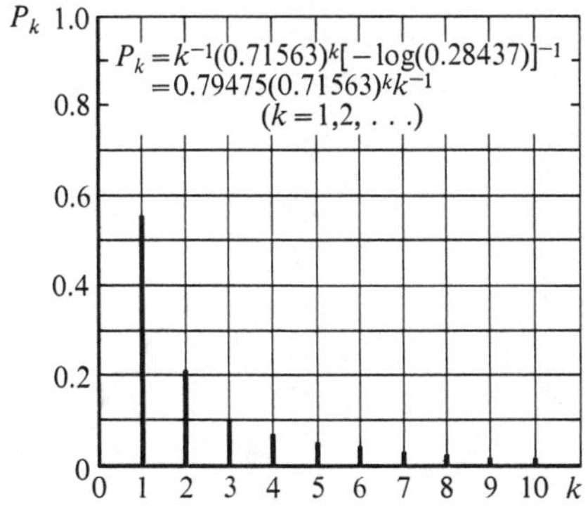

FIGURE 1b

Logarithmic Series Distribution

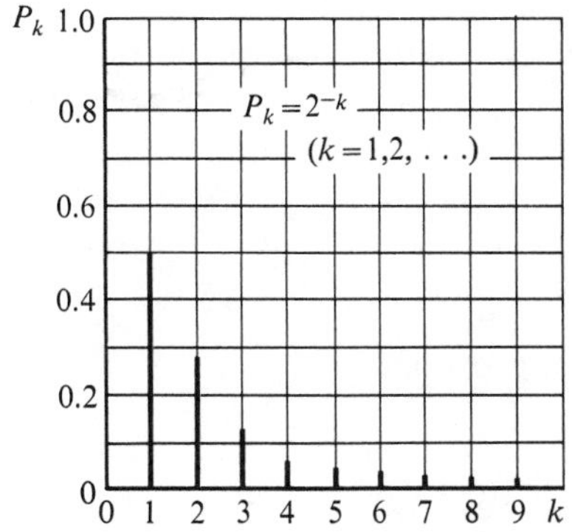

FIGURE 1c

Translated Geometric Distribution

lead one to expect the Zipf distribution to be approached asymptotically for 'sufficiently long' sections of 'stable' text. They also indicate that greater deviations from the Zipf form may be expected at lower, than at higher, frequencies, which does agree with empirical facts in a considerable number of cases.

Given values of n independent random variables, each having the distribution (12), the maximum likelihood estimator, $\hat{\rho}$, of ρ satisfies the equation (Seal [59])

$$n^{-1} \sum_{i=1}^{n} \log \mathbf{x}_i = -\zeta'(\hat{\rho} + 1)/\zeta(\hat{\rho} + 1). \tag{14}$$

Using appropriate tables (see Table 1 below), it is not difficult to obtain an acceptably accurate solution of (14).

TABLE 1

Values of $-\zeta'(\rho + 1)/\zeta(\rho + 1)$

ρ	$-\zeta'(\rho + 1)/\zeta(\rho + 1)$	ρ	$-\zeta'(\rho + 1)/\zeta(\rho + 1)$
0.1	9.441	1.6	0.256
0.2	4.458	1.7	0.228
0.3	2.808	1.8	0.204
0.4	1.990	1.9	0.183
0.5	1.505	2.0	0.164
0.6	1.186	2.2	0.134
0.7	0.961	2.4	0.110
0.8	0.796	2.6	0.0914
0.9	0.669	2.8	0.0761
1.0	0.570	3.0	0.0637
1.1	0.490	3.2	0.0535
1.2	0.425	3.4	0.0451
1.3	0.372	3.6	0.0382
1.4	0.327	3.8	0.0324
1.5	0.289	4.0	0.0276*

*For $\rho > 4$; $\zeta'(\rho + 1)/\zeta(\rho + 1) \doteqdot (1 + 2^{\rho+1})^{-1} \log_e 2$.

An alternative, though usually considerably less accurate on average, method is based on the ratio of the proportions $(\mathbf{f}_1,\mathbf{f}_2)$ of 1's and 2's among $\mathbf{x}_1, \mathbf{x}_2, \ldots, \mathbf{x}_n$. Equating $\mathbf{f}_j$ to $\Pr[\mathbf{x} = j]$ for $j = 1, 2$ we obtain

$$\hat{\rho}^* = \frac{\log (\mathbf{f}_1/\mathbf{f}_2)}{\log 2} - 1. \tag{15}$$

The variance of $\hat{\rho}$ determined from (14) is approximately

$$\left[\frac{d}{d\rho} \frac{\zeta'(\rho + 1)}{\zeta(\rho + 1)}\right]^{-1} n^{-1} = [g(\rho)]^{-1} n^{-1}. \tag{16}$$

Values of $\zeta'(\rho + 1)/\zeta(\rho + 1)$, taken from Walther [71], and values of $g(\rho)$, are given in Tables 1 and 2. (Table 2 was obtained by numerical differentiation of Table 1.)

TABLE 2

Values of $\frac{d}{d\rho}[\zeta'(\rho + 1)/\zeta(\rho + 1)] = g(\rho)$

ρ	$g(\rho)$	ρ	$g(\rho)$
0.5	3.860	1.8	0.225
0.6	2.638	1.9	0.196
0.7	1.909	2.0	0.172
0.8	1.436	2.1	0.152
0.9	1.114	2.2	0.134
1.0	0.904	2.3	0.119
1.1	0.716	2.4	0.106
1.2	0.588	2.5	0.095
1.3	0.490	2.6	0.085
1.4	0.412	2.7	0.076
1.5	0.354	2.8	0.069
1.6	0.300	2.9	0.062
1.7	0.258	3.0	0.056

The variance of $\hat{\rho}^*$, as determined from (15) is approximately (using the approximation to the multinomial distribution, given in Chapter 11 equation (13))

$$[\log 2]^{-2}(1 + 2^{\rho+1})[\zeta(\rho + 1)]^{-1}n^{-1}. \tag{17}$$

(Note that this formula neglects the possibility that $\mathbf{f}_1 = 0$ or $\mathbf{f}_2 = 0$.)

Table 3 gives a few values of the ratio $\text{var}(\hat{\rho}^*)/\text{var}(\hat{\rho})$ calculated from the approximate formulas (16) and (17). It will be observed that according to this table $\text{var}(\hat{\rho}^*)$ is considerably larger than $\text{var}(\hat{\rho})$ when ρ is small. For ρ large, the ratio is approximately $(1 + 2^{-(\rho+1)})^{-1}[\zeta(\rho + 1)]^{-1}(<1)$ and tends to 1 as ρ tends to infinity.

A better estimator than $\hat{\rho}^*$ appears to be obtained by equating population and sample first moments. The estimator $\tilde{\rho}$ satisfies the equation

$$\bar{\mathbf{x}} = \zeta(\tilde{\rho})/\zeta(\tilde{\rho} + 1). \tag{18}$$

Table 4 facilitates solution of this equation (based on a table of Moore [47]). For n large

$$\text{var}(\tilde{\rho}) \doteqdot [\zeta(\rho + 1)]^2[\zeta(\rho - 1)\zeta(\rho + 1) - \{\zeta(\rho)\}^2] \cdot [\zeta(\rho + 1)\zeta'(\rho) - \zeta(\rho)\zeta'(\rho + 1)]^{-2}n^{-1}.$$

When $\rho = 3$, $\text{var}(\tilde{\rho}) \doteqdot 23.1n^{-1}$, while $\text{var}(\hat{\rho}) = 17.9n^{-1}$, a ratio of about 1.3.

TABLE 3

Approximate values of var($\hat{\rho}^*$)/var($\hat{\rho}$)

ρ	Approx. var($\hat{\rho}^*$)/var($\hat{\rho}$)
0.5	(11.8)
1.0	(5.7)
1.5	3.7
2.0	2.7
2.5	2.0
3.0	1.8

TABLE 4

ρ in terms of μ'_1

μ'_1	ρ	μ'_1	ρ	μ'_1	ρ	μ'_1	ρ	μ'_1	ρ
1.01	5.848	1.21	2.419	1.41	1.931	1.61	1.702	1.81	1.566
1.02	4.952	1.22	2.381	1.42	1.915	1.62	1.693	1.82	1.560
1.03	4.449	1.23	2.345	1.43	1.901	1.63	1.685	1.83	1.555
1.04	4.105	1.24	2.311	1.44	1.887	1.64	1.677	1.84	1.550
1.05	3.847	1.25	2.279	1.45	1.873	1.65	1.669	1.85	1.545
1.06	3.642	1.26	2.249	1.46	1.860	1.66	1.662	1.86	1.539
1.07	3.473	1.27	2.220	1.47	1.847	1.67	1.654	1.87	1.535
1.08	3.331	1.28	2.193	1.48	1.834	1.68	1.647	1.88	1.530
1.09	3.208	1.29	2.167	1.49	1.822	1.69	1.640	1.89	1.525
1.10	3.100	1.30	2.143	1.50	1.811	1.70	1.633	1.90	1.520
1.11	3.006	1.31	2.119	1.51	1.799	1.71	1.626	1.91	1.516
1.12	2.921	1.32	2.097	1.52	1.788	1.72	1.619	1.92	1.511
1.13	2.844	1.33	2.075	1.53	1.778	1.73	1.613	1.93	1.507
1.14	2.775	1.34	2.054	1.54	1.767	1.74	1.607	1.94	1.503
1.15	2.711	1.35	2.034	1.55	1.757	1.75	1.600	1.95	1.499
1.16	2.653	1.36	2.015	1.56	1.747	1.76	1.594	1.96	1.495
1.17	2.599	1.37	1.997	1.57	1.737	1.77	1.589	1.97	1.491
1.18	2.549	1.38	1.980	1.58	1.728	1.78	1.583	1.98	1.487
1.19	2.503	1.39	1.963	1.59	1.719	1.79	1.577	1.99	1.483
1.20	2.459	1.40	1.946	1.60	1.710	1.80	1.571	2.00	1.479

3.2 *Yule Distribution*

Simon [62] [63] noted that, for many distributions of word frequencies, a very good fit of distribution (12) could be obtained for the larger (>3) values of r, with $\rho \doteqdot 1$. For $r = 1, 2$, however, this was not the case. If $\rho = 1$, the ratio of relative frequencies of 1's to 2's should be $1:2^{-2} = 4:1$. Observed values of this ratio (in the data studied by Simon) were nearer to 3:1, sometimes even less. Simon suggested consideration be given to distributions of form

(19) $$\Pr[\mathbf{x} = r] = A_\rho B(r,\rho + 1) \qquad (r = 1,2,\ldots)$$

where $A_\rho = \left[\sum_{r=1}^{\infty} B(r,\rho + 1)\right]^{-1}$ and $B(\alpha,\beta)$ is the beta function.

For large values of r, $B(r,\rho + 1)$ is approximately proportional to $r^{-(\rho+1)}$, as in the zeta distribution. The ratio

(20) $$\Pr[\mathbf{x} = 1]/\Pr[\mathbf{x} = 2] = B(1,\rho + 1)/B(2,\rho + 1) = \rho + 2$$

is equal to 3 if $\rho = 1$. Thus taking $\rho = 1$ in (19) gives agreement with both features of the observed phenomena — the ratio 3:1 of frequencies of 1's and 2's, and the approximate proportionality to r^{-2} for large r. In this particular case ($\rho = 1$), the value of A_1 is easily determined. In fact, $A_1 = 1$ and the distribution is defined by

(21) $$\Pr[\mathbf{x} = r] = [r(r + 1)]^{-1} \qquad (r = 1,2,\ldots).$$

The expected value is infinite.

If ρ is a positive integer, then $A_\rho = \rho$ and

(22) $$\Pr[\mathbf{x} = r] = \rho(\rho!)(r^{[\rho+1]})^{-1}.$$

The sth *ascending* factorial moment of $\mathbf{x}$ is

(23) $$\mu_{[s]} = \rho \cdot (\rho!) \sum_{r=1}^{\infty} [(r + s)^{[\rho-s+1]}]^{-1} = \frac{\rho \cdot (\rho!)}{(\rho - s) \cdot (s + 1)^{[\rho-s]}}$$

(for $s < \rho - 1$).

For $s \geq \rho - 1$ the corresponding moment is infinite.

Simon gave the name Yule distribution to (19) because it was obtained by Yule [73] in 1923 as the limiting case of a distribution in mathematical genetics.

Some interesting examples of applications of Zipf and Yule distributions have been given by Kendall [40].

It is noteworthy that Yule obtained the distribution as a compound geometric distribution. In fact, suppose that $\mathbf{y}$ has the geometric distribution

$$\Pr[\mathbf{y} = k] = e^{-\beta\phi}(1 - e^{-\beta\phi})^{k-1} \qquad (k = 1,2,\ldots).$$

and suppose further that ϕ has the exponential distribution, with probability density function

$$p_\phi(\phi) = \theta^{-1} \exp(-\phi\theta^{-1}) \qquad (\phi > 0; \theta > 0).$$

Then the compound distribution

$$\text{Geometric } (e^{-\beta\phi}) \underset{\phi}{\wedge} \text{ Exponential } (\theta)$$

is defined by

$$\text{(24)} \qquad \Pr[\mathbf{y} = k] = \theta^{-1} \int_0^\infty (1 - e^{-\beta\phi})^{k-1} \exp[-\phi(\beta + \theta^{-1})]\, d\phi.$$

Putting $e^{-\beta\phi} = z$, (24) becomes

$$\begin{aligned} \Pr[\mathbf{y} = k] &= (\beta\theta)^{-1} \int_0^1 (1 - z)^{k-1} z^{(\beta\theta)^{-1}}\, dz \\ &= (\beta\theta)^{-1} B(k,(\beta\theta)^{-1} + 1) \qquad (k = 1,2,\ldots) \end{aligned}$$

which is of form (19) with $\rho = (\beta\theta)^{-1}$.

There has been a certain amount of controversy about the suitability of the distributions discussed in this section as representations of distributions of word frequencies. Herdan [34] was doubtful about the fit in the upper tails of the distribution, while Mandelbrot [41] was especially critical of the use of values of ρ less than 1. As mentioned above this gives an infinite expected value and is, in this respect, unrealistic. Judgement as to suitability in particular cases, however, will often be based more on accuracy with which relative frequencies are estimated, rather than specific unrealistic points, though the latter serve as useful warnings.

Prasad [55] has described a generalized form of this distribution, with the following definition

$$\text{(25)} \qquad \Pr[\mathbf{x} = r] = 2\lambda(\lambda + 1)[(\lambda + r - 1)(\lambda + r)(\lambda + r + 1)]^{-1} \qquad (\lambda > 0; r = 1,2\ldots).$$

(Putting $\lambda = 1$, we obtain (19) with $\rho = 2$.)

The expected value of $\mathbf{x}$ is $(1 + \lambda)$ and the mean deviation is

$$4\lambda(\lambda + 1)([\lambda] + 1)(\lambda + [\lambda] + 1)^{-1}(\lambda + [\lambda] + 2)^{-1}.$$

If λ is an integer, this reduces to $2\lambda(\lambda + 1)(2\lambda + 1)^{-1}$. The second and higher moments are infinite. The cumulative distribution function is

$$\text{(26)} \qquad \Pr[\mathbf{x} \le X] = 1 - \lambda(\lambda + 1)(\lambda + X)^{-1}(\lambda + X + 1)^{-1}.$$

The same kind of generalization can be applied to (21) yielding the following expression

$$\text{(27)} \qquad \Pr[\mathbf{x} = r] = \lambda[(\lambda + r - 1)(\lambda + r)]^{-1} \qquad (\lambda > 0;\ r = 1,2,\ldots).$$

Similarly generalizing (19), we have

$$\text{(28)} \qquad \Pr[\mathbf{x} = r] = A_\lambda B(\lambda + r - 1, \rho + 1) \qquad (\lambda > 0;\ r = 1,2,\ldots).$$

Haight [32] has fitted the logarithmic series, Yule and Borel (see Section 6) distributions to four distributions of responses in psychological tests. In each case three methods of fitting were used:

(*a*) equation of sample and population first moment,
(*b*) equation of sample and population proportion of 1's,
(*c*) equation of sample and population proportion of values above some number.

He found that generally the Yule distribution gave the best fit, the logarithmic series the worst fit. Further, fitting by moments (method (*a*)) generally gave a worse fit, for a given distribution, than did methods (*b*) or (*c*).

Another form of modification of the zeta distribution, recently proposed by Haight [32], is the *harmonic distribution*. For this distribution, the probability that the random variable takes the value k is proportional to the number of quantities Zj^{-1}, where j is a positive integer, for which the nearest integer is k. Evidently this distribution depends on the single parameter Z (a positive integer). Since there are $2Z$ values of j for which $Zj^{-1} \geq \frac{1}{2}$, the probability of obtaining a value k is

$$p_k = \frac{1}{2Z}\left\{\left[\frac{2Z}{2k-1}\right] - \left[\frac{2Z}{2k+1}\right]\right\} \tag{29}$$

and the expected value is

$$(2Z)^{-1}\sum_{j=1}^{Z}[2Z(2j-1)^{-1}].$$

As Z tends to infinity, p_k tends to $2(2k-1)^{-1}(2k+1)^{-1}$. (Similarity to the Yule distribution, with $\rho = 1$, may be noted.)

Haight gives tables of the distribution for $Z = 1, 5, 10, 20, 30, 100$ and 1000. It has the unusual feature that p_k is equal to zero for considerable ranges of values of k, interspersed with isolated values $(2Z)^{-1}$.

Generalizing the harmonic distribution in the spirit of Zipf, we replace Zj^{-1} by $Zj^{-\sigma}$ (where σ will usually be near 1). This leads to

$$p_k = (2Z)^{1/\sigma}\left\{\left[\frac{2Z}{2k-1}\right]^{1/\sigma} - \left[\frac{2Z}{2k+1}\right]^{1/\sigma}\right\} \tag{30}$$

with expected value

$$(2Z)^{-1/\sigma}\sum_{j=1}^{Z}[2Z(2j-1)^{-1}]^{1/\sigma}.$$

As Z tends to infinity p_k tends to

$$(2k-1)^{-1/\sigma} - (2k+1)^{-1/\sigma}$$

with mean value $(1 - 2^{-1/\sigma})\zeta(\sigma^{-1})$.

Haight terms this limiting distribution a *zeta* distribution, but it is not the same as the zeta distribution described in this chapter.

4. Hyper-Poisson and Factorial Distributions

4.1 *Hyper-Poisson Distributions*

A generalization similar to that described at the end of the previous section has been applied to the Poisson distribution by Bardwell and Crow [3]. Noting that the Poisson distribution can be defined in the form:

$$\Pr[x = k] = C_\theta \frac{\theta^k}{1(1+1)\dots(1+\overline{k-1})} = C_\theta \theta^k [1^{[k]}]^{-1} \tag{31}$$

(cf. Chapter 4, equation (1)) with $C_\theta = e^{-\theta}(k = 0,1,\dots;\theta > 0)$, the corresponding generalized form

$$\Pr[x = k] = C_{\theta,\lambda} \frac{\theta^k}{\lambda^{[k]}} \qquad (k = 0,1,\dots;\theta > 0,\lambda > 0) \tag{32}$$

is suggested. Bardwell and Crow term this family of distributions *hyper-Poisson*, sub-classified into *sub-Poisson* $(\lambda < 1)$ and *super-Poisson* $(\lambda > 1)$ distributions. Each distribution of the family is defined by values of the two parameters θ and λ.

The value of $C_{\theta,\lambda}$ can be expressed as the reciprocal of a confluent hypergeometric function:

$$C_{\theta,\lambda} = [M(1;\lambda;\theta)]^{-1}. \tag{33}$$

From (32) it is easy to obtain the recurrence relation

$$(\lambda + k)\Pr[x = k + 1] = \theta \Pr[x = k] \qquad (k = 0,1,\dots). \tag{34}$$

By summation (writing $\lambda + k = \lambda - 1 + (k + 1)$) of equations like (34) we obtain

$$(\lambda - 1)(1 - \Pr[x = 0]) + E[x] = \theta$$

or

$$E[x] = \theta + (1 - \lambda)(1 - \Pr[x = 0]) \tag{35}$$
$$= \theta \frac{\partial}{\partial\theta} \log M(1;\lambda;\theta)$$

(noting that

$\{1 - \theta(\lambda - 1)^{-1}\}M(1;\lambda;\theta) + \theta(\lambda - 1)^{-1}\,\partial M(1;\lambda;\theta)/\partial\theta = M(0;\lambda;\theta) = 1$).

Multiplying (34) by $(k + 1)^r$ and summing leads to the following relation among the moments (about zero) of the distribution (32):

$$\lambda\mu'_r + (\mu'_{r+1} - \mu'_r) = \theta\left[\mu'_r + \binom{r}{1}\mu'_{r-1} + \cdots + \binom{r}{r-1}\mu'_1 + 1\right]$$

or

(36) $$\mu'_{r+1} = (\theta - \lambda + 1)\mu'_r + \theta\left[\binom{r}{1}\mu'_{r-1} + \binom{r}{2}\mu'_{r-2} + \cdots + \binom{r}{r-1}\mu'_1 + 1\right] \quad (r = 1,2,\ldots).$$

It follows that all moments of the distribution are finite, and the variance is

(37) $$\begin{aligned}\operatorname{var}(\mathbf{x}) &= (\theta - \lambda + 1)\mu'_1 + \theta - \mu'^2_1 \\ &= \theta + (1 - \lambda)[\theta - \lambda + 1 - (1 - \lambda)\Pr[\mathbf{x} = 0]]\Pr[\mathbf{x} = 0].\end{aligned}$$

The probability generating function is

(38) $$M(1;\lambda;\theta t)/M(1;\lambda;\theta).$$

Bardwell and Crow [3] describe some methods of estimating λ and θ, given values of n independent random variables $\mathbf{x}_1, \mathbf{x}_2, \ldots, \mathbf{x}_n$ each having the distribution (32). Firstly, we suppose λ known.

Since (32) can be regarded as a power series distribution, the maximum likelihood estimator, $\hat{\boldsymbol{\theta}}$, is obtained by equating the expected value (35) with the sample arithmetic mean, $\bar{\mathbf{x}}$, giving

(39) $$\bar{\mathbf{x}} = \hat{\boldsymbol{\theta}} + (1 - \lambda)[1 - \{M(1;\lambda;\hat{\boldsymbol{\theta}})\}^{-1}]$$

or

$$M(1;\lambda;\hat{\boldsymbol{\theta}}) = (1 - \lambda)/(1 - \lambda - \bar{\mathbf{x}} + \hat{\boldsymbol{\theta}}).$$

The variance of $\hat{\boldsymbol{\theta}}$ is approximately

(40) $$\theta^2[n \operatorname{var}(\mathbf{x})]^{-1}$$

where var(**x**) is as given in (37).

The uniformly minimum variance unbiased estimator of θ is *approximately* $\lambda\bar{\mathbf{x}}$ when n is large. Crow and Bardwell [14] obtain the values

$$\begin{array}{ll} 0 & \text{when } \bar{\mathbf{x}} = 0 \\ \lambda n^{-1} & \text{when } \bar{\mathbf{x}} = n^{-1} \\ 2\lambda n^{-1}[1 + n^{-1}(\lambda - 1)(\lambda + 1)^{-1}]^{-1} & \text{when } \bar{\mathbf{x}} = 2n^{-1}. \end{array}$$

Now turning to situations in which both λ and θ are unknown, the equations satisfied by the maximum likelihood estimators $\hat{\boldsymbol{\lambda}}$ and $\hat{\boldsymbol{\theta}}$ are:

(41) $$M(1;\hat{\boldsymbol{\lambda}};\hat{\boldsymbol{\theta}}) = (1 - \hat{\boldsymbol{\lambda}})/(1 - \hat{\boldsymbol{\lambda}} - \bar{\mathbf{x}} + \hat{\boldsymbol{\theta}})$$

(cf. (39)) and

(42) $$\sum_{j=1}^{\infty} \hat{\boldsymbol{\theta}}^j(\hat{\boldsymbol{\lambda}}^{[j]})^{-1}\mathbf{S}_j = M(1;\hat{\boldsymbol{\lambda}};\hat{\boldsymbol{\theta}}) \cdot n^{-1} \sum_{i=1}^{n} \mathbf{S}_{\mathbf{x}_i}$$

where
$$\mathbf{S}_j = \sum_{i=1}^{j} (\hat{\lambda} + i - 1)^{-1}.$$

Crow and Bardwell [14] also describe some other estimators. All of them seem to have rather low relative efficiencies, so we will only mention a pair which use the proportion of 0's ($\mathbf{f}_0$) and the first two sample moments $\bar{\mathbf{x}}$ and $\mathbf{m}'_2 = n^{-1} \sum_{j=1}^{n} \mathbf{x}_j^2$. The estimators are

$$\tilde{\theta} = [(1 - \mathbf{f}_0)\mathbf{m}'_2 - \bar{\mathbf{x}}^2][1 - \mathbf{f}_0(\bar{\mathbf{x}} + 1)]^{-1} \tag{43}$$

$$\tilde{\lambda} = 1 + \frac{(\mathbf{m}'_2 - \bar{\mathbf{x}}^2) - \bar{\mathbf{x}}}{1 - \mathbf{f}_0(\bar{\mathbf{x}} + 1)}. \tag{44}$$

Some queueing theory, with a hyper-Poisson distribution of arrivals, has been worked out by Nishida [50].

4.2 *Factorial Distributions*

Irwin ([35], [36]) has suggested that for discrete distributions with 'extremely long tails' distributions based on inverse factorial series might be useful. Such distributions are defined by equations of form

$$\Pr[\mathbf{x} = k] = a_k(\lambda^{[k+1]})^{-1}\Omega(\lambda) \qquad (k = 0,1,\ldots) \tag{45}$$

where
$$\Omega(\lambda) = \int_0^1 t^{\lambda-1}\phi(t)\,dt$$

and
$$\phi(t) = \sum_{j=0}^{\infty} (a_j/j!)(1 - t)^j,$$

it being supposed that the series converges.

Such distributions may be called *inverse factorial* or *Irwin* distributions. (They should not be confused with '*factorial distributions*' in the sense of Marlow [44], which are forms of Negative Binomial $(N,\mathbf{P}) \underset{\mathbf{P}/\mathbf{Q}}{\wedge}$ Beta distributions.)

The hyper-Poisson (32) is an inverse factorial distribution, with λ replaced by $(\lambda - 1)$, $a_j = 1$ and $\Omega(\lambda - 1) = M(1;\lambda;\theta)/(\lambda - 1)$. The Yule distribution is another member of the family.

Irwin considers in special detail the distribution corresponding to Waring's expansion:
$$(\lambda - a)^{-1} = \sum_{j=0}^{\infty} (\lambda^{[j+1]})^{-1}a^{[j]}.$$

This is

$$\Pr[\mathbf{x} = k] = (\lambda - a)a^{[k]}(\lambda^{[k+1]})^{-1} \qquad (k = 0,1,\ldots) \tag{46}$$

which he calls a *Waring* distribution.

Irwin [35] gives tables of $\Pr[\mathbf{x} = k]$ and $\sum_{j=0}^{k} \Pr[\mathbf{x} = j]$ to 5 decimal places for $a/\lambda = 0.1, 0.5, 0.9$; $\lambda - a = 1(1)10$.

The probability generating function corresponding to equation (46) is the following

$$(47) \qquad (\lambda - a)\lambda^{-1} \sum_{k=0}^{\infty} (a^{[k]}/(\lambda + 1)^{[k]})t^k = (\lambda - a)\lambda^{-1}F(a,1;\lambda + 1;t).$$

and the rth factorial moment is

$$\mu_{(r)} = r!a^{[r]}/(\lambda - a - 1)^{(r)}.$$

The Yule distribution is obtained by putting $a = 1$. It will be remembered that the variance of this distribution is infinite.

If $a, \lambda \to \infty$ with a/λ constant, the limiting distribution is geometric (Chapter 5, Section 2).

5. Occupancy Distributions

These distributions relate to the number of 'occupied' categories (i.e. random variables with non-zero values) when the joint distribution is multinomial (or, occasionally, hypergeometric). (For the multinomial distribution, see Chapter 11.)

We will consider first the symmetrical case when $p_1 = p_2 = \cdots = p_k = k^{-1}$. The probability that in N trials, exactly j categories have non-zero frequencies is (as can be seen from a straightforward application of Boole's formula — Chapter 1, equation (73).)

$$(48) \qquad P(j) = \binom{k}{j} \sum_{i=0}^{j} (-1)^i \binom{j}{i} \left(\frac{j-i}{k}\right)^N$$
$$= k^{-N} \binom{k}{j} \Delta^j 0^N.$$

The cumulative distribution function is

$$F(x) = \sum_{j=0}^{x} P(j).$$

This is sometimes called *Arfwedson's distribution* [1].

Nicholson [49] gives minimum values of N for which $F(x) \leq \gamma$ for $\gamma = 0.01$, 0.25, 0.05 and 0.10; and maximum values of N for which $F(x - 1) \geq \gamma$ for $\gamma = 0.01, 0.025, 0.05$ and 0.10. These tables are also included in Owen [53]. These can be used in connection with tests (of a sequential type) of the hypothesis that the category probabilities are equal.

This distribution was obtained by Stevens [65] in 1937 and Craig [13] in 1953.

The sth factorial moment of the distribution (48) is

(49)
$$\begin{aligned}\mu_{(s)} = E[\mathbf{x}^{(s)}] &= k^{-N} \sum_{x=1}^{k} x^{(s)} \binom{k}{x} \Delta^x 0^N \\ &= k^{-N} \sum_{x=s}^{k} k^{(s)} \binom{k-s}{x-s} \Delta^{x-s}(\Delta^s 0^N) \\ &= k^{-N} k^{(s)} \Delta^s (k-s)^N.\end{aligned}$$

In particular the expected value is

(50)
$$\begin{aligned}E[\mathbf{x}] &= k^{-(N-1)}[k^N - (k-1)^N] \\ &= k[1 - (1 - k^{-1})^N]\end{aligned}$$

and the variance is

(51)

$$\operatorname{var}(\mathbf{x}) = k(1 - k^{-1})^N + k(k-1)(1 - 2k^{-1})^N - k^2(1 - k^{-1})^{2N}$$

Usually N is known. Estimation of k corresponds to estimation of the number of classes in the population. This is discussed in Chapter 11, Section 2.5.

A modified form of this distribution is obtained if it is supposed that there is an upper limit, s, to the value of each of the multinomial variables. This corresponds, for example, to a situation in which there are k boxes, each capable of containing up to s balls, and N balls are placed at random in the available boxes, with the proviso that if any boxes are full, choice must be made among the remaining boxes. The probability that there are exactly x boxes each containing at least one ball is

(52)
$$\frac{\binom{k}{x}}{\binom{ks}{N}} \left[\binom{sx}{N} - \binom{x}{1}\binom{s(x-1)}{N} + \binom{x}{2}\binom{s(x-2)}{N} \cdots + (-1)^t \binom{x}{t}\binom{s(x-t)}{N} \right]$$

where t is the greatest integer such that $s(x - t) \geq N$.

The expected value of the distribution (52) is

$$k(1 - q_1)$$

and the variance is

$$k\{q_1 - q_2 + k(q_2 - q_1^2)\}$$

where

$$q_1 = \binom{s(k-1)}{N} \Big/ \binom{sk}{N}; \quad q_2 = \binom{s(k-2)}{N} \Big/ \binom{sk}{N}.$$

For general values of the p's, use can be made of generating functions.

Denoting by $P_r(N,k)$ the probability that the maximum frequency exceeds r, Richards [56] has shown that

$$\sum_{j=0}^{\infty} P_r(j,k)(t^j/j!) = e^t - \prod_{j=1}^{k} \left\{\sum_{i=1}^{r} [(p_j t)^i/i!]\right\}$$

If $p_j = k^{-1}$ for all j, the right hand side simplifies to

$$e^t - \left[\sum_{i=1}^{r} \{(k^{-1}t)^i/i!\}\right]^k.$$

Barton and David [5] have obtained limiting distributions for this and a number of similar distributions.

A further problem of this kind has been formulated by Tukey [68]. A total of m balls are distributed at random among an *unknown* number of boxes including N specified boxes. The total number of balls falling in the N boxes is m if the total number of boxes is N; otherwise it has a binomial distribution. Suppose that the probability of a specified ball falling into any given one of the N specified boxes is p, then the distribution of the number $(\mathbf{n}_b)$ of (specified) boxes containing b balls each has expected value

$$E(\mathbf{n}_b) = N\binom{m}{b}(1-p)^m[p/(1-p)]^b \tag{53}$$

and variance

$$\text{var}(\mathbf{n}_b) = E(\mathbf{n}_b)\cdot\{1 - [1 - \omega(b,b)]E(\mathbf{n}_b)\} \tag{54}$$

where

$$\omega(b,c) = (1 - N^{-1})\binom{m}{b,c,m-b-c}\binom{m}{b}^{-1}\binom{m}{c}^{-1}$$
$$\times [1 - \{p/(1-p)\}^2]^m\{(1-p)/(1-2p)\}^{b+c}$$

for p small,

$$\text{var}(\mathbf{n}_b) \doteqdot (1 - N^{-1})\binom{m}{b,c,\ m-b-c}\binom{m}{b}^{-1}\binom{m}{c}^{-1}$$

while for $m \gg b + c$, and p small

$$\text{var}(\mathbf{n}_b) \doteqdot (1 - N^{-1}). \tag{55.1}$$

The covariance between $\mathbf{n}_b$ and $\mathbf{n}_c$ is

$$\text{cov}(\mathbf{n}_b,\mathbf{n}_c) = -\{1 - \omega(b,c)\}E(\mathbf{n}_b)E(\mathbf{n}_c). \tag{55.2}$$

6. Distributions Associated with Queues

We are not here concerned with the more general aspects of the "theory of queues", but only with certain distributions arising naturally from simple models of queuing situations.

The *Borel-Tanner* distribution describes the distribution of the total number of customers served before a queue vanishes under condition of a single queue with random arrival times (at constant rate λ) of customers and a constant time (β) occupied in serving each customer. We suppose that the probability of arrival of a customer during the period $(t, t + (\Delta t))$ is $\lambda(\Delta t) + o(\Delta t)$ and that the probability of arrival of two (or more) customers in this period is $o(\Delta t)$. If there are initially M customers in the queue, then the probability that the total number (**x**) of customers served before the queue vanishes is equal to k is

$$\text{(56)} \qquad \Pr[\mathbf{x} = k] = \frac{M}{(k-M)!} k^{k-M-1} (\lambda\beta)^{k-M} e^{-\lambda\beta k} \qquad (k = M, M+1, \ldots).$$

For (56) to represent a proper distribution, it is necessary that $\lambda\beta$ be less than 1. If $\lambda\beta > 1$, $\sum_{k=M}^{\infty} \Pr[\mathbf{x} = k] < 1$.

The parameters λ and β appear only in the form of their product $\lambda\beta$. It is convenient to use a single symbol for this product, and to put $\lambda\beta = \alpha$, say. (As noted above, $\alpha < 1$.) The distribution (56) was obtained in 1942 by Borel [8] for the case $M = 1$, and for general values of M by Tanner [67] in 1953. It is called the *Borel-Tanner distribution* with parameters M and α. For this distribution

$$\text{(57.1)} \qquad E[\mathbf{x}] = M(1-\alpha)^{-1}$$

$$\text{(57.2)} \qquad \text{var}(\mathbf{x}) = M\alpha(1-\alpha)^{-3}.$$

Haight and Breuer [33] have given an expression for the probability generating function of (56) in terms of a function $h(y)$ defined by

$$y = [h(y)] \exp [-h(y)].$$

The expression is

$$\text{(58)} \qquad [h(t\alpha e^{-\alpha})/\alpha]^M = [t \exp \{h(t\alpha e^{-\alpha}) - \alpha\}]^M.$$

Haight and Breuer also give the relationship

$$\text{(59)} \qquad \Psi(t) = M\alpha[\phi(t)]^{M^{-1}} + M(t - \alpha)$$

between the cumulant generating function $\Psi(t)$ and the moment generating function $\phi(t)$.

Tables of the cumulative distribution function $\Pr[\mathbf{x} \le k]$, to five decimal places, are given by Haight and Breuer [33] for $M = 1$; $\alpha = 0.01(0.01)0.62$. They also refer to the existence of (unpublished) tables including values of $\Pr[\mathbf{x} = k]$ for $M = 1$; $\alpha = 0.63(0.01)0.99$; $k \le 34$. Owen [53] gives values of $\Pr[\mathbf{x} \le k]$ to five decimal places for the values $M = 1(1)5$; $\alpha = 0.01(0.01)0.25$.

The distribution has a very long positive tail (unless α is small). The modal value is approximately between $(X - 1)$ and X where

$$\alpha e^{1-\alpha} = (X - M)X^{M+\frac{1}{2}}(X - 1)^{-(M+\frac{3}{2})}$$

and $M + 1 \leq X \leq \frac{2}{3}M^2 + 3M$.

For given M, with n independent random variables each having distribution (56), the maximum likelihood estimator of α is

$$\hat{\boldsymbol{\alpha}} = 1 - M\bar{\mathbf{x}}^{-1}, \quad \text{where } \bar{\mathbf{x}} = n^{-1} \sum_{j=1}^{n} \mathbf{x}_j. \tag{60}$$

By comparison with (57), it can be seen that the same estimator would be obtained by equating population and sample first moments.

If $\alpha > 1$, then there is a non-zero probability of "**x** being infinite". In fact,

$$1 - \sum_{k=M}^{\infty} \Pr[\mathbf{x} = k] = 1 - (\alpha'/\alpha)^M \tag{61}$$

where α' is the root of

$$\alpha' e^{-\alpha'} = \alpha e^{-\alpha}$$

which is less than 1.

Haight [30] has described a distribution, which is 'analogous to the Borel-Tanner' distribution with probabilities:

$$\Pr[\mathbf{x} = k] = \frac{M}{k}\binom{2k - M - 1}{k - 1}\frac{\alpha^{k-M}}{(1 + \alpha)^{2k-M}} \qquad (k = r, r + 1, \ldots). \tag{62}$$

The probability generating function is

$$(2t)^M[1 + \alpha + \{(1 + \alpha)^2 - 4\alpha t\}^{\frac{1}{2}}]^{-M}$$

whence

$$E[\mathbf{x}] = M(1 - \alpha)^{-1} \tag{63.1}$$

$$\text{var}(\mathbf{x}) = M\alpha(1 + \alpha)(1 - \alpha)^{-3}. \tag{63.2}$$

These formulas should be compared with (57.1) and (57.2).

7. Runs

7.1 *Runs of Like Elements*

A considerable variety of forms of distribution have been discovered from the study of 'runs' in sequences of observations. Among the simplest forms of such 'runs' are sequences of identical values when the random variables representing the observed values are independent and take the values 1, 0 with probabilities p, $q(= 1 - p)$ respectively. A concise account of the early history of attempts to find the distribution of the total number of runs in such binomial (Bernoulli) sequences has been given by Mood [46]. In 1917, von Bortkiewicz [10] obtained the expected value and variance of this distribution.

Soon after, in 1921, von Mises [45] demonstrated that the distribution of the number of 'long' runs (of a specified length) can be approximated by a Poisson distribution. (Formulas for the mean and variance of the number of runs of specified length had been obtained in 1906 by Bruns [11].) A further asymptotic result obtained by Wishart and Hirschfeld [72] in 1936 showed that the asymptotic (standardized) distribution of the total number of runs, as the length of the sequence increases, is normal. Results concerning the distribution of runs of like elements when the numbers n_1, n_0 of 1's and 0's in the sequence are fixed (i.e. the conditional distribution given n_1 and n_0) were published by Ising [37] in 1925 and Stevens [66] in 1939.

If the number of runs of i consecutive 1's denoted by r_{1i}, and of j consecutive 0's by r_{0j} then the probability of obtaining such sets of runs of various lengths, conditional on $\sum_i ir_{1i} = n_1$; $\sum_j jr_{0j} = n_0$, is (with $\sum_i r_{1i} = r_1$; $\sum_j r_{0j} = r_0$)

$$(64) \qquad F(n_1,n_2)\binom{r_1}{r_{11},r_{12},\ldots,r_{1N}}\binom{r_0}{r_{01},r_{02},\ldots,r_{0N}}\bigg/\binom{N}{n_1}$$

where

$$F(n_1,n_2) = \begin{cases} 0 & |r_1 - r_0| > 1 \\ 1 & |r_1 - r_0| = 1 \\ 2 & r_1 = r_0 \end{cases}$$

The joint distribution of the total numbers of runs $\mathbf{r}_1$, $\mathbf{r}_0$ is

$$(65) \qquad \Pr[\mathbf{r}_0 = r_0,\mathbf{r}_1 = r_1] = \frac{\binom{n_1 - 1}{r_1 - 1}\binom{n_0 - 1}{r_0 - 1}}{\binom{N}{n_1}} F(r_1,r_0)$$

$$(r_0 + r_1 \geq 2; \quad r_1 \leq n_1, r_0 \leq n_0).$$

Using equation (65) it is straightforward to deduce the marginal distribution of $\mathbf{r}_1$:

$$(66) \qquad \Pr[\mathbf{r}_1 = r_1] = \binom{n_1 - 1}{r_1 - 1}\binom{n_0 + 1}{r_1}\bigg/\binom{N}{n_1} \qquad (r_1 = 1,2,\ldots n_1)$$

(where, of course, $n_0 = N - n_1$).

Distribution (66) is sometimes called the *Ising-Stevens distribution.* It is a special case of the hypergeometric distribution (Chapter 6). The sth factorial moment of $\mathbf{r}_1$ is

$$(67) \qquad \mu_{(s)}(\mathbf{r}_1) = (n_0 + 1)^{(s)}\binom{N - s}{n_1 - s}\bigg/\binom{N}{n_1}$$

whence

$$(68) \qquad E[\mathbf{r}_1] = n_1(n_0 + 1)/N$$

$$(69) \qquad \operatorname{var}(\mathbf{r}_1) = N^{-1}(n_0 + 1)^{(2)}n_1^{(2)}/N^{(2)}.$$

Finally the mean and variance of the *total* number of runs $\mathbf{R}_1 + \mathbf{R}_0$, are

$$E[\mathbf{r}_1 + \mathbf{r}_0] = 2n_1n_0N^{-1} + 1 \tag{70}$$

$$\operatorname{var}(\mathbf{r}_1 + \mathbf{r}_0) = 2n_1n_0(2n_1n_0 - N)N^{-2}(N-1)^{-1}. \tag{71}$$

Asymptotic normality of the distribution of $\mathbf{r}_1$ (or of $\mathbf{r}_1 + \mathbf{r}_0$) as N increases was established by Wald and Wolfowitz [69] in 1940.

A natural generalization of the situation is to consider arrangements of k (instead of 2) kinds of elements, conditional on there being $n_1, n_2, \ldots, n_k$ elements of the 1st, 2nd, ..., kth kind respectively (with $\sum n_j = N$). Letting r_{ij} now denote the number of runs of elements of the i-th kind which are of exact length j, and putting $r_i = \sum_{j=1}^{n_i} r_{ij} (i = 1,2,\ldots,k)$ the probability of obtaining the array $\{r_{ij}\}$ of runs is

$$F(r_1,r_2,\ldots,r_k)\left[\prod_{i=1}^{k}\binom{r_i}{r_{i1},r_{i2},\ldots,r_{in_i}}\right]\Big/\binom{N}{n_1,n_2,\ldots,n_k} \tag{72}$$

where $F(r_1,r_2,\ldots,r_k)$ is the coefficient of $\prod_{i=1}^{k} x_i^{r_i}$ in the expansion of

$$\left(\sum_{i=1}^{k} x_i\right)^{k} \prod_{j=1}^{k}\left[\sum_{i=1}^{k} x_i - x_j\right]^{r_j-1}.$$

Various additional theoretical aspects of the 'runs' distributions are discussed in Part IV of the Introduction to David, Kendall and Barton's *Symmetric Functions and Allied Tables*, (1966), pp. 46–50 (reference [3] of Chapter 1).

7.2 *Runs Up and Down*

Another kind of 'run' which has received attention is a run of increasing or decreasing values in a sequence of N independent random variables each having the same distribution, usually assumed to be continuous (obviating the need to consider tied values). Evidently the numbers of such runs are unaltered if each value is replaced by its 'rank order' — that is 1 for the smallest value, 2 for the next smallest, and so on up to N for the largest value. Accordingly, it is only necessary to study the occurrence of runs in rearrangements of the integers 1, 2, ..., N. The results will be valid, whatever the common continuous distribution of the random variables representing the observations.

Wallis and Moore [70] in constructing a test of randomness of a sequence, considered the distribution of the number of 'turning-points' in a sequence of N values. The jth observation constitutes a 'turning-point' if either $\mathbf{x}_j = \min(\mathbf{x}_{j-1},\mathbf{x}_j,\mathbf{x}_{j+1})$ or $\mathbf{x}_j = \max(\mathbf{x}_{j-1},\mathbf{x}_j,\mathbf{x}_{j+1})$. Neither $\mathbf{x}_1$ nor $\mathbf{x}_n$ can be turning-points, but for each of the other $(N-2)$ observations, the probability that it is a turning-point is $\frac{2}{3}$. Hence the expected value of the number of turning-points is $\frac{2}{3}(N-2)$. The variance, and third and fourth moments of this statistic are

$$\begin{aligned}\mu_2 &= (16N - 29)/90 \\ \mu_3 &= -16(N + 1)/945 \\ \mu_4 &= (448N^2 - 1976N + 2301)/4725.\end{aligned} \tag{73}$$

For $N > 12$, the distribution can be assumed to be normal.

The number of runs (up or down) is equal to

$$(\text{number of turning-points}) + 1.$$

Wallis and Moore [70] defined *phases* as runs excluding those beginning with $\mathbf{x}_1$ or ending with $\mathbf{x}_N$. There are no phases if there are no turning-points; otherwise the number of phases is one *less* than the number of turning-points.

Defining the *duration* of a phase between turning-points $\mathbf{x}_j$ and $\mathbf{x}_{j'}(j < j')$ to be $(j - j')$, they showed that the expected number of phases of duration d is $2(d^2 + 3d + 1)(N - d - 2)/(d + 3)!$

In [70] the authors trace the history of these distributions back to two papers by Bienaymé in 1874 (*Bulletin de la Société Mathématique de France*, *2*, 153–154). They ascribe credit for the formula for distribution of phase durations to Besson in 1920 (*Monthly Weather Review*, *48*, 89–94).

7.3 '*Records*'

Chandler [12] studied the distribution of the serial numbers of items in a sequence (of the kind considered in Sections 7.1 and 7.2) constituting records, i.e. having the smallest, or greatest, of all values so far recorded. For example, if the rank orders in a sequence of 10 observations are 3, 5, 2, 1, 9, 10, 7, 6, 4, 8 there are low records at the 1st, 3rd and 4th observations, and high records at the 1st, 2nd, 5th and 6th observations. (By convention, the first observation is regarded as both a low and high record!)

Chandler obtained the distribution of the serial number $\mathbf{u}_n$ at which the nth high record is observed. (Clearly, the same distribution applies if low records are being considered.) The joint probability of obtaining the values $\mathbf{u}_2, \mathbf{u}_3, \ldots \mathbf{u}_n$ is

$$\mathbf{u}_n^{-1} \prod_{j=2}^{n} (\mathbf{u}_j - 1)^{-1} \tag{74}$$

whence the distribution of $\mathbf{u}_n$ is

$$\Pr[\mathbf{u}_n = k] = k^{-1}K_{n-2}(k - 1) \qquad (k = n, n + 1 \ldots) \tag{75}$$

where

$$K_0(z) = z^{-1} \qquad (z \geq 1)$$

and

$$K_{r+1}(z) = z^{-1} \sum_{j=r+1}^{z-1} K_r(j) \qquad (z \geq r + 2).$$

The expected value of distribution (75) is infinite.

The distribution of $(\mathbf{u}_n - \mathbf{u}_{n-1})$, the number of observations between the $(n-1)$-th and n-th high records is of especial interest. Chandler [12] gave tables of the cumulative distribution functions of $\mathbf{u}_n$ and $(\mathbf{u}_n - \mathbf{u}_{n-1})$ to six decimal places for $n = 3(1)9$. He also showed that if the common probability density function of the observed random variables in the sequence is $p_{\mathbf{x}}(x)$, the probability that the n-th *low* record is less than X is

(76) $$\frac{1}{(n-1)!}\int_0^{-\log F} t^{n-1}e^{-t}\,dt \quad \text{where} \quad F = \int_{-\infty}^{X} p_{\mathbf{x}}(x)\,dx.$$

Foster and Stuart [20] showed that the *total* numbers of records in a sequence of length N has expected value $2\sum_{j=2}^{N} j^{-1}$ and variance $2\sum_{j=2}^{N} j^{-1} - 4\sum_{j=2}^{N} j^{-2}$. The probability generating function is $(N!)^{-1}\prod_{j=0}^{N-2}(j+2t)$. The distribution is asymptotically normal as N increases. Table 5 (from [20]) compares exact and (asymptotic) approximate values of the cumulative distribution function (a continuity correction was used in calculating the latter).

A related distribution is that of the number of *local* records. These are values which are the largest in *some* sequence of k successive values. Thus $\mathbf{x}_i$ is a local record if it is the greatest value among $\{\mathbf{x}_{i-k+1},\mathbf{x}_{i-k+2},\ldots,\mathbf{x}_i\}$ or among $\{\mathbf{x}_{i-k+2},\ldots,\mathbf{x}_{i+1}\}$ or among $\{\mathbf{x}_{i-k+3},\ldots,\mathbf{x}_{i+2}\}$ ... or among $\{\mathbf{x}_i,\mathbf{x}_{i+1},\ldots,\mathbf{x}_{i+k-1}\}$. The distribution of the number $\mathbf{r}$ of such local maxima in a sequence of n values $\mathbf{x}_1, \mathbf{x}_2 \ldots \mathbf{x}_n$ is sometimes called a *Morse distribution* (Freimer et al. [21]). The name is also applied to the distribution of $(n - \mathbf{r}) = \mathbf{s}$, and we will consider the distribution of this latter quantity, following Freimer et al. [21]. The distribution depends on n and k, and we denote the probability that $\mathbf{s}$ takes the value y by $P_{n,k}(y)$. When $n < k$, there are not, in fact, any sequences of length k. For this case we *define*

$$\begin{cases} P_{n,k}(n) = 1 \\ P_{n,k}(y) = 0 \qquad (y < n) \end{cases}$$

(i.e. in effect it is assumed that if the series were completed there would be a value included which is greater than any of those in $x_1, x_2, \ldots, x_n$). The recurrence relation (see Austin et al. [2]) below is valid for $n \geq k$.

(77.1) $$P_{n,k}(y) = n^{-1}\sum_{i=0}^{n-1}\sum_{j=0}^{y} P_{i,k}(j)P_{n-i-1,k}(y-j)$$

There is a corresponding relationship between the probability generating functions $g_{n,k}(t) = \sum_{y=0}^{n} t^y P_{n,k}(y)$.

(77.2) $$g_{n,k}(t) = n^{-1}\sum_{i=0}^{n-1} g_{i,k}(t)g_{n-i-1,k}(t) \qquad (n \geq k).$$

TABLE 5

Cumulative Distribution of Number of Records Pr[s ≤ *S*]

	N = Length of Sequence													
S	3	4	5	6	7	8	9	10	11	12	13	14	15	Normal approximation $N = 15$
2	.333	.167	.100	.067	.048	.036	.028	.022	.018	.015	.013	.011	.010	.020
3	1.000	.667	.467	.345	.265	.211	.172	.143	.121	.104	.090	.079	.070	.080
4		1.0000	.867	.733	.622	.533	.461	.403	.356	.317	.284	.256	.233	.228
5			1.0000	.956	.892	.825	.760	.700	.646	.598	.555	.516	.481	.464
6				1.0000	.987	.964	.933	.898	.862	.826	.791	.757	.725	.715
7					1.0000	.9968	.989	.978	.964	.947	.928	.908	.888	.890
8						1.0000	.9993	.9974	.9938	.989	.982	.975	.966	.970
9							1.0000	.9999	.9994	.9985	.9970	.9949	.9922	.9945
10								1.0000	1.0000	.9999	.9997	.9993	.9987	.9993
11										1.0000	1.0000	.9999	.9998	.9999
12												1.0000	1.0000	1.0000

The mean and the variance of the distribution are

$$(78.1) \qquad E(\mathbf{s}) = \begin{cases} n & (0 \le n \le k-1) \\ (k-1)(k+1)^{-1}(n+1) & (n \ge k). \end{cases}$$

$$(78{\cdot}2) \qquad \operatorname{var}(\mathbf{s}) = (n+1)\left[8(k+1)^{-1}\sum_{j=k}^{2k-1}(j+2)^{-1} - 2k(5k+3)(2k+1)^{-1}(k+1)^{-2}\right] \qquad (n \ge 2k).$$

$$\operatorname{var}(\mathbf{s}) \doteqdot (n+1)[k^{-1}(8\log 2 - 5)] \qquad \text{(for large } k\text{)}.$$

The limiting standardized distribution, as n increases to infinity, is normal.

The name of this distribution (*Morse*) arises from some relevance to a method of machine decoding hand-keyed Morse code, based upon the identification of the largest and smallest of each successive sequence of six spaces with a 'long' space (separating adjacent characters — 'letters') and a 'short' space (separating two elements — 'dot' or 'dash') of the same character.

David and Barton [15] give many interesting examples of distributions of the kind discussed in Sections 7.1–7.3.

8. Distributions Based on Partial Sums

From a given "parent" distribution, in which values 0, 1, 2, . . . are taken with probabilities $P_0, P_1, P_2, \ldots$ (with $\sum_{j=0}^{\infty} P_j = 1$) a distribution can be derived with probabilities proportional to the complement of the parent cumulative distribution function. This distribution is

$$(79) \qquad Pr[\mathbf{x} = k] = (\mu_1'^*)^{-1} \sum_{j=k+1}^{\infty} P_j \qquad (k = 0,1,\ldots).$$

(We use $\mu_r'^*$, $\mu_{(r)}'^*$ to denote the rth moment about zero and the rth factorial moment respectively of the parent distribution.)

The rth moment about zero ($r \ge 1$) of this distribution is

$$\begin{aligned} \mu_r' &= (\mu_1'^*)^{-1} \sum_{k=0}^{\infty} k^r \sum_{j=k+1}^{\infty} P_j \\ &= (\mu_1'^*)^{-k} \sum_{j=1}^{\infty} P_j \sum_{k=0}^{j-1} k^r \\ &= (\mu_1'^*)^{-1} \sum_{j=1}^{\infty} P_j \sum_{k=0}^{j-1} \left\{ \sum_{i=1}^{r} (\Delta^i 0^r / i!) k^{(i)} \right\} \\ &= (\mu_1'^*)^{-1} \sum_{j=1}^{\infty} P_j \sum_{i=1}^{r} (\Delta^i 0^r/(i+1)!) j^{(i+1)} \\ &= (\mu_1'^*)^{-1} \sum_{i=1} (\Delta^i 0^r/(i+1)!) \mu_{(i+1)}'^*. \end{aligned}$$

The moment generating function is

$$(\mu_1'^*)^{-1}(1-e^t)^{-1}(1-\phi^*(t))$$

where

$$\phi^*(t) = \sum_{j=0}^{\infty} P_j e^{jt}.$$

The distribution obtained when the parent distribution is Poisson, with

$$P_j = e^{-\theta}(\theta^j/j!)$$

is called *Poisson's exponential binomial limit.* For this distribution $\mu_j'^* = \theta^j$ and

(80.1) $$E[\mathbf{x}] = \tfrac{1}{2}\theta$$

(80.2) $$\operatorname{var}(\mathbf{x}) = \tfrac{1}{2}\theta + \tfrac{1}{12}\theta^2$$

(80.3) $$\mu_3(\mathbf{x}) = \tfrac{1}{2}\theta + \tfrac{1}{4}\theta^2$$

(80.4) $$\mu_4(\mathbf{x}) = \tfrac{1}{2}\theta + \tfrac{4}{3}\theta^2 + \tfrac{1}{4}\theta^3 + \tfrac{1}{80}\theta^4.$$

As θ tends to infinity, the moment-ratios $\sqrt{\beta_1}$ and β_2 tend to 0 and 1.8 respectively, the values for rectangular distributions (Chapter 25).

Gold [25] defines Poisson's exponential binomial limit distribution by the formula

(79)′ $$\Pr[\mathbf{x} = k] = (\mu_1'^* + 1)^{-1} \sum_{j=k}^{\infty} P_j$$

rather than (79). In this case

$$\mu_r' = (\mu_1'^* + 1)^{-1}\left(\sum_{i=1}^{r} (\Delta^i 0^r/(i+1)!)\mu_{(i+1)}'^* + \mu_r'^*\right).$$

For (79)′

$$E[\mathbf{x}] = \theta(\tfrac{1}{2}\theta + 1)(\theta + 1)^{-1}$$

and

$$\operatorname{var}(\mathbf{x}) = (\tfrac{1}{12}\theta^3 + \tfrac{5}{6}\theta^2 + \tfrac{3}{2}\theta + 1)(\theta + 1)^{-2}.$$

The moment generating function is

$$E[e^{t\mathbf{x}}] = (\mu_1'^* + 1)^{-1}(1-e^t)^{-1}(1 - e^t\phi^*(t)).$$

Gerstenkorn [22] has given formulas for the moments of this distribution and also for another distribution introduced by Gold [25]. This is the *general Poisson* defined by

(81) $$\Pr[\mathbf{x} = k] = Kq^k e^{-q\theta} \sum_{j=k}^{\infty} (\theta^j/j!) \qquad (k = 0,1,2,\ldots)$$

with K suitably chosen. For $q = 1$ this reduces to distribution (79)′. The moment generating function is

$$\begin{aligned} E[e^{tu}] &= Ke^{-q\theta} \sum_{k=0}^{\infty} (qe^t)^k \sum_{j=k}^{\infty} (\theta^j/j!) \\ &= Ke^{-q\theta} \sum_{j=0}^{\infty} (\theta^j/j!) \sum_{k=0}^{j} (qe^t)^k \\ &= Ke^{-q\theta} \sum_{j=0}^{\infty} (\theta^j/j!)(1 - (ge^t)^{j+1}(1 - ge^t)^{-1} \\ &= K(1 - ge^t)^{-1} \{\exp[(1 - q)\theta] - ge^t \exp[q(e^t - 1)\theta]\}. \end{aligned}$$

Putting $t = 0$ we find

$$K = (1 - q)(e^{(1-q)\theta} - q)^{-1}.$$

This distribution should not be confused with the "generalized Poisson" introduced in Chapter 4, Section 10.5 which has been studied (under that name) by Goodman [27] and Haight [29].

Bissinger [6] has constructed systems of distributions derived from a parent distribution by another summation process. The 'Bissinger system' distribution is defined by

$$\text{(82)} \qquad \Pr[\mathbf{x} = k] = (1 - P_0)^{-1} \sum_{j=k+1}^{\infty} j^{-1}P_j \qquad (k = 0,1,2,\ldots).$$

It is easily verified that $\sum_{k=0}^{\infty} \Pr[\mathbf{x} = k] = 1$, so (82) defines a proper distribution. Bissinger gave this class of distributions the name *STER*, from the phrase "*s*ums of *t*runcated forms of the *e*xpected value of the *r*eciprocal" (of a variable having the 'parent' distribution).

From (82)

$$\text{(83)} \qquad \Pr[\mathbf{x} = k - 1] - \Pr[\mathbf{x} = k] = (1 - P_0)^{-1}k^{-1}P_k \geq 0.$$

Hence the successive values of $\Pr[\mathbf{x} = k]$ are non-increasing as k increases.

The rth moment about zero ($r > 0$) is

$$\begin{aligned} \mu'_r &= (1 - P_0)^{-1} \sum_{k=1}^{\infty} k^r \sum_{j=k+1}^{\infty} j^{-1}P_j \\ &= (1 - P_0)^{-1} \sum_{j=2}^{\infty} P_j j^{-1} \sum_{k=1}^{j-1} k^r \\ &= (1 - P_0)^{-1} \sum_{i=1}^{r} \frac{\Delta^i 0^r}{(i + 1)!} \sum_{j=2}^{\infty} (j - 1)^{(i-1)}P_j \end{aligned}$$

Using the formulas

$$\sum_{k=1}^{j-1} k = \tfrac{1}{2}j(j - 1); \quad \sum_{k=1}^{j-1} k^2 = \tfrac{1}{6}j(j - 1)(2j - 1)$$

we find

$$\mu_1' = \tfrac{1}{2}[(1 - P_0)^{-1}\mu_2'^* - 1]$$
$$\mu_2' = \tfrac{1}{6}[(1 - P_0)^{-1}(2\mu_2'^* - 3\mu_1'^*) + 1]$$

where $\mu_r'^*$ denotes the rth moment about zero of the parent distribution. The probability generating function of distribution (82) is

$$(84)\quad \sum_{k=0}^{\infty} t^k \sum_{j=k+1}^{\infty} j^{-1}P_j = \sum_{j=1}^{\infty} j^{-1}P_j(1 + t + t^2 + \cdots + t^{j-1})(1 - P_0)^{-1}$$
$$= (1 - t)^{-1} \sum_{j=1}^{\infty} j^{-1}P_j(1 - t^j)(1-P_0)^{-1}$$
$$= (1 - t)^{-1}\{E^*[\mathbf{x}^{-1}] - E^*[t^{\mathbf{x}}\mathbf{x}^{-1}]\}(1-P_0)^{-1}$$

where E^* denotes expectation taken with respect to the parent distribution. Bissinger [6] obtains the following relation with the moments ($\mu_r'^*$) about zero of the parent distribution:

$$(85)\qquad \mu_r'^* = (1 - P_0)^{-1} \sum_{j=0}^{r} \binom{r+1}{j} \mu_j'.$$

We further note that

$$\Pr[\mathbf{x} = k] = \sum_{j=k}^{\infty} (j + 1)^{-1}P_j \qquad (k = 0,1,2,\ldots)$$

defines a proper distribution. For this distribution

$$\mu_r' = \sum_{i=1}^{r} \frac{\Delta^i 0^r}{(i + 1)!} \mu_{(i)}'^*$$

which it is interesting to compare with

$$\mu_r'^* = \sum_{i=1}^{r} \frac{\Delta^i 0^r}{i!} \mu_{(i)}'^*.$$

9. Matching Distributions

The classical example of a matching distribution arises in the following way. Suppose that a set of n subjects, each numbered 1, 2, . . . n respectively, are arranged in random order. We seek to find the distribution of the number of subjects for which position in the random order equals the number originally assigned to them. Applying Boole's formula, it is found that this distribution is defined by

$$(86)\quad \Pr[\mathbf{x} = k] = (k!)^{-1}[1 - (1!)^{-1} + (2!)^{-1} \ldots + (-1)^{n-k}((n - k)!)^{-1}] \qquad (k = 0,1,\ldots n).$$

(Note that $\Pr[\mathbf{x} = n - 1] = 0$.)

The expected value is

$$E(\mathbf{x}) = 1, \tag{87}$$

and the variance is also

$$\operatorname{var}(\mathbf{x}) = 1. \tag{88}$$

From (86) it can be seen that, except when $(n - k)$ is small,

$$\Pr[\mathbf{x} = k] \doteqdot (k!)^{-1}e^{-1} \tag{89}$$

to a close degree of approximation. The successive probabilities are, thus, approximated by the corresponding probabilities for a Poisson distribution with expected value 1. The close connection between the two distributions is brought out by considering the expressions for their moment generating functions.

The moment generation function of distribution (86) is given in the following expression

$$\sum_{j=0}^{n} (-1)^j (j!)^{-1}(1 - e^t)^j \tag{90}$$

while that of the Poisson distribution with expected value 1 is (Chapter 4, equation (9))

$$\exp\left[-(1 - e^t)\right] = \sum_{j=0}^{\infty} (-1)^j (j!)^{-1}(1 - e^t)^j. \tag{91}$$

From comparison of (90) and (91), it can be seen that the two distributions have identical 1st, 2nd, 3rd, . . . , nth moments. (Their first n cumulants are, of course, also identical.)

There are many variants on this simple classical problem. (An interesting brief historical discussion has been given by Barton [4].) We now describe a simple extension. Suppose, for example, that we have two identical decks of cards, and that each deck of cards has s suits of c (distinct) cards per suit (so that each deck contains $cs = n$ cards). The two decks are then dealt out, in random order, at the same rate and a *match* is said to occur if the cards dealt from the two decks at the same time are of the same suit. If $c = 1$, then distribution (86) is obtained. For general values of c, by an application of Boole's formula,

$$\Pr[\mathbf{x} = k] = (n!)^{-1} \sum_{j=k}^{n} (-1)^{j-k} \binom{j}{k} (n - j)!\,H_j \tag{92}$$

where H_j is the coefficient of z^j in the expansion of

$$\left[\sum_{i=0}^{c} \frac{(c!)^2}{i![(c - i)!]^2} z^i\right]^s.$$

Tables of $\Pr[\mathbf{x} \geq k]$ to 5 decimal places, for

$$\begin{aligned} c &= 1, 2 \text{ and } s = 2(1)11; \\ c &= 3 \quad \text{ and } s = 2(1)8; \\ c &= 4, 5 \text{ and } s = 2(1)5; \end{aligned}$$

and also

$$\begin{aligned} s &= 2 \quad \text{and } c = 6(1)12; \\ s &= 3 \quad \text{and } c = 6, 7 \end{aligned}$$

are given by Gilbert [23]. These tables are also included in Owen [53].

Silva [61] has shown that for s large, there are the following approximations for $\Pr[\mathbf{x} = k]$:

$$\frac{c^k e^{-c}}{k!}\left[1 - \frac{c-1}{2cs} \cdot \frac{(c-k)^2 - k}{c}\right] \tag{92)'}$$

$$\frac{c^k}{k!}\left(1 - \frac{1}{s}\right)^{cs}\left[1 + \frac{1}{2s} + \frac{(c-1)(2c+1-k)k}{2c^2 s}\right] \tag{92)''}$$

For the more general problem, in which there are $c_{11}, \ldots, c_{1s}$ cards per suit in the first deck and $c_{21}, \ldots, c_{2s}$ cards per suit in the second deck (with $\sum_{i=1}^{s} c_{1i} = \sum_{i=1}^{s} c_{2i} = n$), H_j is the coefficient of z^j in

$$\prod_{i=1}^{s}\left[\sum_{k=1}^{c_i} \frac{c_{1i}!c_{2i}!}{k!(c_{1i}-k)!(c_{2i}-k)!} z^k\right]$$

where $c_i = \min(c_{1i}, c_{2i})$. (Greville [28]).

Limiting forms of these and related distributions have been studied by Barton [4]. Joseph and Bizley [39] give some formulas from which values of $\Pr[\mathbf{x} = k]$ can be calculated, for the case $c_{1i} = c_{2i}$ for all i.

10. Absorption Distributions

This name may be given to distributions which represent the number of 'individuals' (e.g. particles) which fail to cross a specified region containing hazards of various kinds. Possible distributions of this kind are as unlimited as are the possible variations in the specifications of the regions and the hazards therein. We will discuss some of the simpler cases (Blomquist [7], Zacks and Goldfarb [74]).

Firstly, we consider a 'region' consisting of a straight line with a number of 'absorption' points on it. When a particle meets such a point there is a probability p that it will be absorbed. If it is absorbed it fails to make any further progress, but also the point is incapable of absorbing any more particles. When there are M active absorption points, the probability of a particle being absorbed is $(1 - (1 - p)^M)$. If there are initially M absorption points, and n particles attempt to cross the (linear) region, the probability that k particles are absorbed is

(93)

$$P_n(k) = \begin{cases} (1-p)^{nM} & (k = 0) \\ \prod_{j=0}^{k-1} [1-(1-p)^{M-j}](1-p)^{(M-k)(n-k)}K_n(1-p,k) & (1 \le k \le n) \\ 0 & (k > n) \end{cases}$$

where $K_n(1-p,k)$ is a polynomial in $(1-p)$ of degree $k(n-k)$.

Putting $K_n(1-p,k) = \sum_{j=0}^{k(n-k)} a_j(n,k)(1-p)^j$ the coefficients $a_j(n,k)$ can be determined from the recurrence relations

(94)
$$a_j(n,1) = 1 \qquad (j = 0,1,\ldots,(n-1))$$
$$a_j(n,k) = \sum_{i=0}^{[j/k]} a_{j-ik}(n-i-1,k-1)$$
$$\left(j = 0,1,\ldots,\left[\frac{k(n-k)}{2}\right];k = 2,\ldots,[\tfrac{1}{2}n]\right).$$

The problem may be modified and generalized in many ways. We consider here a region of two dimensions, an infinite 'slab' of width b. We suppose that absorption points are distributed at random over the region in such a way that the initial expected number of points encountered by a particle in crossing the region is θ. Further we suppose the number (n) of particles and the total number (M) of absorption points to be large. The probability of exactly k absorptions was found by Blomquist [7] to be

(95)
$$P(k) = q^{(M-k)(n-k)} \frac{\prod_{j=n-k+1}^{n} (1-q^j) \prod_{j=M-k+1}^{M} (1-q^j)}{\prod_{j=1}^{k} (1-q^j)}$$
$$(k = 0,1,\ldots \min (M,n))$$

where $q = 1 - \theta/M$.

Borenius [9] states that the expected value of the distribution is

$$(1-q^M)(1-q^n)$$

and the variance is approximately

$$(1-q)^{-1}[1-(1-q^M)(1-q^n)]^{-2}(1-q^M)q^M(1-q^n)q^n.$$

The standardized distribution tends rapidly to normality as M and n increase. The following diagrams, taken from [9], show how closely the distribution (95) (represented by the step curve) is approximated by the normal distribution with the mean and variance stated above.

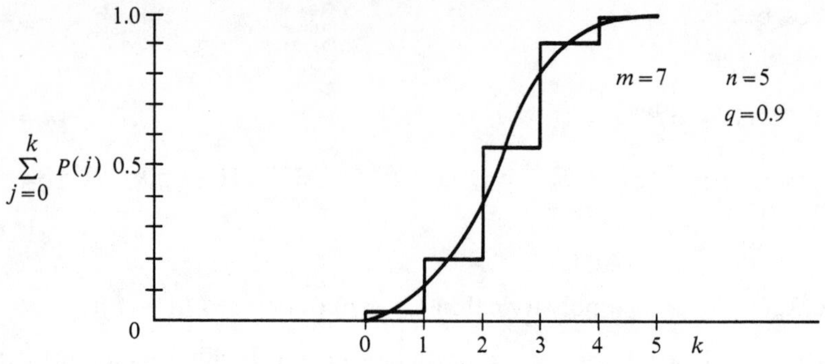

FIGURE 2a

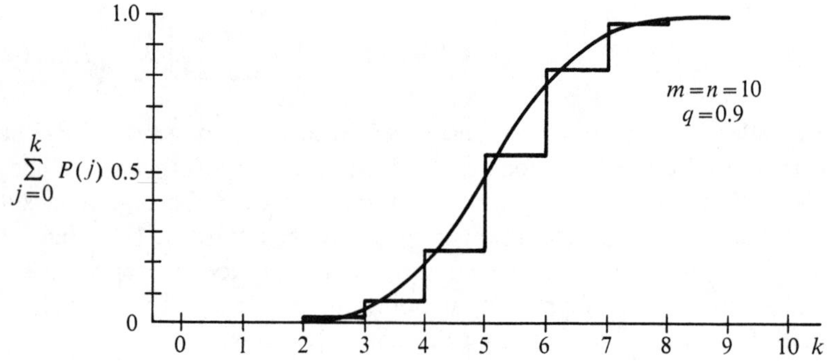

FIGURE 2b

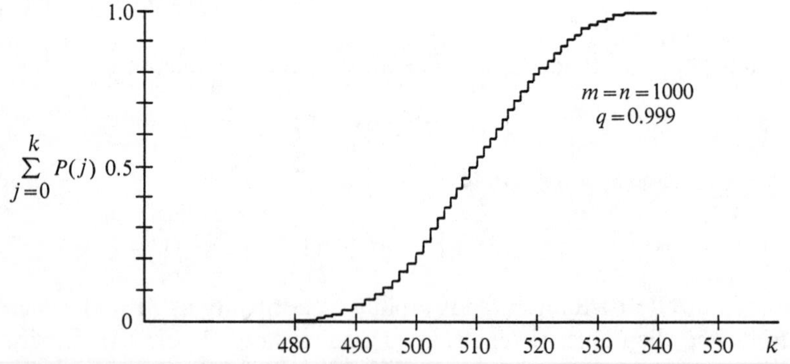

FIGURE 2c

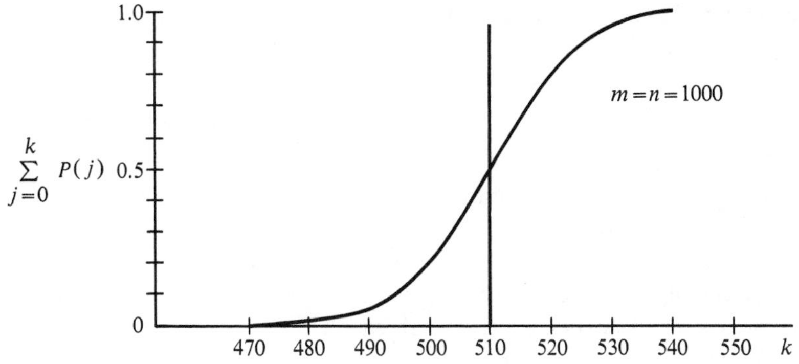

FIGURE 2d

Normal Cumulative Distribution Function with $\mu = 510.1$ *and* $\sigma = 12.24$

11. 'Interrupted' Distributions

This name may be given to distributions arising from circumstances in which some values of a random variable are not observed, because of their juxtaposition in time or space to some other value. Consider, for example, a Geiger-Muller counter recording the arrival of radioactive particles. Suppose there is a constant resolving ('dead') time, D say, succeeding the recording of a particle, during which the counter is unable to record any arriving particles. The distribution of the number of arrivals *recorded* in a fixed period of time (of length T, say) will not be the same as that of the *actual* number of arrivals. We will work out the theory for the case when the actual arrivals occur randomly in time (as described in Chapter 4, Section 1 — see also Chapter 18, Section 2) at a rate λ so that the actual number of arrivals in time period T has a Poisson distribution with expected value λT, and times (**t**) between successive arrivals are independent, and each has the same probability density function:

$$p_{\mathbf{t}}(t) = \lambda e^{-\lambda t} \quad (t \geq 0). \tag{96}$$

This case is considered in [18], [19] and [24], and, in connection with telephone service, in [16]. In the analysis we shall need to work with continuous random variables such as t. Reference to Chapter 12 may help readers unfamiliar as yet with the appropriate methods.

We will consider an especially simple case in which no particle has been recorded in time D preceding the beginning of the time period of length T. More complicated problems of this kind can easily be imagined — by supposing that the initial point of the interval of length T is chosen at random; that no arrival within time D after another arrival (whether recorded or not) can be recorded; that resolving time is variable, etc. Our simple case makes it possible to set out the method of solution in a relatively simple manner.

Given that there are K actual arrivals (the probability of this is $e^{-\lambda T}(\lambda T)^K/K!$), what is the probability, $P_{k,K}$ that exactly k will be recorded? A few values can be written down immediately. Evidently

$$P_{k,K} = 0 \qquad \text{for } (k-1)D \geq T$$
$$\text{(i.e. for } k \geq [T/D] + 1)$$

and also $P_{0,K} = 0$ for $K > 0$ (since the first arrival must be recorded) while $P_{0,0} = 1$.

In order to obtain a general formula we classify the event "exactly k out of K arrivals are observed, given there are exactly K arrivals," according to the time $\mathbf{t}_k^*$ of the last recorded arrival. Given $\mathbf{t}_k^*$, the probability that all the remaining $(K-k)$ values fall within a time D after one of the values $\mathbf{t}_1^*, \mathbf{t}_2^*, \ldots, \mathbf{t}_k^*$ is

$$(97) \qquad [T^{-1}\{(k-1)D + T - \mathbf{t}_k^*\}]^{K-k}$$
$$= \begin{cases} [1 - \{\mathbf{t}_k^* - (k-1)D\{T^{-1}]^{K-k} & (\mathbf{t}_k^* \geq T - D) \\ [kDT^{-1}]^{K-k} & (\mathbf{t}_k^* \leq T - D). \end{cases}$$

Suppose, for the moment, that the set $(\mathbf{t}_1^*, \ldots, \mathbf{t}_k^*)$ is a rearrangement of the set $(\mathbf{t}_1, \ldots, \mathbf{t}_k)$. (Actually it can be a rearrangement of any one of $\binom{K}{k}$ possible subsets of size k from $\mathbf{t}_1, \mathbf{t}_2, \ldots, \mathbf{t}_K$.) If $\mathbf{t}'_1 \leq \mathbf{t}'_2 \leq \cdot \leq \mathbf{t}'_k$ are the order statistics corresponding to $(\mathbf{t}_1, \mathbf{t}_2, \ldots, \mathbf{t}_k)$ the joint probability density function of $\mathbf{t}'_1, \ldots, \mathbf{t}'_k$ is

$$p_{\mathbf{t}'_1, \ldots \mathbf{t}'_k}(t'_1, \ldots, t'_k) = k! T^{-k} \qquad (0 \leq t'_1 \leq t'_2 \leq \cdots \leq t'_k \leq T).$$

The probability that all are recorded is

$$(98) \qquad k! T^{-k} \int_{(k-1)D}^{T} \cdots \int_{D}^{t'_3 - D} \int_{0}^{t'_2 - D} dt'_1\, dt'_2 \ldots dt'_k = [1 - (k-1)D/T]^k.$$

The conditional probability (given that all are recorded) that $\mathbf{t}_k^* (\equiv \mathbf{t}'_k)$ is less than τ is

$$[\tau - (k-1)D]^k [T - (k-1)D]^{-k} \quad (\text{if } (k-1)D \leq \tau \leq T).$$

Hence the probability density function of $\mathbf{t}_k^*$ is

(99)
$$p_{\mathbf{t}_k^*}(t_k^*) = k[t_k^* - (k-1)D]^{k-1}[T - (k-1)D]^{-k} \qquad ((k-1)D \leq t_k^* \leq T).$$

From (98) and (99) (remembering that there are $\binom{K}{k}$ ways of selecting the recorded $\mathbf{t}_j^*$'s)

(100)

$$P_{k,K} = \binom{K}{k}[1-(k-1)D/T]^k$$
$$\times\left[\left(\frac{kD}{T}\right)^{K-k}\Pr[\mathbf{t}_k^* \le T-D] + k[T-(k-1)D]^{-k}\right.$$
$$\left.\times\int_{T-D}^{T}[t-(k-1)D]^{k-1}[1-\{t-(k-1)D\}T^{-1}]^{K-k}\,dt\right]$$
$$= \binom{K}{k}\left[(kD/T)^{K-k}g^k + k\int_g^{1-(k-1)D/T} y^{k-1}(1-y)^{K-k}\,dy\right]$$

where
$$g = \begin{cases} 0 & \text{if } k \ge T/D \\ 1-kD/T & \text{if } k \le T/D. \end{cases}$$

Finally, the overall probability of recording exactly $k(\ge 1)$ arrivals is

(101)

$$P_k = e^{-\lambda T}\sum_{K=k}^{\infty}\{(\lambda T)^K/K!\}P_{k,K}$$
$$= e^{-\lambda T}\{(\lambda T)^k/k!\}\left[g^k\sum_{K=k}^{\infty}\{(\lambda kD)^{K-k}/(K-k)!\} + k\int_g^{1-(k-1)D/T} y^{k-1}\right.$$
$$\left.\times\sum_{K=k}^{\infty}\{(\lambda T[1-y])^{K-k}/(K-k)!\}\,dy\right]$$
$$= e^{-\lambda(T-kD)}(\lambda Tg)^k/k! + \frac{1}{(k-1)!}\int_{\lambda Tg}^{\lambda(T-(k-1)D)} z^{k-1}e^{-z}\,dz.$$

Inserting the value of g, we obtain

(102)

$$P_k = \begin{cases} 0 & (T \le (k-1)D) \\ \dfrac{1}{(k-1)!}\displaystyle\int_0^{\lambda[T-(k-1)D]} z^{k-1}e^{-z}\,dz & ((k-1)D \le T \le kD) \\ e^{-\lambda(T-kD)}[\lambda(T-kD)]^k/k! & \\ \quad + \dfrac{1}{(k-1)!}\displaystyle\int_{\lambda(T-kD}^{\lambda[T-(k-1)D]} z^{k-1}e^{-z}\,dz & (0 < kD \le T) \\ e^{-\lambda T} & (k=0). \end{cases}$$

Oliver [51] has used a model leading to the same distribution to represent traffic counts over a fixed interval of time T. In this model the distribution of times between successive arrivals of vehicles is supposed to have probability density function

$$(103) \qquad p_t(t) = \begin{cases} 0 & (t \leq D) \\ \lambda \exp[-\lambda(t - D)] & (t \geq D). \end{cases}$$

The number of arrivals in time T will have the distribution given by (102), if the time period starts later than time D after the last vehicle to pass. Oliver [51] considers the situation in which the interval starts immediately after the arrival of a vehicle, and so obtains a different distribution, which can, however be obtained from (102) by replacing T by $(T - D)$.

Oliver [51] shows (in effect) that the expected value of the distribution (102) is

$$\mu_1' = \sum_{j=1}^{[T/D]+1} [(j-1)!]^{-1} \int_0^{\lambda[T-(j-1)D]} z^{j-1} e^{-z}\, dz,$$

and the variance is

$$\sum_{j=1}^{[T/D]+1} [(2j-1)/(j-1)!] \int_0^{\lambda[T-(j-1)D]} z^{j-1} e^{-z}\, dz.$$

Note that the integrals in these expressions are incomplete gamma functions. Figure 3, based on [51], shows the distribution (102) for several values of μ_1' with a fixed value of $[T/D]$.

A model of a similar kind has been used by Singh [64] to represent the

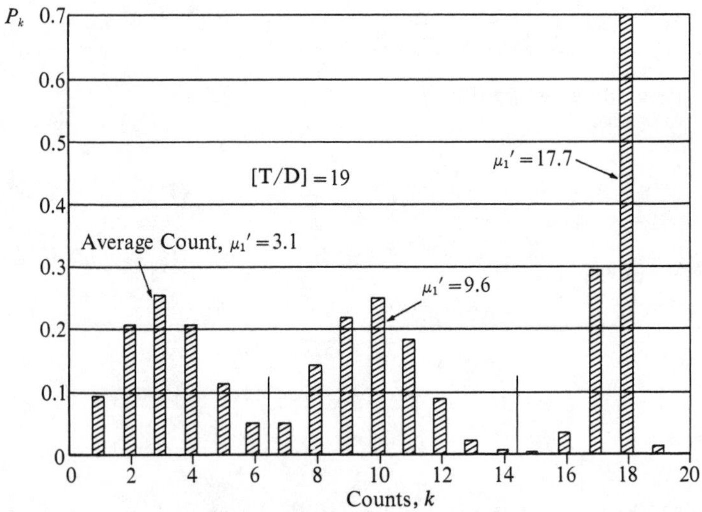

FIGURE 3

Three 'Interrupted' Distributions

distribution of the numbers of conceptions, over a fixed interval of time (T) to married couples in a specified population. Singh assumes that for a time D following a conception no further conception is possible. Instead of representing the time between conceptions as a continuous random variable, however, he divides T into m subintervals of equal length (T/m) and supposes that provided at least ($mDT^{-1} - 1$) subintervals have elapsed since a subinterval in which there was a conception, the probability of a conception in a subinterval is p. (mDT^{-1} is taken to be an integer.) Then, by an argument similar to that used earlier in this section, the probability of k conceptions in the interval T can be made up of cases when the last conception is in the $(m - j + 1)$th subinterval ($j = 1,2,\ldots,(mDT^{-1} - 1)$) for which the probability is

$$\binom{m - j - (k - 1)(mDT^{-1} - 1)}{k - 1} p^k(1 - p)^{m-j-(k-1)mD/T}$$

and cases when the last conception is in some earlier subinterval, for which the probability is

$$\binom{m - k(mDT^{-1} - 1)}{k} p^k(1 - p)^{m-kmD/T}.$$

Combining these values the distribution is defined by

(104)

$$P_k = \left[\binom{m - k(mDT^{-1} - 1)}{k} + \sum_{j=1}^{mDT^{-1}-1} \binom{m - j - (k - 1)(mDT^{-1} - 1)}{k - 1} \times (1 - p)^{mDT^{-1}-j}\right] \cdot p^k(1 - p)^{m(1-kD/T)}.$$

(Singh, in fact, divides the population into two parts, one with distribution (104) and the other with $P_0 = 1$, $P_k = 0$ for $k > 0$, to represent infertile couples. The resultant distribution is then (104) 'modified' in the sense described in Chapter 8, Section 4.)

Distribution (104) tends to distribution (102) as m increases, and p decreases, with $mp = \lambda T$.

12. Discrete Student's Distribution

We have already described, in Section 4 of Chapter 2, a system of discrete distributions studied by Ord [52]. Here, we wish to devote special attention to those members of the system roughly analogous to Type IV of Pearson's system of continuous distributions (see Chapter 12). For these distributions, the recurrence relation

$$P_j = \frac{(j + a)^2 + d^2}{(j + m + a)^2 + b^2} P_{j-1} \qquad (m > 0) \tag{105}$$

is satisfied for all integer j (positive, negative or zero). The expected value, variance and third central moment of the distribution are respectively (with $M = m - 1$)

$$\mu_1' = -1 - \tfrac{1}{2}M - a + \frac{d^2 - b^2}{2M} \tag{106}$$

$$\mu_2 = \frac{1}{4(2M-1)}\left[M^2 + 2(b^2 + d^2) + \frac{(b^2 - d^2)^2}{M^2}\right] \tag{107}$$

and

$$\mu_3 = \frac{d^2 - b^2}{2M^3(2M-1)(2M-2)}[(M^2 + b^2 - d^2)^2 + 4M^2d^2]. \tag{108}$$

For $r \geq 2m - 1$, μ_r is infinite.

From (105)

$$\begin{aligned} j^2P_j &+ 2(m+a)jP_j + [(m+a)^2 + b^2]P_j \\ &= (j-1)^2P_{j-1} + 2(a+1)(j-1)P_{j-1} + [(a+1)^2 + d^2]P_{j-1}. \end{aligned} \tag{109}$$

Multiplying both sides of (109) by j^r, and writing

$$j^r = \sum_{i=0}^{r} \binom{r}{i} (j-1)^i$$

on the right hand side, and then summing with respect to j, we obtain the following relationship between moments about zero (provided these are finite):

(110)

$$\begin{aligned} 2(m+a)\mu_{r+1}' &+ [(m+a)^2 + b^2]\mu_r' \\ &= \sum_{i=0}^{r-1}\binom{r}{i}\mu_{i+2}' + 2(a+1)\sum_{i=0}^{r}\binom{r}{i}\mu_{i+1}' + [(a+1)^2 + d^2]\sum_{i=0}^{r}\binom{r}{i}\mu_i'. \end{aligned}$$

$$\left(\text{Remember that } \sum_{j=-\infty}^{\infty} j^rP_j = \mu_r' = \sum_{j=-\infty}^{\infty}(j-1)^rP_{j-1}\right).$$

On solving these equations we find that μ_r' and so μ_r, must be expressible as the ratio of polynomials (of finite degree) in a. Now it is clear that the value of μ_r is unchanged if a be increased or decreased by an integer. (If a be increased by 1, for example, the only effect is to move the entire distribution one unit in the negative direction.) Therefore μ_r cannot, in fact, depend on a at all, and so the moment ratio $\mu_r/\mu_2^{\frac{1}{2}r}$ does not depend on a. However, the *shape* of the distribution does depend on a — more precisely, on the fractional part of a. As already noted, the shape *is* the same for all pairs of values of a differing by an integer.

If b is equal to d, and also m is an integer, then multipliers and divisors in the ratio P_j/P_{j+l} cancel for $l \geq m$, and

$$P_j = K \prod_{i=1}^{m} \{(j + a + i)^2 + b^2\}^{-1} \qquad (-\infty < j < \infty)$$

where K depends on a and b, and is such that $\sum_{j=-\infty}^{\infty} P_j = 1$.

For these distributions, all finite odd order central moments are zero. However, they are symmetrical only if the fractional part of a is 0 or $\frac{1}{2}$. Ord [52] has called these 'nearly symmetrical' distributions *discrete Student's distributions*.

REFERENCES

[1] Arfwedson, G. (1951). A probability distribution connected with Stirling's second class numbers, *Skandinavisk Aktuarietidskrift*, **34,** 121–132.

[2] Austin, T., Fagen, R., Lehrer, T., and Penney, W. (1957). The distribution of the number of locally maximal elements in a random sample from a continuous distribution. *Annals of Mathematical Statistics*, **28,** 786–790.

[3] Bardwell, G. E. and Crow, E. L. (1964). A two-parameter family of hyper-Poisson distributions, *Journal of the American Statistical Association*, **59,** 133–141.

[4] Barton, D. E. (1958). The matching distributions: Poisson limiting forms and derived methods of approximation, *Journal of the Royal Statistical Society, Series B*, **20,** 73–92.

[5] Barton, D. E. and David, F. N. (1959). Combinatorial extreme-value distributions, *Mathematika*, **6,** 63–76.

[6] Bissinger, B. H. (1963). A type-resisting distribution generated from considerations of an inventory decision model, *Proceedings of the International Symposium on Discrete Distributions, Montreal*, 14–17.

[7] Blomquist, N. (1952). On an exhaustion process, *Skandinavisk Aktuarietidskrift*, **35,** 201–210.

[8] Borel, E. (1942). Sur l'emploi du théorème de Bernoulli pour faciliter le calcul d'un infinité de coefficients. Application au problème de l'attente à un guichet, *Comptes Rendus de l'Académie des Sciences, Paris*, **214,** 452–456.

[9] Borenius, G. (1953). On the statistical distribution of mine explosions, *Skandinavisk Aktuarietidskrift*, **36,** 151–157.

[10] Bortkiewicz, L. von, (1917). *Die Iterationen*, Berlin: Springer.

[11] Bruns, H. (1906). *Wahrscheinlichkeitsrechnung und Kollektivmasslehre*, Leipzig & Berlin: Teubner.

[12] Chandler, K. N. (1952). The distribution and frequency of record values, *Journal of the Royal Statistical Society, Series B*, **14,** 220–228.

[13] Craig, C. C. (1953). On the utilization of marked specimens in estimating populations of flying insects, *Biometrika*, **40,** 170–176.

[14] Crow, E. L. and Bardwell, G. E. (1963). Estimation of the parameters of the hyper-Poisson distributions, *Proceedings of the International Symposium on Discrete Distributions, Montreal*, 127–140.

[15] David, F. N. and Barton, D. E. (1962). *Combinatorial Chance*, London: Griffin.

[16] Erlang, A. K. (1918). Solutions of some problems in the theory of probabilities. *Post Office Electrical Engineering Journal*, **10,** 189. (See also *Transactions of the Danish Academy of Technology and Science* (1948). No. 2: The Life and Works of A. K. Erlang.)

[17] Estoup, J. B. (1916). *Les Gammes Sténographiques*, Paris: Institut Sténographique.

[18] Feix, M. (1955). Théorie d'enregistrement d'évenements aléatoires, *Journal Physique Radium*, **16,** 719–727.

[19] Feller, W. (1948). On probability problems in the theory of counters. *Studies presented to R. Courant*, Interscience, 105–115.

[20] Foster, F. G. and Stuart, A. (1954). Distribution-free tests in time-series based on the breaking of records, *Journal of the Royal Statistical Society, Series B*, **16,** 1–13.

[21] Freimer, M., Gold, B., and Tritter, A. L. (1959). The Morse distribution, *Transactions of IRE—Information Theory*, **5,** 25–31.

[22] Gerstenkorn, T. (1962). On the generalized Poisson distribution, *Lodzkie Towarzystwo Naukowe, Wydzial III*, No. 85, 1–40.

[23] Gilbert, E. J. (1956). The matching problem, *Psychometrika*, **21,** 253–266.

[24] Giltay, J. (1943). A counter arrangement with constant resolving time, *Physica*, **10,** 725–734.

[25] Gold, L. (1957). Generalized Poisson Distribution, *Annals of the Institute of Statistical Mathematics, Tokyo*, **9,** 43–48.

[26] Good, I. J. (1957). Distribution of word frequencies, *Nature, London*, **179,** 595.

[27] Goodman, L. A. (1952). On the Poisson-gamma distribution, *Annals of the Institute of Statistical Mathematics, Tokyo*, **3,** 123–125.

[28] Greville, T. N. E. (1941). The frequency distribution of a general matching problem, *Annals of Mathematical Statistics*, **12,** 350–354.

[29] Haight, F. A. (1959). The generalized Poisson distribution, *Annals of the Institute of Statistical Mathematics, Tokyo*, **11,** 101–105.

[30] Haight, F. A. (1961). A distribution analogous to the Borel-Tanner, *Biometrika*, **48,** 167–173.

[31] Haight, F. A. (1961). Index to the distributions of mathematical statistics, *Journal of Research of the National Bureau of Standards, Section B*, **65,** 23–60.

[32] Haight, F. A. (1966). Some statistical problems in connection with word association data. *Journal of Mathematical Psychology*, **3,** 217–233.

[33] Haight, F. A. and Breuer, M. A. (1960). The Borel-Tanner distribution, *Biometrika*, **47,** 143–150.

[34] Herdan, G. (1961). A critical examination of Simon's model of certain distribution functions in linguistics, *Applied Statistics*, **10,** 65–76.

[35] Irwin, J. O. (1963). Inverse factorial series as frequency distributions, *Proceedings of the International Symposium on Discrete Distributions, Montreal*, 159–174.

[36] Irwin, J. O. (1965). A unified derivation of some well-known frequency distributions, *Journal of the Royal Statistical Society, Series A*, **118,** 394–404.

[37] Ising, E. (1925). Beitrag zür Theorie des Ferromagnetismus. *Zeitschrift für Physik*, **31,** 253–258.

[38] Iyer, P. V. K. (1954). Some distributions arising in matching problems, *Journal of the Indian Society for Agricultural Statistics*, **6,** 5–29.

[39] Joseph, A. W. and Bizley, M. T. L. (1960). The two-pack matching problem, *Journal of the Royal Statistical Society, Series B*, **22,** 114–130.

[40] Kendall, M. G. (1961). Natural law in the social sciences, *Journal of the Royal Statistical Society, Series A*, **124,** 1–16.

[41] Mandelbrot, B. (1959). A note on a class of skew distribution functions. Analysis and critique of a paper by H. A. Simon, *Information and Control*, **2,** 90–99.

[42] Mandelbrot, B. (1961). On the theory of word frequencies and on related Markovian models of discourse. *Proceedings of Symposia on Applied Mathematics*, **1,** 190–219. (American Mathematical Society)

[43] Mandelbrot, B. (1965). Information Theory and Psycholinguistics, (Chapter 29 in *Scientific Psychology* (Ed. B. W. Wolman). Basic Books Publishing Co.)

[44] Marlow, W. H. (1965). Factorial distributions, *Annals of Mathematical Statistics*, **36,** 1066–1068.

[45] Mises, R. von (1921). Das problem der Iterationen, *Zeitschrift für angewandte Mathematik und Mechanik*, **1,** 298–307.

[46] Mood, A. M. (1940). The distribution theory of runs, *Annals of Mathematical Statistics*, **11,** 367–392.

[47] Moore, P. G. (1956). The geometric, logarithmic and discrete Pareto forms of series, *Journal of the Institute of Actuaries*, **82,** 130–136.

[48] Mosteller, F. (1941). Note on an application of runs to quality control charts, *Annals of Mathematical Statistics*, **12,** 228–232.

[49] Nicholson, W. L. (1961). Occupancy probability distribution critical points, *Biometrika*, **48,** 175–180.

[50] Nishida, T. (1962). On the multiple exponential channel queueing system with hyper-Poisson arrivals, *Journal of the Operations Research Society, Japan*, **5,** 57–66.

[51] Oliver, R. M. (1961). A traffic counting distribution, *Operations Research*, **9,** 802–810.

[52] Ord, J. K. (1968). The discrete Student's *t* distribution, *Annals of Mathematical Statistics*, **39,** 1513–1516.

[53] Owen, D. B. (1962). *Handbook of Statistical Tables*, Reading, Mass.: Addison-Wesley Publishing Co., Inc.

[54] Pierce, J. A. (1940). A study of a universe of *n* finite populations, with applications to moment-function adjustments for grouped data. *Annals of Mathematical Statistics*, **11,** 311–334.

[55] Prasad, A. (1957). A new discrete distribution, *Sankhyā*, **17,** 353–354.

[56] Richards, P. I. (1968). A generating function. (Problem 67–18), *SIAM Review*, **10,** 455–456.

[57] Rider, P. R. (1951). The distribution of the range in samples from a discrete rectangular population, *Journal of the American Statistical Association*, **46,** 375–378.

[58] Seal, H. L. (1947). A probability distribution of deaths at age *x* when policies are counted instead of lives, *Skandivisk Aktuarietidskrift*, **30,** 18–43.

[59] Seal, H. L. (1952). The maximum likelihood fitting of the discrete Pareto law, *Journal of the Institute of Actuaries*, **78,** 115–121.

[60] Shreider, Yu. A. (1967). On the possibility of a theoretical model of statistical laws for texts (towards a basis for Zipf's law). *Problemy Peredachi Informatsii*, **3,** 57–63. (In Russian.)

[61] Silva, G. (1941). Una generalizzazione del problema delle concordanze, *Instituto Veneto de Scienze, Lettere ed Arti, Venezia Classe de Scienze Matematiche e Naturali*, **100,** 689–709.

[62] Simon, H. A. (1954). On a class of skew distribution functions, *Biometrika*, **42,** 425–440.

[63] Simon, H. A. (1960). Some further notes on a class of skew distribution functions, *Information and Control*, **3,** 90–88.

[64] Singh, S. N. (1964). A probability model for couple fertility, *Sankhyā, Series B*, 89–94.

[65] Stevens, W. L. (1937). Significance of grouping, *Annals of Eugenics*, **8,** 57–69.

[66] Stevens, W. L. (1939). Distribution of groups in a sequence of alternatives, *Annals of Eugenics*, **9,** 10–17.

[67] Tanner, J. C. (1951). A derivation of the Borel distribution, *Biometrika*, **38,** 383–392.

[68] Tukey, J. W. (1949). Moments of random group size distributions, *Annals of Mathematical Statistics*, **20,** 523–539.

[69] Wald, A. and Wolfowitz, J. (1940). On a test whether two samples are from the same population, *Annals of Mathematical Statistics*, **11,** 147–162.

[70] Wallis, W. A. and Moore, G. H. (1941). *A Significance Test for Time Series*, Technical Paper No. 1, National Bureau of Economic Research, New York.

[71] Walther, A. (1926). Anschauliches zür Riemannschen Zetafunktion, *Acta Mathematica*, **48,** 393–400.

[72] Wishart, J. and Hirschfeld, H. O. (1936). A theorem concerning the distribution of joins between line segments, *Journal of the London Mathematical Society*, **11,** 227–235.

[73] Yule, G. U. (1923). A mathematical theory of evolution, based on the conclusions of Dr. J. C. Willis, F.R.S., *Philosophical Transactions of the Royal Society of London, Series B*, **213,** 21–87.

[74] Zacks, S. and Goldfarb, D. (1966). Survival probabilities in crossing a field containing absorption points, *Naval Research Logistics Quarterly*, **13,** 35–48.

[75] Zipf, G. K. (1949). *Human Behaviour and the Principle of Least Effort*, Reading, Mass.: Addison-Wesley Publishing Co., Inc.

11

Multivariate Discrete Distributions

1. Introduction

Most of the distributions discussed in this chapter are closely related to univariate distributions discussed in earlier Chapters. In particular, the 'marginal distributions' of individual variables are for the most part simple binomial, Poisson, negative binomial, hypergeometric or logarithmic series distributions, and in other cases, distributions obtained by modifying or compounding one of these. The appropriate chapters on univariate distributions and also Chapter 17 of Volume 2 should be used as references when reading the present chapter. Since we are using a great deal of prior material, it has been possible to condense the treatment, and include all the multivariate distributions in a single chapter. Some of the more important distributions (particularly the multinomial and negative multinomial) merit an entire chapter, but we preferred to present all the practically useful distributions together.

In passing from univariate to multivariate distributions, some essentially new features require attention. These are connected with relations among sets of variables and include correlation, regression and, generally, conditional distributions. These aspects are given special attention in this chapter. There is less emphasis on estimation techniques, not because estimation is unimportant but because there has been slower historical growth of estimation techniques for multivariate distributions. This may be ascribed to (a) the greater expense of obtaining multivariate data, and (b) the heavier calculations involved in applying techniques of estimation. The importance of these two factors will decrease, and there may be a rapid development of multivariate estimation techniques in the near future.

Each distribution (or family of distributions) has a section in this chapter.

2. Multinomial Distribution

2.1 *Genesis and Definition*

Consider a series of independent trials, in each of which just one of k mutually exclusive events $E_1, E_2, \ldots, E_k$ must be observed, and in which the probability of occurrence of event E_j is equal to p_j for each trial (with, of course, $\sum_{j=1}^{n} p_j = 1$). Then the joint distribution of the random variables $\mathbf{n}_1, \mathbf{n}_2, \ldots, \mathbf{n}_k$ representing the numbers of occurrences of the events $E_1, E_2, \ldots, E_k$ respectively, in N trials (with $\sum_{j=1}^{k} \mathbf{n}_j = N$) is defined by

$$(1) \qquad P(\mathbf{n}_1,\mathbf{n}_2,\ldots,\mathbf{n}_k) = N! \prod_{j=1}^{k} (p_j^{\mathbf{n}_j}/\mathbf{n}_j!) \qquad \left(0 \le \mathbf{n}_j; \sum_{j=1}^{k} \mathbf{n}_j = N\right).$$

This is often termed a *multinomial* distribution, with parameters $N, p_1, p_2, \ldots, p_k$.

The expression for $P(\mathbf{n}_1,\mathbf{n}_2,\ldots,\mathbf{n}_k)$ shows that it can be regarded as the coefficient of $\prod_{j=1}^{k} t_j^{\mathbf{n}_j}$ in the multinomial expansion of

$$(t_1p_1 + t_2p_2 + \cdots + t_kp_k)^N.$$

Note that if $k = 2$, this reduces to a binomial distribution (for either $\mathbf{n}_1$ or $\mathbf{n}_2$). Also the marginal distribution of any $\mathbf{n}_j$ is binomial with

$$(2) \qquad P(\mathbf{n}_j) = \binom{N}{\mathbf{n}_j} p_j^{\mathbf{n}_j}(1 - p_j)^{N-\mathbf{n}_j} \qquad (\mathbf{n}_j = 0,1,\ldots,N).$$

The multinomial is thus a natural extension of the binomial distribution. J. Bernoulli, who worked with the binomial distribution (see Chapter 3), also used the multinomial distribution.

2.2 *Properties*

The joint distribution of any subset $\mathbf{n}_{a_1}, \mathbf{n}_{a_2}, \ldots, \mathbf{n}_{a_s}$ of the $\mathbf{n}_j$'s is also multinomial (with an $(s + 1)$th variable equal to $N - \sum_{j=1}^{s} \mathbf{n}_{a_j}$). In fact

$$(3) \qquad P(\mathbf{n}_{a_1},\ldots,\mathbf{n}_{a_s}) = N!\left[\left(N - \sum_{j=1}^{s} \mathbf{n}_{a_j}\right)! \prod_{j=1}^{s} \mathbf{n}_{a_j}!\right]^{-1}$$

$$\times \left(1 - \sum_{j=1}^{s} p_{a_j}\right)^{N-\sum_1^s \mathbf{n}_{a_j}} \prod_{j=1}^{s} p_{a_j}^{\mathbf{n}_{a_j}}.$$

Formula (2) is a special case of (3), with $s = 1$.

The *conditional* joint distribution of $\mathbf{n}_{a_1}, \ldots, \mathbf{n}_{a_s}$, *given* the remaining $\mathbf{n}_j$'s is also multinomial. It depends on the values of the remaining $\mathbf{n}_j$'s only through their sum $\mathbf{S}$. In fact

$$P(\mathbf{n}_{a_1},\ldots,\mathbf{n}_{a_s} \mid \mathbf{S}) = (N - \mathbf{S})! \prod_{j=1}^{s} \left[\left(p_{a_j} \Big/ \sum_{j=1}^{s} p_{a_j} \right)^{\mathbf{n}_{a_j}} \Big/ \mathbf{n}_{a_j}! \right]. \tag{4}$$

Combining (3) and (4), it can be seen that the joint distribution of $\mathbf{n}_{a_1}, \ldots, \mathbf{n}_{a_s}$, given any *subset* of the remaining $\mathbf{n}_j$'s, is also multinomial, with

$$(N - \mathbf{S})! \prod_{j=1}^{s} \left[\left(p_{a_j} \Big/ \sum_{j=1}^{s} p_{a_j} \right)^{\mathbf{n}_{a_j}} \Big/ \mathbf{n}_{a_j}! \right].$$

(5)

$$P(\mathbf{n}_{a_1},\ldots,\mathbf{n}_{a_s} \mid \mathbf{n}_{b_1},\ldots,\mathbf{n}_{b_r}) = (N - \mathbf{S}_b)! \left[\left(N - \mathbf{S}_b - \sum_{j=1}^{s} \mathbf{n}_{a_j} \right)! \prod_{j=1}^{s} \mathbf{n}_{a_j}! \right]^{-1}$$
$$\cdot \left[1 - \sum_{j=1}^{s} p_{a_j} \Big/ \left(1 - \sum_{j=1}^{r} p_{b_j} \right) \right]^{N - \mathbf{S}_b - \sum_1^s \mathbf{n}_{a_j}}$$
$$\cdot \prod_{j=1}^{s} \left[p_{a_j} \Big/ \left(1 - \sum_{j=1}^{r} p_{b_j} \right) \right]^{\mathbf{n}_{a_j}}$$

where

$$\mathbf{S}_b = \sum_{j=1}^{r} \mathbf{n}_{b_j}.$$

The *regression* of $\mathbf{n}_i$ on $\mathbf{n}_j$ is linear:

$$E[\mathbf{n}_i \mid \mathbf{n}_j] = (N - \mathbf{n}_j) p_i (1 - p_j)^{-1}. \tag{6}$$

The *multiple regression* of $\mathbf{n}_i$ on $\mathbf{n}_{b_1}, \mathbf{n}_{b_2}, \ldots, \mathbf{n}_{b_r} (b_j \neq i)$ is also linear.

$$E[\mathbf{n}_i \mid \mathbf{n}_{b_1},\ldots,\mathbf{n}_{b_r}] = \left(N - \sum_{j=1}^{r} \mathbf{n}_{b_j} \right) p_i \left(1 - \sum_{j=1}^{r} p_{b_j} \right)^{-1}. \tag{7}$$

The *mode*(*s*) of the distribution is (are) located, broadly speaking, near the expected value point, i.e. $\mathbf{n}_1 \doteqdot Np_1$, $\mathbf{n}_2 \doteqdot Np_2, \ldots, \mathbf{n}_k \doteqdot Np_k$. More precisely, Moran has shown (see Feller [21]) that the values of $\mathbf{n}_1, \mathbf{n}_2, \ldots, \mathbf{n}_k$ satisfy the inequalities

$$Np_i - 1 < \mathbf{n}_i \leq (N + k - 1)p_i \qquad (i = 1,2,\ldots,k). \tag{8}$$

These inequalities (together with the condition $\sum_{j=1}^{k} \mathbf{n}_j = N$) restrict the possible modes to a relatively few points. If k is small, there may indeed be only one combination of values of $\mathbf{n}_1, \mathbf{n}_2, \ldots, \mathbf{n}_k$ satisfying these conditions.

Finucan [22] has devised a procedure which locates the possible modes more efficiently, and also indicates possible configurations of joint modes. He also showed that there is only one 'modal region' in the sense that there is no local maximum remote from the set of equiprobable points, each giving the maximum value of $P(\mathbf{n}_1,\mathbf{n}_2,\ldots,\mathbf{n}_k)$.

Olbrich [56] has suggested an interesting geometrical representation of the multinomial distribution. By the transformation $\mathbf{m}_j = \cos^{-1}(\mathbf{n}_j/N)^{\frac{1}{2}}$, points $(\mathbf{m}_1,\mathbf{m}_2,\ldots,\mathbf{m}_k)$ on a k dimensional sphere are obtained. The angle between the vectors joining two such independently obtained points (with $N = N_1, N_2$ respectively) to the origin has a variance approximately equal to $(N_1^{-1} + N_2^{-1})$. (Note that the transformation is approximately variance equalizing — see Chapter 1, Section 7.) Olbrich suggests this statistic might be used in testing for identity of two multinomial distributions.

We have already noted, in Chapter 3, equation (34), that sums of binomial distribution probabilities can be expressed in terms of incomplete beta function ratios. Olkin and Sobel [57] and Stoka [73] have generalized this to obtain expressions for sums of multinomial distribution probabilities as sums of multiple integrals. We give one example.

Writing

$$W(h_1,\ldots,h_{k-1};p_1,\ldots,p_{k-1}) = \sum_{\mathbf{n}_1=0}^{h_1}\cdots\sum_{\mathbf{n}_{k-1}=0}^{h_{k-1}} P(\mathbf{n}_1,\mathbf{n}_2,\ldots,\mathbf{n}_k) \quad \left(0 \le h_j, \sum_{j=1}^{k-1} h_j \le N\right),$$

we have

$$\begin{aligned} W(h_1,\ldots,h_{k-1};p_1,\ldots,p_{k-1}) &= 1 - \sum_{j=1}^{k-1} I_{p_j}(h_j - 1, N - h_j) \\ &\quad + \sum_{0\le j_1<j_2}^{k-1}\sum^{k-1} I_{p_{j_1},p_{j_2}}(h_{j_1} - 1, N - h_{j_1}; h_{j_2} - 1, N - h_{j_2}) \\ &\quad - \cdots\cdots\cdots\cdots \\ &\quad + (-1)^{k-1} I_{p_1,\ldots p_{k-1}}(h_1 - 1, N - h_1;\ldots;h_{k-1} - 1, N - h_{k-1}), \end{aligned}$$

where $I_{j_1,j_2\ldots j_l}(h_{j_1} - 1, N - h_{j_1};\ldots;h_{j_l} - 1, N - h_{j_l})$

$$= \frac{N!}{\left(N - l - \sum_{i=1}^{l} h_{j_i}\right)! \prod_{i=1}^{l} h_{j_i}!} \times \int_0^{p_{j_1}}\cdots\int_0^{p_{j_l}} \left(1 - \sum_{j=1}^{l} x_j\right)^{N-l-\sum_{i=1}^{l} h_{j_i}} \prod_{i=1}^{l} x_i^{h_{j_i}}\, dx_1\ldots dx_l.$$

Mallows [51] has established the interesting inequalities

$$\Pr\left[\bigcap_{j=1}^{k} (\mathbf{n}_j \le a_j)\right] \le \prod_{j=1}^{k} \Pr[\mathbf{n}_j \le a_j] \tag{9}$$

which are valid for any multinomial distribution, and any values of $a_1, a_2, \ldots, a_k$.

2.3 *Moments*

The *mixed factorial moment*

$$\mu_{(r_1,r_2,\ldots,r_k)} = E[\mathbf{n}_1^{(r_1)}\mathbf{n}_2^{(r_2)}\ldots\mathbf{n}_k^{(r_k)}]$$

is equal to

$$N^{\left(\sum_1^k r_j\right)} p_1^{r_1}p_2^{r_2}\ldots p_k^{r_k}. \tag{10}$$

From this general formula we find

$$\operatorname{cov}(\mathbf{n}_i,\mathbf{n}_j) = -Np_ip_j \tag{11.1}$$

whence the correlation between $\mathbf{n}_i$ and $\mathbf{n}_j$ is

$$\operatorname{corr}(\mathbf{n}_i,\mathbf{n}_j) = -\sqrt{\frac{p_ip_j}{q_iq_j}}. \tag{11.2}$$

(The moments of $\mathbf{n}_i$ are those of a binomial distribution with parameters N, p_i.)

The probability generating function (as has already been mentioned in Section 2.1) is

$$(t_1p_1 + t_2p_2 + \cdots + t_kp_k)^N.$$

The moment generating function is

$$(p_1e^{t_1} + p_2e^{t_2} + \cdots + p_ke^{t_k})^N.$$

It follows that if $(\mathbf{n}'_1,\mathbf{n}'_2,\ldots,\mathbf{n}'_k)$ have a joint multinomial distribution with parameters N', $p_1, p_2, \ldots p_k$ and $(\mathbf{n}''_1,\mathbf{n}''_2,\ldots,\mathbf{n}''_k)$ have a joint multinomial distribution with parameters N'', $p_1, p_2, \ldots, p_k$, and the two sets of variables are mutually independent of each other, then $(\mathbf{n}'_1 + \mathbf{n}''_1,\mathbf{n}'_2 + \mathbf{n}''_2,\ldots,\mathbf{n}'_k + \mathbf{n}''_k)$ have a joint multinomial distribution with parameters $(N' + N'')$, p_1, p_2, $\ldots, p_k$. This property may be described by saying that the multinomial distribution is *reproductive* with respect to N.

The *cumulant generating function* is

$$N\log\left(\sum_{j=1}^k p_je^{t_j}\right).$$

From this formula Guldberg's [32] recurrence relations can be derived:

$$\kappa_{r_1,r_2,\ldots,r_{i-1},r_i+1,r_{i+1},\ldots,r_k} = (p_i/p_k)\frac{\partial}{\partial(p_i/p_k)}[\kappa_{r_1,r_2,\ldots,r_k}].$$

Explicit formulas for several lower cumulants are given by Wishart [82]. They include (in an obvious notation).

$$\kappa_{..21} = -Np_ip_j(q_i - p_i);$$

$$\kappa_{.111.} = 2Np_ip_jp_k$$

$$\kappa_{..31} = -Np_ip_j(1 - 6p_iq_i)$$

$$\kappa_{..22} = -Np_ip_j\{(q_i - p_i)(q_j - p_j) + 2p_ip_j\}$$

$$\kappa_{..211.} = 2Np_ip_jp_k(q_i - 2p_i)$$

$$\kappa_{.1111.} = -6Np_ip_jp_kp_l.$$

2.4 *Approximations*

Fisz [23] has shown that if the set of random variables $\{A_j^{(N)}\mathbf{n}_j + B_j^{(N)}\}$, with appropriately chosen constants $A_j^{(N)}$, $B_j^{(N)}$ has a proper $(k - 1)$-dimensional limiting distribution, this must be the joint distribution of m independent Poisson variables and $(r - 1 - m)$ multinormally distributed variables, independent of the Poisson variables. (See also Rjauba [62].) Khatri and Mitra [44] have constructed approximations based on this result.

The routine computation of exact probabilities connected with the multinomial distribution is usually almost prohibitively difficult. However, there are useful approximate methods. The most commonly used is the 'X^2-approximation'. This is obtained in a natural way (see Pearson [59], Johnson and Tetley [41]) by considering the limiting form of $P(\mathbf{n}_1,\mathbf{n}_2,\ldots,\mathbf{n}_k)$ as N tends to infinity, $p_1, p_2, \ldots, p_k$ remaining constant. Since $\mathbf{n}_j$ is distributed binomially with parameters N, p_j, the ratio $(n_j - Np_j)(Np_jq_j)^{-\frac{1}{2}}$ will, with high probability, be of order 1 (see Chapter 3, Section 8.1). Retaining terms in the exponent of e up to and including $N^{-\frac{1}{2}}$, we have

(12)

$$P(\mathbf{n}_1,\mathbf{n}_2,\ldots,\mathbf{n}_k) \doteqdot (2\pi N)^{-\frac{1}{2}}\left(\prod_{j=1}^{k} p_j\right)^{-\frac{1}{2}} \exp\left[-\frac{1}{2}\sum_{j=1}^{k}\frac{(\mathbf{n}_j - Np_j)^2}{Np_j} - \frac{1}{2}\sum_{j=1}^{k}\frac{\mathbf{n}_j - Np_j}{Np_j} + \frac{1}{6}\sum_{j=1}^{k}\frac{(\mathbf{n}_j - Np_j)^3}{(Np_j)^2}\right].$$

Neglecting the terms in $N^{-\frac{1}{2}}$ we have

$$(13)\qquad P(\mathbf{n}_1,\mathbf{n}_2,\ldots,\mathbf{n}_k) \doteqdot (2\pi N)^{-\frac{1}{2}}\left(\prod_{j=1}^{k} p_j\right)^{-\frac{1}{2}} \exp\left[-\frac{1}{2}\sum_{j=1}^{k}\frac{(\mathbf{n}_j - Np_j)^2}{Np_j}\right].$$

The exponent $-\frac{1}{2}\sum_{j=1}^{k}\frac{(\mathbf{n}_j - Np_j)^2}{Np_j}$ can be written $-\frac{1}{2}\mathbf{X}^2$ where $\mathbf{X}^2$ is the standard quantity

$$\sum \frac{(\text{observed value} - \text{expected value})^2}{\text{expected value}}$$

often referred to as 'chi-squared'.

There have been a number of numerical investigations of the accuracy of approximation (13) (e.g. Argentiero et al. [5], Shanawany [64], Wise [80], [81], Lancaster [47], Bennett [11], Lancaster and Brown [48] and, from a rather more analytical standpoint, Studer [74]), but it is not easy to give a simple summary of the results. The accuracy increases as min $(Np_1, Np_2, \ldots, Np_k)$ increases and decreases with increasing k. We note that

$$(14)\qquad \begin{cases} E(\mathbf{X}^2) = k - 1 = E(\chi^2_{k-1}) \\ \operatorname{var}(\mathbf{X}^2) = 2(k-1)(1 - N^{-1}) + N^{-1}\left(\sum_{j=1}^{k} p_j^{-1} - k^2\right) \end{cases}$$

$$(\text{while } \operatorname{var}(\chi^2_{k-1}) = 2(k-1)).$$

Recently Rüst [63] obtained an explicit but rather complicated expression for the moment generating function of the random variable $\mathbf{X}^2$.

Wise [81] suggested replacing $\mathbf{X}^2$, in the exponent in (13) by

$$(15)\qquad \mathbf{X}'^2 = \sum_{j=1}^{k} \frac{(\mathbf{n}_j - Np_j)^2}{Np_j + \frac{1}{2}}.$$

Pearson [59] showed that $\mathbf{X}^2$ is approximately distributed as χ^2 with $(k - 1)$ degrees of freedom. The following improvement on this approximation has been suggested by Hoel [36]. He expanded the moment generating function of the multinomial and expressed its leading terms as linear functions of moment generating functions of χ^2 distributions. This leads to the addition of a corrective term $n^{-1}[R_1S_1 + R_2S_2]$ in the formula

$$(16)\qquad \Pr[\mathbf{X}^2 \leq c] \doteqdot \Pr[\chi^2_{k-1} \leq c] + N^{-1}(R_1S_1 + R_2S_2) \qquad \text{where}$$

(17)

$$S_1 = \frac{1}{8}\left[\sum_{i=1}^{k} p_i^{-1} - (k^2 + 2k - 2)\right];\ S_2 = \frac{1}{24}\left[S\sum_{i=1}^{k} p_i^{-1} - (3k^2 + 6k - 4)\right],$$

and (for k odd)

$$(18.1)\qquad R_1 = \frac{c^{\frac{1}{2}(k-1)}e^{-\frac{1}{2}c}}{2.4\ldots(k-1)(k+1)}[c - (k+1)]$$

$$(18.2)\qquad R_2 = \frac{c^{\frac{1}{2}(k-1)}e^{-\frac{1}{2}c}}{2.4\ldots(k+1)(k+3)}[c^2 - 2(k+3)c + (k+3)(k+1)],$$

while for k even

$$(19.1)\qquad R_1 = \sqrt{\frac{2}{\pi}}\cdot\frac{c^{\frac{1}{2}(k-1)}e^{-\frac{1}{2}c}}{1.3\ldots(k-1)(k+1)}[c - (k+1)]$$

(19.2)

$$R_2 = \sqrt{\frac{2}{\pi}}\cdot\frac{c^{\frac{1}{2}(k-1)}e^{-\frac{1}{2}c}}{1.3\ldots(k+1)(k+3)}[c^2 - 2(k+3)c + (k+3)(k+1)].$$

Vora [79] has obtained upper and lower bounds for $\Pr[\mathbf{X}^2 \leq c]$. These are not given explicitly here, as they are rather complicated, and also are not always very close. They are of form

$$\tau_2 \Pr[\chi_{k-1}'^2(\delta_2^2) \leq c_2] \leq \Pr[\mathbf{X}^2 \leq c] \leq \tau_1 \Pr[\chi_{k-1}'^2(\delta_1^2) \leq c_1] \tag{20}$$

where τ_1, τ_2, δ_1^2, δ_2^2, c_1, c_2 depend on c, N, $p_1, p_2, \ldots, p_k$ and $\chi'^2(\cdot)$ denotes a noncentral χ^2 variable (as in Chapter 28). Vora obtained bounds (in terms of *central* χ^2 distributions) for $\Pr[\mathbf{X}''^2 \leq c]$ with

$$\mathbf{X}''^2 = \sum_{j=1}^{k} (\mathbf{n}_j + \tfrac{1}{2} - Np_j)^2(Np_j)^{-1}.$$

It is important to keep in mind that there are *two* approximations to be considered: (*i*) the accuracy of (13) and (*ii*) the accuracy of the χ^2 distribution as an approximation to the distribution of $\mathbf{X}^2$.

Johnson [40] suggested approximating to the whole multinomial distribution by using a joint probability density function for the variables

$$\mathbf{y}_j = \mathbf{n}_j/N (j = 1,2,\ldots,k).$$

The function proposed was

$$f_{\mathbf{y}_1,\ldots,\mathbf{y}_k}(y_1,y_2,\ldots,y_k) = \Gamma(N-1) \cdot \prod_{j=1}^{k} \left[\frac{y_j^{(N-1)p_j-1}}{\Gamma(N-1)p_j}\right] \left(0 \leq y_j; \sum_{j=1}^{k} y_k = 1\right) \tag{21}$$

which is to be interpreted as giving the joint probability density function of any $(k-1)$ of the $\mathbf{y}_j$'s, if the remaining $\mathbf{y}_j$ is expressed in terms of the other $(k-1)$ $\mathbf{y}_j$'s.

This approximation gives the correct first and second order moments and product moments of the joint distribution of the $\mathbf{y}_j$'s, and therefore of the $\mathbf{n}_j$'s.

It is particularly useful to note the identity between this approximate distribution and the Dirichlet distribution (Chapter 42). This implies that the approximation is equivalent to taking

$$\mathbf{y}_j = \mathbf{v}_j \bigg/ \left(\sum_{i=1}^{k} \mathbf{v}_i\right)$$

where $\mathbf{v}_1, \mathbf{v}_2, \ldots, \mathbf{v}_k$ are independent random variables, and $\mathbf{v}_j$ is distributed as χ^2 with $2(N-1)p_j$ degrees of freedom $(j = 1,2,\ldots,k)$. In turn this indicates that the distributions of various functions of the $\mathbf{y}_j$'s would be approximated by those of corresponding functions of independent χ^2's. Johnson and Young [42] have used this result to obtain (in the case $p_1 = p_2 = \cdots = p_k = k^{-1}$) approximations to the distribution of $\max_j \mathbf{n}_j$ and $(\max_j \mathbf{n}_j)/(\min_j \mathbf{n}_j)$. (See also Young [84].)

Unpublished tables of the distributions of $\max \mathbf{n}_j$ and $\min \mathbf{n}_j$ for $5 \leq k \leq 50$, $10 \leq N \leq 200$ by Rappeport are reported by Mallows [51].

Tables of the mean and variance (to four decimal places) of $\max \mathbf{n}_j$ and $\min \mathbf{n}_j$ for the case

$$p_1 = p_2 = p_{k-1} = (k - 1 + a)^{-1}; \quad p_k = a(k - 1 + a)^{-1}$$

for $k = 2(1)10$; $N = 2(1)15$; $a = 1(1)5$, $a^{-1} = 2(1)5$ have been provided by Gupta and Nagel [33].

Young [85] proposed an approximation to the distribution of range

$$\mathbf{W} = \max \mathbf{n}_j - \min \mathbf{n}_j$$

according to which

$$\Pr[\mathbf{W} \leq w] \doteqdot \Pr[\mathbf{W}' \leq w(k/N)^{\frac{1}{2}}]$$

where $\mathbf{W}'$ is distributed as the range of k independent unit normal variables. Bennett and Nakamura [10] have given values of the largest integer N such that

$$\Pr[\mathbf{W} \geq w] \leq \alpha$$

for $\alpha = 0.01, 0.025, 0.05$; and (i) $k = 2(1)4$; $w = 4(1)15$:

(ii) $k = 5$; $w = 3(1)13$:

(iii) $k = 6, 7$; $w = 3(1)10$:

(iv) $k = 8(1)10$; $w = 3(1)9$.

For the mean deviation

$$\mathbf{M} = k^{-1} \sum_{j=1}^{k} |\mathbf{n}_j - Nk^{-1}|,$$

Young [86] proposed the approximation

$$\Pr[\mathbf{M} \leq m] \doteqdot \Pr[\mathbf{M}' \leq m(k/N)^{\frac{1}{2}}]$$

where $\mathbf{M}'$ is distributed as the mean deviation of k independent unit normal variates.

Some aspects of the joint distribution of the $\mathbf{n}_j$'s are discussed in Section 5 of Chapter 10.

2.5 *Estimation of Parameters*

We will consider first situations in which N and k are known, and it is necessary to estimate $p_1, p_2, \ldots, p_k$.

If $\mathbf{n}_1, \mathbf{n}_2, \ldots, \mathbf{n}_k$ represent an observed set of frequencies of $E_1, E_2, \ldots, E_k$ respectively, then the maximum likelihood estimators of $p_1, p_2, \ldots, p_k$ are the relative frequencies

$$\hat{\mathbf{p}}_j = \mathbf{n}_j/N \qquad (j = 1,2,\ldots,k). \tag{22}$$

The variances and covariances of these estimators can be written down immediately from (9) and (11).

In 1964, Quesenberry and Hurst [60] gave the following formula for the end-points of simultaneous confidence intervals for $p_1, p_2, \ldots, p_k$, the joint confidence coefficient being approximately $1 - \alpha$:

(23) Limits for p_i are:

$$\frac{\chi^2_{k-1,1-\alpha} + 2n_i \pm [\chi^2_{k-1,1-\alpha}\{\chi^2_{k-1,1-\alpha} + 4n_iN^{-1}(N - n_i)\}]^{\frac{1}{2}}}{2(N + \chi^2_{k-1,1-\alpha})}.$$

In 1965, Goodman [30] improved on these results (for $k > 2$), replacing the upper $100\alpha\%$ point of χ^2_{k-1}, which is $\chi^2_{k-1,1-\alpha}$, by $\chi^2_{1,1-\alpha/k}$. (Note that $\chi^2_{1,1-\alpha/k}$ can be found from tables of percentage points of the unit normal distribution.) Goodman also gave a system of intervals for the differences $\{p_i - p_j\}$. The limits for $p_i - p_j$ are

$$(24)\quad (n_i - n_j)N^{-1} \pm [\chi^2_{1,1-2\alpha k^{-1}(k-1)^{-1}}\{(n_i + n_j)N^{-1} - (n_i - n_j)^2N^{-2}\}N^{-1}]^{\frac{1}{2}}.$$

A number of methods of *sequential estimation* of parameters of the multinomial distribution have been developed. Bhat and Kulkarni [12], applying Wolfowitz' [83] general result, have shown that for any unbiased estimator,

$$\mathbf{T}(\mathbf{n}_1,\mathbf{n}_2,\ldots,\mathbf{n}_k)$$

of a differentiable function $g(p_1,p_2,\ldots,p_k)$

$$(25)\qquad \operatorname{var}(\mathbf{T}) \geq [E[\mathbf{N}]]^{-1}\left[\sum_{j=1}^{k} p_j\left(\frac{\partial g}{\partial p_j}\right)^2 - \left(\sum_{j=1}^{k} p_j\frac{\partial g}{\partial p_j}\right)^2\right].$$

(The inequality becomes an equality if and only if $\mathbf{T}$ is a linear function of the variables $\mathbf{n}_jp_j^{-1} - \mathbf{n}_{j'}p_{j'}^{-1}$.) In (25), $E[\mathbf{N}]$ denotes the expected value of the 'number of trials', $\mathbf{N}$.

There is an infinite number of possible systems of sequential sampling. Among these are *inverse multinomial sampling*, in which sampling is terminated as soon as one or other of a specified subset of 'classes' has occurred k times (see Section 3, on the negative multinomial distribution).

In the special case of a *trinomial distribution* (i.e., multinomial with $k = 3$), a sequential method of estimation of p_1, p_2 and p_3 has been developed by Muhamedhanova and Suleimanova [54]. The method is a natural generalization of that of Girshick et al. [26] for the binomial distribution. It is based on the enumeration of the paths followed by the point $(\mathbf{n}_1,\mathbf{n}_2,\mathbf{n}_3)$, as N increases, which conclude at an observed termination point. (The termination points are points on the boundary of the continuation region, and this boundary defines the sampling procedure.) If the termination point be $(\mathbf{n}_1,\mathbf{n}_2,\mathbf{n}_3)$ and $K(\mathbf{n}_1,\mathbf{n}_2,\mathbf{n}_3)$ be the number of possible paths from (0,0,0) ending at $(\mathbf{n}_1,\mathbf{n}_2,\mathbf{n}_3)$, while $K_i(\mathbf{n}_1,\mathbf{n}_2,\mathbf{n}_3)$ is the number of such paths starting from (1,0,0)(i = 1), (0,1,0)(i = 2), or (0,0,1)(i = 3), then

$$K_i(\mathbf{n}_1,\mathbf{n}_2,\mathbf{n}_3)/K(\mathbf{n}_1,\mathbf{n}_2,\mathbf{n}_3) \tag{26}$$

is an unbiased estimator of p_i.

The method can evidently be extended to values of k greater than 3, but evaluation of K_i and K rapidly increases in complexity.

Lewontin and Prout [49] describe a more unusual situation, in which it is known that $p_j = k^{-1}(j = 1,\ldots,k)$ but k is unknown, and must be estimated. The maximum likelihood estimator, $\hat{\mathbf{k}}$, of k satisfies the equation

$$N\hat{\mathbf{k}}^{-1} = \sum_{j=\hat{\mathbf{k}}-\mathbf{K}+1}^{\hat{\mathbf{k}}} j^{-1} \tag{27}$$

where $\mathbf{K}$ is the number of different 'classes' represented in a random sample of size N. Equation (27) is approximately equivalent to

$$N\hat{\mathbf{k}}^{-1} \doteqdot \log\,[\hat{\mathbf{k}}/(\hat{\mathbf{k}} - \mathbf{K} + 1)].$$

For large values of N

$$\text{var}(\hat{\mathbf{k}}) \doteqdot k[e^{N/k} - (1 + N/k)]^{-1}. \tag{28}$$

2.6 *Applications of the Multinomial Distribution*

This distribution is employed in so many diverse fields of statistical analysis that an exhaustive catalogue would be very lengthy. It is used in the same circumstances as those in which a binomial distribution might be used, when there are multiple categories of events instead of a simple dichotomy.

An important field of application is in the kinetic theory of classical physics. 'Particles' (e.g. molecules) are considered to occupy cells in phase space. That is, each particle is assigned to a 'cell' in a six-dimensional space (three dimensions for position, three for velocity). The 'cells' are formed by a fine subdivision of the space. Each allocation of N particles among the K cells available constitutes a 'microstate'. If the particles are not distinguishable each configuration corresponds to a 'macrostate'. The *thermodynamic probability* of a macrostate is proportional to the number of ways it can be realized by different 'microstates'. Thus for the macrostate consisting of n_j particles in state j $(j = 1,\ldots,k)$ the probability is

$$\frac{N!}{\prod_1^k n_j!}\Bigg/\left(\sum_{n_1}\cdots\sum_{n_k}\frac{N!}{\prod_1^k n_j!}\right) = \frac{N!}{\prod_{j=1}^k n_j!}\cdot k^{-N}. \tag{29}$$

This corresponds to "Maxwell-Boltzmann" statistical thermodynamics.

The multinomial distribution may be applicable whenever data obtained by random sampling are grouped in a finite number of groups. Provided successive observations are independent (e.g. effects of finite sample size are negligible) conditions for a multinomial distribution are satisfied. Therefore, one might apply the multinomial distribution to many problems requiring estimation of a population distribution, since we are, in effect, estimating the 'cell' probabilities of such a distribution. (In fact approximations, such as the use of continuous distributions, are often introduced.)

A situation where multinomial distributions are commonly used, is the analysis of contingency tables (e.g. two-, three- (or M-) way tables representing the joint incidence of two-, three- (or M-) factors each at a number of different levels).

2.7 *'Truncated' Multinomial Distributions*

Suppose that $\mathbf{n}_{11}, \mathbf{n}_{12}, \ldots, \mathbf{n}_{1k}$ have a joint multinomial distribution with parameters $N_1, p_1, p_2, \ldots, p_k$. Suppose further that $\mathbf{n}_{21}, \mathbf{n}_{22}, \ldots, \mathbf{n}_{2k'} (k' < k)$ have a joint multinomial distribution, independent of $\mathbf{n}_{11}, \mathbf{n}_{12}, \ldots, \mathbf{n}_{1k}$, with parameters

$$N_2, p_j\left(\sum_{j=1}^{k'} p_j\right)^{-1} \qquad (j = 1,2,\ldots,k').$$

This corresponds to a situation in which the data represented by $\mathbf{n}_{11}, \mathbf{n}_{12}, \ldots, \mathbf{n}_{1k}$ are supplemented by a further 'incomplete' experiment in which cells $(k'+1), (k'+2), \ldots, k$ are not observed.

Asano [6] has shown that the maximum likelihood estimators of the p_i's are

$$\text{(30)} \qquad \begin{cases} \hat{\mathbf{p}}_i = \mathbf{n}_{1i}/N_1 & \text{if } i > k' \\ \hat{\mathbf{p}}_i = \dfrac{\mathbf{n}_{1i} + \mathbf{n}_{2i}}{N_1\left[1 + N_2\left(\sum_{j=1}^{k'} \mathbf{n}_{1j}\right)^{-1}\right]} & \text{if } i \leq k'. \end{cases}$$

These formulas apply to the 'simple truncated' sampling described above. They are special cases of more general results established by Batschelet [9] and Geppert [25], corresponding to 'compound truncation', in which there are R sets $(\mathbf{n}_{ra_{r_1}}, \mathbf{n}_{ra_{r_2}}, \ldots, \mathbf{n}_{ra_{rk_r}})$ with cells $a_{r_1}, a_{r_2}, \ldots, a_{rk_r}$, only, observed in the rth set $(r = 1, \ldots, R)$, and $\sum_{j=1}^{k_r} \mathbf{n}_{ra_{rj}} = N_r$.

Asano [6] has shown that for the p_i's to be estimable, a necessary and sufficient condition is that (a) it appears in at least one set $\{a_{rj}\}$ and (b) every set $\{a_{rj}\}$ has at least one member in common with at least one other set

$$\{a_{r'j}\} \qquad (r \neq r').$$

3. Negative Multinomial (Multivariate Negative Binomial)

3.1 *Definition*

Recall that the binomial distribution has the probability generating function $(q + pt)^n$ with $q + p = 1$, ($p > 0$ and $n > 0$), but the negative binomial distribution has the probability generating function $(Q - Pt)^{-N}$, with $Q - P = 1$ ($P > 0$ and $N > 0$). Also, while it is possible to define *two* random variables ($\mathbf{x}$ and $(n - \mathbf{x})$) in the binomial case, there is only a single random variable for the negative binomial.

Coming to consider the construction of a k-variate distribution analogous to the (univariate) negative binomial distribution it is, therefore, natural to expect the k-variate distribution to be related to the multinomial distribution with $(k + 1)$ (and not k) variables. Since the probability generating function of this last distribution is $\left(\sum_{i=1}^{k+1} p_i t_i\right)^N \left(\text{with } \sum_{j=1}^{k} p_i = 1\right)$, we (by analogy) define a k-variate *negative multinomial distribution* to correspond to the probability generating function

$$\left(Q - \sum_{i=1}^{k} P_i t_i\right)^{-N} \tag{31}$$

(with $P_i > 0$ for all i, $N > 0$, and $Q - \sum_{i=1}^{k} P_i = 1$).

From (31) we obtain the distribution function

$$P(\mathbf{n}_1,\mathbf{n}_2,\ldots,\mathbf{n}_k) = \frac{\Gamma\left(N + \sum_{i=1}^{k} \mathbf{n}_i\right)}{\left(\prod_{i=1}^{k} \mathbf{n}_i!\right)\Gamma(N)} Q^{-N} \prod_{i=1}^{k} (P_i/Q)^{\mathbf{n}_i} \qquad (\mathbf{n}_j \geq 0). \tag{32}$$

This is termed a *negative multinomial* distribution with parameters N, P_1, $P_2, \ldots, P_k$ (since $Q = 1 + \sum_{j=1}^{k} P_j$ it need not be mentioned, though sometimes it also is included).

There is clearly some similarity in formal expression between the negative multinomial and multinomial distributions (as there is, also, between the negative binomial and binomial distributions). This is, perhaps, more easily seen if (32) is expressed in terms of new parameters $\theta_0 = Q^{-1}$; $\theta_j = -P_j/Q$ ($j = 1,2,\ldots,k$)

$$P(\mathbf{n}_1,\mathbf{n}_2,\ldots,\mathbf{n}_k) = (-N)^{(\Sigma \mathbf{n}_j)} \theta_0^{-N-\Sigma \mathbf{n}_j} \prod_{j=1}^{k} (\theta_j^{\mathbf{n}_j}/\mathbf{n}_j!). \tag{32'}$$

However, it should be noted that N can be fractional for the negative multinomial distribution.

3.2 *Genesis and Historical Remarks*

The negative binomial distribution is a compound Poisson distribution. In fact (see Section 4, Chapter 5) the distribution

$$\text{Poisson}(\boldsymbol{\theta}) \underset{\boldsymbol{\theta}}{\wedge} \text{Gamma}(\alpha,\beta) \tag{33}$$

is a negative binomial distribution. If

$$p_{\boldsymbol{\theta}}(\theta) = [\Gamma(\alpha)]^{-1}\beta^{\alpha}\theta^{\alpha-1}e^{-\beta\theta} \qquad (0 \leq \theta; \alpha > 0, \beta > 0)$$

then the probability generating function of (33) is

$$[\Gamma(\alpha)]^{-1}\beta^{\alpha}\int_0^{\infty} \theta^{\alpha-1}\exp[-\theta(\beta + 1 - t)]\,d\theta = \beta^{\alpha}(\beta + 1 - t)^{-\alpha}$$
$$= (1 + \beta^{-1} - \beta^{-1}t)^{-\alpha}.$$

(Note that we use β^{-1} in place of β here.)

Suppose that we now consider the joint distribution of k random variables $\mathbf{n}_1$, $\mathbf{n}_2, \ldots, \mathbf{n}_k$ which correspond to the same value of $\boldsymbol{\theta}$. For a specific value of $\boldsymbol{\theta}$, the joint probability generating function is given by the following expression

$$\prod_{j=1}^{k} \exp[-\theta(1 - t_i)] = \exp\left[-\theta\left(k - \sum_{i=1}^{k} t_i\right)\right].$$

The joint probability generating function of $\mathbf{n}_1, \mathbf{n}_2, \ldots, \mathbf{n}_k$ over the distribution of $\boldsymbol{\theta}$ is

$$[\Gamma(\alpha)]^{-1}\beta^{\alpha}\int_0^{\infty} \theta^{\alpha-1}\exp\left[-\theta\left(\beta + k - \sum_{i=1}^{k} t_i\right)\right]d\theta \tag{34}$$
$$= \left(1 + k\beta^{-1} - \beta^{-1}\sum_{i=1}^{k} t_i\right)^{-\alpha}$$

which corresponds to a special form of the negative multinomial distribution with $P_1 = P_2 = \cdots = P_k$. If the values of $\boldsymbol{\theta}$ for successive $\mathbf{n}$'s are in known ratios, the more general form (32) is obtained.

This model has been applied (see, in particular, Bates and Neyman [8]) to represent the joint distribution of numbers of accidents suffered by the same individual in k separate periods. There are evidently many other physical situations where this kind of model is appropriate (Sibuya et al. [65] give seven examples).

Neyman [55] has given a lucid account of the historical development of the negative multinomial distribution, with particular reference to its application in models representing accident proneness, but also describing applications to spatial distribution of galaxies, military aviation, and the theory of epidemics.

3.3 *Properties*

The distribution is sometimes called the *multivariate negative binomial distribution;* the name *negative multinomial* seems to be more accurately descriptive, however. The former name refers particularly to the fact that the marginal distributions (i.e. distributions of $\mathbf{n}_1, \mathbf{n}_2, \ldots, \mathbf{n}_k$ separately) are each negative binomial.

The negative multinomial has a number of properties, relating to conditional distributions, which are similar to properties of the multinomial distribution. Putting $t_j = 1$ (for $j \neq a_1, a_2, \ldots, a_s$) in (31) we obtain, for the joint probability generating function of $\mathbf{n}_{a_1}, \mathbf{n}_{a_2}, \ldots, \mathbf{n}_{a_s}$

$$(35) \qquad \left(Q - \sum_{j \neq a_i} P_j - \sum_{i=1}^{s} P_{a_i} t_{a_i}\right)^{-N}.$$

The joint distribution of $\mathbf{n}_{a_1}, \mathbf{n}_{a_2}, \ldots, \mathbf{n}_{a_s}$ is thus negative multinomial with parameters $N, P_{a_1}, \ldots, P_{a_s}$. The distribution is defined by

$$(36) \qquad P(\mathbf{n}_{a_1}, \ldots, \mathbf{n}_{a_s}) = \frac{\Gamma\left(N + \sum_{i=1}^{s} \mathbf{n}_{a_i}\right)}{\left(\prod_{i=1}^{s} \mathbf{n}_{a_i}!\right) \Gamma(N)} Q'^{-N} \prod_{i=1}^{s} (P_{a_i}/Q')^{\mathbf{n}_{a_i}}$$

with $Q' = Q - \sum_{j \neq a_i} P_j = 1 + \sum_{i=1}^{s} P_{a_i}$.

Dividing (32) by (36), we obtain the conditional distribution of

$$\{\mathbf{n}_j\} (j \neq a_1, \ldots, a_s) \text{ given } \mathbf{n}_{a_1}, \ldots, \mathbf{n}_{a_s}:$$

(37)

$$P(\{\mathbf{n}_j\} (j \neq a_1, \ldots, a_s) \mid \mathbf{n}_{a_1}, \ldots, \mathbf{n}_{a_s})$$

$$= \frac{\Gamma\left(N + \sum_{i=1}^{s} \mathbf{n}_{a_i} + \sum_{j \neq a_i} \mathbf{n}_j\right)}{\Gamma\left(N + \sum_{i=1}^{s} \mathbf{n}_{a_i}\right) \prod_{j \neq a_i} \mathbf{n}_j!} (Q/Q')^{-\left(N + \sum_{i=1}^{s} \mathbf{n}_{a_i}\right)} \prod_{j \neq a_i} [(P_j/Q')(Q/Q')^{-1}]^{\mathbf{n}_j}.$$

The conditional distribution is a negative multinomial with parameters

$$\left(N + \sum_{i=1}^{s} \mathbf{n}_{a_i}\right), (P_j/Q') \qquad (j = 1, 2, \ldots, k \text{ excluding } a_1, a_2, \ldots, a_s).$$

Combining (35) and (37) we see that the conditional joint distribution of any subset of the $\mathbf{n}_j$'s, given any other (distinct) subset, is also a negative multinomial distribution.

Khatri and Mitra [44] have obtained expressions for sums of negative multinomial probabilities similar to those described for multinomial probabilities in the last part of Section 2.2.

3.4 *Moments and Cumulants*

The moments of individual **n**'s are those of negative binomial variables with parameters N, P_j. The joint factorial moment

$$\mu_{(r_1,r_2,\ldots r_k)} = E\left[\prod_{j=1}^{k} \mathbf{n}_j^{(r_j)}\right]$$

is equal to

$$N^{\left[\sum_1^k r_j\right]} \prod_{j=1}^{k} P_j^{r_j}. \tag{38}*$$

From (38) it follows that the correlation between $\mathbf{n}_i$ and $\mathbf{n}_j$ ($i \neq j$) is

$$\rho_{ij} = \sqrt{P_iP_j(1 + P_i)^{-1}(1 + P_j)^{-1}}. \tag{39}$$

Note that the correlation is positive, while for the multinomial distribution it is negative.

Also, from the results in Section 2.2, the multiple regression of $\mathbf{n}_j$ on $\mathbf{n}_{a_1}$, $\mathbf{n}_{a_2}, \ldots, \mathbf{n}_{a_s}(j \neq a_1,a_2 \ldots, a_s)$ is

$$E[\mathbf{n}_j \mid n_{a_1},\ldots,n_{a_s}] = \left(N + \sum_{i=1}^{s} n_{a_i}\right)(P_j/Q'). \tag{40}$$

In particular the regression $\mathbf{n}_j$ on $\mathbf{n}_i$ ($i \neq j$) is

$$E[\mathbf{n}_j \mid n_i] = (N + n_i)P_j(1 + P_i)^{-1}. \tag{41}$$

The regressions are all linear. Note that the (common) regression coefficient

$$P_j/Q' = P_j\left(1 + \sum_{i=1}^{s} P_{a_i}\right)^{-1}$$

decreases as the number (s) of given values $n_{a_1}, \ldots, n_{a_s}$ increases.

The moment generating function of the negative multinomial distribution (32) is

$$\left(Q - \sum_{i=1}^{k} Pe^{t_i}\right)^{-N}, \tag{42}$$

and the cumulant generating function is

$$-N\log\left(Q - \sum_{i=1}^{k} P_ie^{t_i}\right).$$

The factorial moment generating function is

$$\left(1 - \sum_{i=1}^{k} P_it_i\right)^{-N}, \tag{43}$$

*$N^{[\;]}$ denotes an *ascending* factorial.

and the factorial cumulant generating function is

$$-N \log\left(1 - \sum_{i=1}^{k} P_i t_i\right).$$

Wishart [82] gives formulas for several mixed (ordinary) cumulants including (in an obvious notation)

$$\kappa_{...21.} = NP_iP_j(1 + 2P_i) \qquad \kappa_{..31.} = NP_iP_j[1 + 6P_i(1 + P_i)]$$
$$\kappa_{.111.} = 2NP_iP_jP_k \qquad \kappa_{..22.} = NP_iP_j[(1 + 2P_i)(1 + 2P_j) + 2P_iP_j]$$
$$\kappa_{.211.} = 2NP_iP_jP_k(1 + 3P_i)$$
$$\kappa_{.1111.} = 6NP_iP_jP_kP_l.$$

These values can be obtained from Guldberg's [32] formula

$$\kappa_{r_1,r_2,\ldots,r_i+1,\ldots,r_k} = P_i \frac{\partial}{\partial P_i} \kappa_{r_1,r_2,\ldots,r_i,\ldots,r_k}.$$

3.5 *Estimation*

Given a set of observed values of $\mathbf{n}_1, \mathbf{n}_2, \ldots, \mathbf{n}_k$ the equations formed by differentiating the likelihood function (32) with respect to P_j, and to N respectively, and then equating to zero (and replacing N, P_j by $\hat{\mathbf{N}}$, $\hat{\mathbf{P}}_j$)

$$\mathbf{n}_j/\hat{\mathbf{P}}_j = \left(\hat{\mathbf{N}} + \sum_{j=1}^{k} \mathbf{n}_j\right) \Big/ \left(1 + \sum_{j=1}^{k} \hat{\mathbf{P}}_j\right) \tag{44}$$

$$\sum_{j=0}^{\mathbf{S}-1} (\hat{\mathbf{N}} + j)^{-1} = \log\left(1 + \sum_{j=1}^{k} \hat{\mathbf{P}}_j\right) \tag{45}$$

where
$$\mathbf{S} = \sum_{j=1}^{k} \mathbf{n}_j.$$

From (44) it is clear that

$$\hat{\mathbf{P}}_1 : \hat{\mathbf{P}}_2 : \ldots : \hat{\mathbf{P}}_k = \mathbf{n}_1 : \mathbf{n}_2 : \ldots : \mathbf{n}_k.$$

Putting $\mathbf{c} = \mathbf{n}_j/\hat{\mathbf{P}}_j$ we find from (44) $\mathbf{c} = \hat{\mathbf{N}}$, where (45) can be rewritten

$$\sum_{j=0}^{\mathbf{S}-1} (\hat{\mathbf{N}} + j)^{-1} = \log(1 + \hat{\mathbf{N}}^{-1}\mathbf{S}). \tag{46}$$

It is impossible to solve this equation with a positive value for $\hat{\mathbf{N}}$ (the left hand side is always greater than the right hand side). The reason can be appreciated on studying the comments on the method of maximum likelihood applied to the (univariate) negative binomial distribution. It was pointed out there (Chapter 5, Section 8, Method (3)) that if $\mathbf{s}^2 \leq \bar{\mathbf{x}}$ the equation for $\hat{\mathbf{N}}$ might not be soluble. Here we are, essentially, attempting to estimate the value of N, for a negative binomial distribution with parameters N, $\sum_{i=1}^{k} P_i$, using only *one* observed value ($\mathbf{S}$), so that necessarily $\mathbf{s}^2 = 0$.

In order to obtain proper estimators, more than one set of observed values of $\mathbf{n}_1, \mathbf{n}_2, \ldots, \mathbf{n}_k$ must be available. If m sets

$$\{\mathbf{n}_{1t}, \mathbf{n}_{2t}, \ldots, \mathbf{n}_{kt}\} \qquad (t = 1,2, \ldots m)$$

are available, equations for maximum likelihood estimators are

$$\mathbf{n}_j/\hat{\mathbf{P}}_j = \left(m\hat{\mathbf{N}} + \sum_{t=1}^{m} \mathbf{S}_t\right) \Big/ \left(1 + \sum_{j=1}^{k} \hat{\mathbf{P}}_j\right) \tag{47}$$

$$\sum_{t=1}^{m} \sum_{j=0}^{\mathbf{S}_t - 1} (\hat{\mathbf{N}} + j)^{-1} = m \log\left(1 + \sum_{j=1}^{k} \hat{\mathbf{P}}_j\right) \tag{48}$$

where
$$\mathbf{S}_t = \sum_{j=1}^{k} \mathbf{n}_{jt}; \quad \mathbf{n}_j = \sum_{t=1}^{m} \mathbf{n}_{jt}.$$

Arguments similar to those used in analyzing the consequences of (44) and (45) lead to the formulas

$$\hat{\mathbf{P}}_j = \mathbf{n}_j/(m\hat{\mathbf{N}}) \tag{49}$$

$$\sum_{j=1}^{\infty} (\hat{\mathbf{N}} + j - 1)^{-1}\mathbf{F}_j = \log\left(1 + \frac{\mathbf{S}_t}{m\hat{\mathbf{N}}}\right) \tag{50}$$

where $\mathbf{F}_j$ = proportion of $\mathbf{S}_t$'s which are greater than or equal to j.

For a discussion of the existence of a solution (in $\mathbf{N}$) of (50) see Section 8, Chapter 5. Compare (50) with equation (31.1) of Chapter 5.

4. Multivariate Poisson Distributions

If, in (1), $p_1, p_2, \ldots p_{k-1}$ are assumed to tend to zero (and so $p_k \to 1$) and N to tend to infinity in such a way that $Np_j = \theta_j (j = 1,2 \ldots k - 1)$ then the probability of obtaining the set of values $n_1, \ldots, n_{k-1}$ tends to

$$\lim_{\substack{N\to\infty \\ \{p_j\}\to 0}} P(n_1,n_2,\ldots,n_{k-1}) = \prod_{j=1}^{k-1} [e^{-\theta_j}\theta_j^{n_j}/n_j!] \tag{51}$$

This is simply the joint distribution of $(k - 1)$ independent variables each having a Poisson distribution. As such, it does not require any special analysis. However, the name of *multiple Poisson* has been given to such distributions (see e.g. Patil and Bildikar [58]) and formulas for the moments have been given by Banerjee [7] and also Sibuya et al. [65].

The distribution can also be derived as a limiting form of the negative multinomial.

A non-trivial form of *bivariate Poisson distribution* has been constructed by Holgate [37] (see also Campbell [14]). It is the joint distribution of

$$\mathbf{x}_1 = \mathbf{u} + \mathbf{v}_1 \quad \text{and} \quad \mathbf{x}_2 = \mathbf{u} + \mathbf{v}_2$$

where $\mathbf{u}$, $\mathbf{v}_1$ and $\mathbf{v}_2$ are independent Poisson variables with expected values ξ, θ_1 and θ_2 respectively. For this distribution $\Pr[\mathbf{x}_1 = x_1, \mathbf{x}_2 = x_2]$ is

(52)
$$P(x_1,x_2) = e^{-(\xi+\theta_1+\theta_2)} \sum_{j=0}^{\min(x_1,x_2)} \frac{\xi^j}{j!} \frac{\theta_1^{x_1-j}}{(x_1-j)!} \frac{\theta_2^{x_2-j}}{(x_2-j)!}.$$

Clearly the marginal distributions of $\mathbf{x}_1$ and $\mathbf{x}_2$ are Poisson with expected values $\xi + \theta_1$, $\xi + \theta_2$ respectively. In the computation of numerical values of $P(x_1,x_2)$ the following recurrence relations (Teicher [78]) are useful.

(53)
$$x_1P(x_1,x_2) = \theta_1P(x_1-1,x_2) + \xi P(x_1-1,x_2-1)$$
$$x_2P(x_1,x_2) = \theta_2P(x_1,x_2-1) + \xi P(x_1-1,x_2-1)$$
(If either x_1 or x_2 is negative $P(x_1,x_2) = 0$).

The moments of the joint distribution are most easily evaluated by expressing $\mathbf{x}_1$, $\mathbf{x}_2$ in terms of the variables $\mathbf{u}$, $\mathbf{v}_1$ and $\mathbf{v}_2$. Thus, the moment generating function is

(54)
$$E[\exp(t_1\mathbf{x}_1 + t_2\mathbf{x}_2)] = E[\exp\{(t_1+t_2)\mathbf{u} + t_1\mathbf{v}_1 + t_2\mathbf{v}_2\}]$$
$$= \exp[-\xi(1 - e^{t_1+t_2}) - \theta_1(1 - e^{t_1}) - \theta_2(1 - e^{t_2})]$$

and the cumulant generating function is

(55)
$$\xi(e^{t_1+t_2} - 1) + \theta_1(e^{t_1} - 1) + \theta_2(e^{t_2} - 1).$$

The covariance between $\mathbf{x}_1$ and $\mathbf{x}_2$ is

(56)
$$\operatorname{cov}(\mathbf{x}_1,\mathbf{x}_2) = \operatorname{cov}(\mathbf{u}+\mathbf{v}_1,\mathbf{u}+\mathbf{v}_2) = \operatorname{var}(\mathbf{u}) = \xi.$$

Hence the correlation between $\mathbf{x}_1$ and $\mathbf{x}_2$ is

(57)
$$\operatorname{corr}(\mathbf{x}_1,\mathbf{x}_2) = \xi(\xi+\theta_1)^{-\frac{1}{2}}(\xi+\theta_2)^{-\frac{1}{2}}.$$

This cannot exceed $\xi\{\xi + \min(\theta_1,\theta_2)\}^{-1}$, or, as Holgate [37] points out $\min[(\xi+\theta_1)^{\frac{1}{2}}/(\xi+\theta_2)^{\frac{1}{2}}, (\xi+\theta_2)^{\frac{1}{2}}/(\xi+\theta_1)^{\frac{1}{2}}]$, i.e. the correlation cannot exceed the square root of the ratio of the smaller to the larger of the means of the two marginal distributions. This is a limitation on the applicability of the joint distribution. It also provides a check on whether there is a possibility that the model ($\mathbf{x}_j = \mathbf{u} + \mathbf{v}_j$, $j = 1,2$) is not justified.

The conditional distribution of $\mathbf{x}_1$, given $\mathbf{x}_2$, is

$$(58)\qquad P(x_1 \mid x_2) = P(x_1,x_2)\cdot[e^{-(\xi+\theta_2)}(\xi+\theta_2)^{x_2}/x_2!]^{-1}$$
$$= e^{-\theta_1}\sum_{j=0}^{\min(x_1,x_2)}\binom{x_2}{j}\left(\frac{\xi}{\xi+\theta_2}\right)^{j}\left(\frac{\theta_2}{\xi+\theta_2}\right)^{x_2-j}\frac{\theta_1^{x_1-j}}{(x_1-j)!}.$$

This is the distribution of the sum of two independent variables, one distributed as $\mathbf{v}_1$ and the other having a binomial distribution with parameters x_2, $\xi/(\xi+\theta_2)$ (distribution of $\mathbf{u}$, given $\mathbf{x}_2$). Hence

$$(59)\qquad E[\mathbf{x}_1 \mid x_2] = \theta_1 + x_2\xi/(\xi+\theta_2)$$

$$(60)\qquad \operatorname{var}(\mathbf{x}_1 \mid x_2) = \theta_1 + x_2\xi\theta_2/(\xi+\theta_2)^2.$$

Holgate [37] considered methods of estimating ξ, θ_1 and θ_2, given n independent pairs of observed values $(x_{1i},x_{2i})(i = 1,2,\ldots,n)$. From (52)

$$(61)\qquad \frac{\partial P(x_1,x_2)}{\partial \xi} = P(x_1-1,x_2-1) - P(x_1,x_2)$$

$$(62)\qquad \frac{\partial P(x_1,x_2)}{\partial \theta_1} = P(x_1-1,x_2) - P(x_1,x_2)$$

$$(63)\qquad \frac{\partial P(x_1,x_2)}{\partial \theta_2} = P(x_1,x_2-1) - P(x_1,x_2).$$

The equations satisfied by maximum likelihood estimators $\hat{\xi}$, $\hat{\theta}_1$, $\hat{\theta}_2$ of ξ, θ_1, θ_2 respectively are

$$\sum_{i=1}^{n}\frac{1}{P(\mathbf{x}_{1i},\mathbf{x}_{2i})}\frac{\partial P(\mathbf{x}_{1i},\mathbf{x}_{2i})}{\partial \phi} = 0 \qquad (\phi = \hat{\xi},\hat{\theta}_1,\hat{\theta}_2).$$

Using (61), (62), (63) and also (53), these equations can be put in the form

$$(64)\qquad \bar{\mathbf{x}}_t - \hat{\xi}\bar{R} = \hat{\theta}_t \qquad (t = 1,2)$$

$$(65)\qquad \hat{\theta}_1^{-1}(\bar{\mathbf{x}}_1 - \hat{\xi}\bar{R}) + \hat{\theta}_2^{-1}(\bar{\mathbf{x}}_2 - \hat{\xi}\bar{R}) = 1 + \bar{R}$$

where

$$\bar{\mathbf{x}}_t = n^{-1}\sum_{i=1}^{n}\mathbf{x}_{ti} \qquad (t = 1,2)$$

$$\bar{\mathbf{R}} = n^{-1}\sum_{i=1}^{n}\frac{P(\mathbf{x}_{1i}-1,\mathbf{y}_{1i}-1)}{P(\mathbf{x}_{1i},\mathbf{y}_{1i})}.$$

From (64) and (65) it follows that

$$\hat{\xi} + \hat{\theta}_t = \bar{\mathbf{x}}_t \qquad (t = 1,2)$$

and

$$\bar{\mathbf{R}} = 1.$$

The second equation is a polynomial in $\hat{\xi}$ which must be solved numerically

Evidently, $\text{var}(\hat{\boldsymbol{\xi}} + \hat{\boldsymbol{\theta}}_t) = (\xi + \theta_t)n_t^{-1}$

and
$$\begin{aligned}\text{corr}(\hat{\boldsymbol{\xi}} + \hat{\boldsymbol{\theta}}_1, \hat{\boldsymbol{\xi}} + \hat{\boldsymbol{\theta}}_2) &= \text{corr}(\mathbf{x}_{1i}, \mathbf{x}_{2i}) \\ &= \xi(\xi + \theta_1)^{-\frac{1}{2}}(\xi + \theta_2)^{-\frac{1}{2}}.\end{aligned}$$

Also (Holgate [37])

$$\text{var}(\hat{\boldsymbol{\xi}}) \doteqdot \frac{\xi^2(\theta_1 + \theta_2 - 1) + (\theta_1 - \xi)(\theta_2 - \xi)}{[(\theta_1 + \theta_2)\xi + \theta_1\theta_2](Q - 1) - (\theta_1 + \theta_2)} \tag{66}$$

where
$$Q = \sum_{\mathbf{x}_1, \mathbf{x}_2 = 1}^{\infty} \frac{[P(\mathbf{x}_1 - 1, \mathbf{x}_2 - 1)]^2}{P(\mathbf{x}_1, \mathbf{x}_2)}.$$

Alternatively ξ may be estimated by the sample covariance (see (56)). The variance of this estimator is approximately

$$[\xi(1 + \xi) + (\xi + \theta_1)(\xi + \theta_2)]n^{-1}. \tag{67}$$

It is uncorrelated with $\hat{\boldsymbol{\theta}}_1$ and $\hat{\boldsymbol{\theta}}_2$. The efficiency of this method of estimating ξ approaches 100% as ξ tends to zero, but decreases as ξ/θ_t increases.

The bivariate Poisson distribution (termed the bivariate Poisson *correlation* function) as described above was introduced in 1934 by Campbell [14] and further discussed in 1935 by Aitken and Gonin [2] and in 1936 by Aitken [1]. Campbell derived the distribution as a limiting case of the bivariate binomial distribution (see Section 9 below). The method of definition we have used gives, in our opinion, a clearer conception of the nature of the distribution, though the earlier approaches provide possible fields of application for it, at least as an approximation.

It is straightforward to generalize the definition to the joint distribution of k variables $\mathbf{x}_1, \mathbf{x}_2, \ldots, \mathbf{x}_k$, putting

$$\mathbf{x}_j = \mathbf{u} + \mathbf{v}_j \qquad (j = 1, \ldots, k)$$

with $\mathbf{u}, \mathbf{v}_1, \ldots, \mathbf{v}_k$ independent Poisson variables with expected values ξ, $\theta_1, \ldots, \theta_k$ respectively. More elaborate structures can easily be constructed, still retaining the expression of each $\mathbf{x}_j$ as the sum of a number of independent Poisson variables. Teicher [78] and Dwass and Teicher [19] have shown that the probability generating functions of such distributions are of the form

$$\exp\left[\sum_{j=1}^{k} A_j t_j + \sum\sum_{i<j} A_{ij} t_i t_j + \cdots + A_{12\ldots k} t_1 t_2 \ldots t_k - A\right] \tag{68}$$

where
$$\sum_{j=1}^{k} A_j + \sum\sum_{i<j} A_{ij} + \cdots + A_{12\ldots k} = A.$$

5. Multivariate Hypergeometric Distribution

Consider a population of M individuals, of which M_1 are of type 1, M_2 of type 2, ..., M_k of type k, with $\sum_{j=1}^{k} M_j = M$. Suppose a sample of size N is chosen, *without replacement*, from among these M individuals. Then the joint distribution of the random variables $\mathbf{n}_1, \mathbf{n}_2, \ldots, \mathbf{n}_k$ representing the numbers of individuals of types $1, 2, \ldots, k$ respectively in the sample, is defined by

$$P(\mathbf{n}_1,\mathbf{n}_2,\ldots,\mathbf{n}_k) = \left[\prod_{j=1}^{k} \binom{M_j}{\mathbf{n}_j}\right] \Big/ \binom{M}{N} \tag{69}$$

with
$$\sum_{j=1}^{k} \mathbf{n}_j = N;\ 0 \leq \mathbf{n}_j \leq M_j \qquad (j = 1,2,\ldots,k).$$

This distribution is called the *multivariate hypergeometric distribution* with parameters $N, M_1, M_2, \ldots, M_k$. When $k = 2$ it reduces to the ordinary hypergeometric distribution (note that there are really only $(k - 1)$ distinct variables, since $\mathbf{n}_k = N - \sum_{j=1}^{k-1} \mathbf{n}_j$).

A number of relationships satisfied by the multinomial distribution also hold, with appropriate modifications in detail, for the multivariate hypergeometric distribution. This is not unexpected, in view of the similarity of genesis of the two distributions. In fact as M tends to infinity with $M_j/M = p_j$ constant the distribution (69) tends to the multinomial distribution with parameters $N, p_1, p_2, \ldots, p_k$ (see (1)).

The marginal distribution of $\mathbf{n}_j$ is hypergeometric with parameters N, M_j, M, i.e. $P(\mathbf{n}_j) = \binom{M_j}{\mathbf{n}_j}\binom{M - M_j}{N - \mathbf{n}_j} \Big/ \binom{M}{N}$. The joint distribution of

$$\mathbf{n}_{a_1}, \ldots, \mathbf{n}_{a_s}, N - \sum_{i=1}^{s} \mathbf{n}_{a_i} \quad (s \leq k - 1)$$

is multivariate hypergeometric with parameters $N, M_{a_1}, \ldots, M_{a_s}, M - \sum_{j=1}^{s} M_{a_j}$. The conditional joint distribution of the remaining $\mathbf{n}$'s, given $\mathbf{n}_{a_1}, \ldots, \mathbf{n}_{a_s}$ depends only on $\mathbf{S} = \sum_{j=1}^{s} \mathbf{n}_{aj}$ and is also multivariate hypergeometric, with parameters $N - \mathbf{S}, M_1, M_2, \ldots, M_s$.

The joint factorial moment

$$\mu_{(r_1,r_2,\ldots,r_k)} = E\left[\prod_{j=1}^{k} \mathbf{n}_j^{(r_j)}\right]$$

is equal to

$$[N^{\left(\sum_1^k r_j\right)}/M^{\left(\sum_1^k r_j\right)}] \prod_{j=1}^{k} M_j^{(r_j)}. \tag{70}$$

In particular the correlation between $\mathbf{n}_i$ and $\mathbf{n}_j$ is

$$\operatorname{corr}(\mathbf{n}_i,\mathbf{n}_j) = -\sqrt{M_i M_j (M - M_i)^{-1}(M - M_j)^{-1}}. \tag{71}$$

The similarity with (11) is apparent.

Steyn [66] has given a unified treatment of a number of distributions which have probability generating functions which can be expressed as multivariate hypogeometric series

$$F(a;b_1,b_2,\ldots,b_k;c;t_1,t_2,\ldots,t_k) = \sum_{j_1=0}^{\infty}\cdots\sum_{j_k=0}^{\infty}\frac{a^{[\Sigma j_\ell]}}{c^{[\Sigma j_\ell]}}\prod_{\ell=1}^{k}\left[\frac{b_{j_\ell}^{[j_\ell]}t^{j_\ell}}{j_\ell!}\right]$$

where $\sum j_\ell$ denotes $\sum_{\ell=1}^{k} j_\ell$. By giving a, c and the b's special values the multivariate hypergeometric, negative multinomial and negative multivariate hypergeometric (see Section 8.1) distributions can be obtained (Steyn [67]).

6. Multivariate Logarithmic Series Distribution

It will be remembered (Chapter 7) that the univariate logarithmic series distribution can be derived as a limiting form of the truncated (by omission of zero class) negative binomial distribution. The limit is taken as the value of the parameter N tends to zero. A similar limiting operation can be performed on the multivariate negative binomial (negative multinomial) distribution. From (32), the distribution truncated by omission of $n_1 = n_2 = \ldots n_k = 0$ is

$$P^*(\mathbf{n}_1,\mathbf{n}_2,\ldots,\mathbf{n}_k) = \frac{\Gamma\left(N+\sum_1^k \mathbf{n}_j\right)}{\left(\prod_{j=1}^{k}\mathbf{n}_j!\right)\Gamma(N)\cdot(1-Q^{-N})}Q^{-N}\prod_{j=1}^{k}(P_i/Q)^{\mathbf{n}_j} \tag{72}$$

As N tends to zero

$$\lim_{N\to0}\frac{\Gamma\left(N+\sum_1^k \mathbf{n}_j\right)}{\Gamma(N)\cdot(1-Q^{-N})} = \lim_{N\to0}\frac{N(N+1)\cdots\left(N+\sum_1^k \mathbf{n}_j - 1\right)}{1-Q^{-N}}$$

$$= \left(\sum_{j=1}^{k}\mathbf{n}_j - 1\right)!\lim_{N\to0} N(1-Q^{-N})^{-1}$$

$$= \frac{\left(\sum_{j=1}^{k}\mathbf{n}_j - 1\right)!}{\log Q}.$$

Hence

$$\lim_{N\to0} P^*(\mathbf{n}_1,\mathbf{n}_2,\ldots\mathbf{n}_k) = \frac{\left(\sum_{j=1}^{k}\mathbf{n}_j - 1\right)!}{\left(\prod_{j=1}^{k}\mathbf{n}_j!\right)\log Q}\prod_{j=1}^{k}(P_i/Q)^{\mathbf{n}_j}. \tag{73}$$

Putting $P_i/Q = \theta_i$ $(0 < \sum_{i=1}^{k}\theta_i < 1)$ and noting that $Q = (1 - \sum_{i=1}^{k}\theta_i)^{-1}$, the right hand side of equation (73) can be rearranged to give a formula defining the *multivariate logarithmic series distribution with parameters* $\theta_1, \theta_2, \ldots, \theta_k$:

$$(74) \qquad P(\mathbf{n}_1,\mathbf{n}_2,\ldots,\mathbf{n}_k) = \frac{\left(\sum_{j=1}^{k} \mathbf{n}_j - 1\right)!}{\left(\prod_{j=1}^{k} \mathbf{n}_j!\right)\left[-\log\left(1 - \sum_{i=1}^{k} \theta_i\right)\right]} \prod_{j=1}^{k} \theta_j^{\mathbf{n}_j}$$

$$\left(\mathbf{n}_j \geq 0; \sum_{i=1}^{k} \mathbf{n}_i > 0\right).$$

This distribution has been studied by Patil and Bildikar [58]. They applied the bivariate logarithmic series distribution to the observed joint distribution of numbers of males and females in different occupations in a city. Use of the multivariate logarithmic series distribution has also been considered by Chatfield et al. [16]. They came to the conclusion that the multivariate negative binomial has some general advantages (particularly in flexibility) though the multivariate logarithmic series distribution can be valuable when marginal distributions are very skew.

Summing (74), keeping $\sum \mathbf{n}_i = N > 0$, we have

$$\sum \prod_{j=1}^{n} (\theta_j^{\mathbf{n}_j}/\mathbf{n}_j!) = \left(\sum_{j=1}^{k} \theta_j\right)^N \Big/ N!$$

and it follows that $\sum_{j=1}^{k} \mathbf{n}_j$ has a logarithmic series distribution with parameter $\sum_{1}^{k} \theta_j$.

One point of special importance is to note that the marginal distributions of (74) are *not* logarithmic series distributions. This is evident, since the value $\mathbf{n}_1 = 0$ is not excluded (all that is excluded is the *combination* of values $\mathbf{n}_1 = \mathbf{n}_2 = \cdots = \mathbf{n}_k = 0$).

In fact, the joint probability generating function is (see (31))

$$(75) \qquad \lim_{N\to 0} \frac{\left(Q - \sum_{i=1}^{k} P_i t_i\right)^{-N} - Q^{-N}}{1 - Q^{-N}} = \frac{-\log\left[1 - \sum_{i=1}^{k} P_i t_i/Q\right]}{\log Q}$$

$$= \frac{\log\left(1 - \sum_{i=1}^{k} \theta_i t_i\right)}{\log\left(1 - \sum_{i=1}^{k} \theta_i\right)}.$$

Putting $t_2 = t_3 = \cdots = t_k = 1$ we obtain the probability generating function of $\mathbf{n}_1$:

$$(76) \qquad \frac{\log\left(1 - \sum_{i=2}^{k} \theta_i - \theta_1 t_1\right)}{\log\left(1 - \sum_{i=1}^{k} \theta_i\right)}$$

whence

$$\Pr[\mathbf{n}_1 = 0] = \frac{\log\left(1 - \sum_{i=2}^{k} \theta_i\right)}{\log\left(1 - \sum_{i=1}^{k} \theta_i\right)} \tag{77}$$

$$\Pr[\mathbf{n}_1 = r] = \frac{r^{-1}}{\left[-\log\left(1 - \sum_{i=1}^{k} \theta_i\right)\right]}\left[\frac{\theta_1}{1 - \sum_{i=2}^{k} \theta_i}\right]^r \qquad (r \geq 1). \tag{78}$$

This will be recognized as a *modified* logarithmic series distribution (in the sense of Chapter 8, Section 4). (Note that (78) can be written

$$\Pr[\mathbf{n}_1 = r] = \left\{1 - \frac{\log\left(1 - \sum_{i=2}^{k} \theta_i\right)}{\log\left(1 - \sum_{i=1}^{k} \theta_i\right)}\right\} \frac{r^{-1}}{\left[-\log\left(1 - \dfrac{\theta_1}{1 - \sum_{i=2}^{k} \theta_i}\right)\right]} \times \left[\frac{\theta_1}{1 - \sum_{i=2}^{k} \theta_i}\right]^r.$$

Similarly the marginal distribution of any $\mathbf{n}_j$ is a modified logarithmic series distribution. Also the joint distribution of any subset $\mathbf{n}_{a_1}, \mathbf{n}_{a_2}, \cdots \mathbf{n}_{a_s} (1 < s < k)$ is a modified multivariate logarithmic series distribution with joint probability generating function

(79)

$$\frac{\log\left(1 - \sum_{j \neq a_i} \theta_j - \sum_{i=1}^{s} \theta_{a_i} t_{a_i}\right)}{\log\left(1 - \sum_{j=1}^{k} \theta_j\right)}$$

$$= \frac{\log\left(1 - \sum_{j \neq a_i} \theta_j\right) + \left[\log 1 - \left(\sum_{i=1}^{s} \theta_{a_i} t_{a_i}\right)\left(1 - \sum_{j \neq a_i} \theta_j\right)^{-1}\right]}{\log\left(1 - \sum_{j=1}^{k} \theta_j\right)}$$

$$= \frac{\log\left(1 - \sum_{j \neq a_i} \theta_j\right)}{\log\left(1 - \sum_{j=1}^{k} \theta_j\right)} + \frac{\log\left[1 - \left(\sum_{i=1}^{s} \dfrac{\theta_{a_i}}{1 - \sum_{j=1}^{k} \theta_j} t_{a_i}\right)\right]}{\log\left[1 - \sum_{i=1}^{s} \dfrac{\theta_{a_i}}{1 - \sum_{j=1}^{k} \theta_j}\right]}$$

$$\times\left[1-\frac{\log\left(1-\sum_{j\neq a_i}\theta_j\right)}{\log\left(1-\sum_{j=1}^{k}\theta_j\right)}\right].$$

The parameters of the unmodified multivariate logarithmic series distribution are $\theta_{a_i}(1-\sum_{j\neq a_i}\theta_j)^{-1}$. The modification consists of assigning a probability of $[\log(1-\sum_{j\neq a_i}\theta_j)][\log(1-\sum_1^k\theta_j)]^{-1}$ to the event $\mathbf{n}_{a_1}=\mathbf{n}_{a_2}=\cdots=\mathbf{n}_{a_s}=0$.

As is to be expected from the nature of the joint distribution of $\mathbf{n}_{a_1}$, $\mathbf{n}_{a_2}$, $\ldots$, $\mathbf{n}_{a_s}$ the conditional distribution of the remaining $\mathbf{n}_j$'s, given $n_{a_1}\ldots, n_{a_s}$ takes a different form when $n_{a_1}=n_{a_2}=\cdots=n_{a_s}=0$. If this event occurs then the conditional distribution of the remaining $\mathbf{n}$'s is a multivariate logarithmic series distribution with parameters $\{\theta_j\}$, $(j\neq a_1,a_2,\ldots,a_s)$. In all other cases (i.e. when $\sum_{i=1}^{s}n_{a_i}>0$) the conditional distribution is not of logarithmic series type at all, but is a multivariate *negative binomial* distribution with parameters $\sum_{i=1}^{s}n_{a_s}$, $\{\theta_j\}$, $(j\neq a_1,a_2,\ldots,a_s)$. Chatfield et al. [16] seem to regard this as implying some kind of lack of homogeneity in the multivariate logarithmic series distribution. While there may not always be good reason to expect that all conditional distributions should be of the same form as the overall distribution, it is undoubtedly true that it is easier to visualize a distribution with conditional distributions of common form. It makes theoretical analysis tidier, and may correspond to practical requirements if the different variables represent similar kinds of physical quantities. From these points of view, the multivariate logarithmic series distribution is at a disadvantage compared with the distributions considered in Sections 2–5.

Remembering that the mode of the univariate logarithmic series distribution is at 1, it is, perhaps, not surprising that the mode of the multivariate logarithmic series distribution is at

$$n_j=1,\ n_i=0(i\neq j)\quad\text{where}\quad\theta_j=\max(\theta_1,\theta_2,\ldots,\theta_k). \tag{80}$$

If two or more θ's are equal maxima, then there are a corresponding number of equal modal values.

From (74) we calculate the factorial moment

$$\mu_{(r_1,r_2,\ldots,r_k)}=E\left[\prod_{j=1}^{k}\mathbf{n}_j^{(r_j)}\right] \tag{81}$$

$$=\left\{\frac{\left(\sum_{j=1}^{k}r_j-1\right)!}{-\log\left(1-\sum_{j=1}^{k}\theta_j\right)}+E\left[\left(\sum_{j=1}^{k}\mathbf{n}_j\right)^{\left[\sum_1^k r_j\right]}\right]\right\}\prod_{j=1}^{k}\theta_j^{r_j}$$

(the first term in { } on the right hand side is necessary because the summation formally includes $\sum_{j=1}^{k} n_j = 0$). Since $\sum_{j=1}^{k} \mathbf{n}_j$ has a logarithmic series distribution with parameter $\sum_{j=1}^{k} \theta_j$ we have

(82)
$$\mu_{(r_1,r_2,\ldots,r_k)} = \frac{\left(\sum_{j=1}^{k} r_j - 1\right)! \prod_{j=1}^{k} \theta_j^{r_j}}{\left[-\log\left(1 - \sum_{j=1}^{k} \theta_j\right)\right]\left(1 - \sum_{j=1}^{k} \theta_j\right)^{\sum_1^k r_j}} \qquad (r_j\text{'s not all zero}).$$

Hence

(83)
$$E[\mathbf{n}_j] = \frac{\theta_j}{-\left(1 - \sum_{i=1}^{k} \theta_i\right) \log\left(1 - \sum_{i=1}^{k} \theta_i\right)}$$

$$\operatorname{var}(\mathbf{n}_j) = -\frac{\theta_j}{\left(1 - \sum_{i=1}^{k} \theta_i\right) \log\left(1 - \sum_{i=1}^{k} \theta_i\right)} \times \left[\frac{\theta_j}{1 - \sum_{i=1}^{k} \theta_i}\left\{1 - \frac{1}{-\log\left(1 - \sum_{i=1}^{k} \theta_i\right)}\right\}^{-1}\right]$$

$$\operatorname{cov}(\mathbf{n}_i,\mathbf{n}_j) = -\frac{\theta_i\theta_j}{\left(1 - \sum_{i=1}^{k} \theta_i\right)^2\left[-\log\left(1 - \sum_{i=1}^{k} \theta_1\right)\right]}\left[\frac{1}{-\log\left(1 - \sum_{i=1}^{k} \theta_i\right)} - 1\right].$$

7. Bivariate Neyman Type A Distributions

Holgate [38] has described three ways in which bivariate distributions with Neyman Type A marginal distributions can be constructed. The first two are compound distributions — a bivariate Poisson (as described in Section 4) compounded by a (univariate) Poisson distribution, and vice-versa. Thus there are

Type I: Bivariate Poisson $(\xi',\theta_1',\theta_2') \bigwedge_{\xi'/\theta_1,\theta_1'/\theta_1,\theta_2'/\theta_2}$ Poisson (λ)

and *Type II:* $\left.\begin{array}{l}\text{Poisson } (\phi_1') \\ \text{Poisson } (\phi_2')\end{array}\right\} \bigwedge_{\phi_1'/\theta_1,\phi_2'/\theta_2}$ Bivariate Poisson $(\lambda,\lambda_1,\lambda_2)$.

For *Type I*, we have

(84) $$E[\mathbf{x}_j] = \lambda(\xi + \theta_j) \qquad (j = 1,2)$$

(85) $$\operatorname{var}(\mathbf{x}_j) = \lambda(\xi + \theta_j) \qquad (j = 1,2)$$

(86) $$\operatorname{cov}(\mathbf{x}_1,\mathbf{x}_2) = \lambda\xi$$

whence

(87) $$\operatorname{corr}(\mathbf{x}_1,\mathbf{x}_2) = \xi[(\xi + \theta_1)(\xi + \theta_2)]^{-\frac{1}{2}}$$

(as for the original bivariate Poisson distribution — see (57)).

The probability generating function is

$$\exp\,[\lambda \exp(\theta_1 t_1 + \theta_2 t_2 + \xi t_1 t_2 - \theta_1 - \theta_2 - \xi)].$$

Holgate [38] obtained an expression for individual probabilities, but says that for computational purposes it is better to use the formula

(88) $$P(x_1,x_2) = e^{-\lambda} \sum_{j=0}^{\infty} (\lambda^j/j!) P(x_1,x_2 \mid \lambda\xi,\lambda\theta_1,\lambda\theta_2),$$

where $P(x_1,x_2 \mid \lambda\xi,\lambda\theta_1,\lambda\theta_2)$ is the right hand side of (52) with ξ, θ_1, θ_2 replaced by $\lambda\xi$, θ_1, θ_2 respectively.

For *Type II*, the variable $\mathbf{x}_j$ is distributed as Neyman Type A with parameters $\lambda + \lambda_j$, ϕ_j and

(89) $$E[\mathbf{x}_j] = (\lambda + \lambda_j)\phi_j \qquad (j = 1,2)$$

(90) $$\operatorname{var}(\mathbf{x}_j) = (\lambda + \lambda_j)\phi_j(\phi_j + 1) \qquad (j = 1,2)$$

(91) $$\operatorname{cov}(\mathbf{x}_1,\mathbf{x}_2) = \lambda\phi_1\phi_2,$$

whence

(92) $$\operatorname{corr}(\mathbf{x}_1,\mathbf{x}_2) = \lambda\sqrt{\phi_1\phi_2[(\phi_1 + 1)(\phi_2 + 1)(\lambda + \lambda_1)(\lambda + \lambda_2)]^{-1}}$$

(If $\phi_1 = \phi_2 = \phi$, the correlation is $[\phi/(\phi + 1)]\ \lambda(\lambda + \lambda_1)^{-\frac{1}{2}}(\lambda + \lambda_2)^{-\frac{1}{2}}$.)

The probability generating function is

(93) $$\exp\,[\lambda_1 e^{\phi_1(t_1-1)} + \lambda_2 e^{\phi_2(t_2-1)} + \lambda e^{\phi_1(t_1-1)+\phi_2(t_2-1)} - \lambda_1 - \lambda_2 - \lambda].$$

Holgate [38] obtained the recurrence relation

(94) $$(x_1 + 1)P(x_1 + 1,x_2) = \lambda\phi_1 e^{-\phi_1-\phi_2} \sum_{i=0}^{x_1} \sum_{j=0}^{x_2} \frac{\phi_1^i\phi_2^j}{i!\,j!} P(x_1 - i,x_2 - j)$$
$$+ \lambda_1\phi_1 e^{-\phi_1} \sum_{i=0}^{x_1} \frac{\phi_1^i}{i!} P(x_1 - i,x_2).$$

The third type (*Type III*) of bivariate Neyman Type A proposed by Holgate is directly analogous to the bivariate Poisson. The two variables $\mathbf{x}_1$ and $\mathbf{x}_2$ are formed as sums $\mathbf{x}_1 = \mathbf{u} + \mathbf{v}_1$, $\mathbf{x}_2 = \mathbf{u} + \mathbf{v}_2$ where $\mathbf{u}$, $\mathbf{v}_1$ and $\mathbf{v}_2$ are mutually independent random variables having Neyman Type A distributions with parameters (λ,ϕ), (λ_1,ϕ), (λ_2,ϕ) respectively. (Note that the value of ϕ is the same for all three of the distributions.) Then $\mathbf{x}_j$ also has a Neyman Type A distribution with parameters $(\lambda + \lambda_j, \phi)$, and

$$E[\mathbf{x}_j] = (\lambda + \lambda_j)\phi \tag{95}$$

$$\text{var}(\mathbf{x}_j) = (\lambda + \lambda_j)\phi(1 + \phi) \tag{96}$$

$$\text{cov}(\mathbf{x}_1, \mathbf{x}_2) = \lambda\phi(1 + \phi) \tag{97}$$

whence

$$\text{corr}(\mathbf{x}_1, \mathbf{x}_2) = \lambda[(\lambda + \lambda_1)(\lambda + \lambda_2)]^{-\frac{1}{2}} \tag{98}$$

(Note that the correlation does not depend on ϕ).

The probability generating function is

$$\exp\,(\lambda_1 e^{\phi(t_1-1)} + \lambda_2 e^{\phi(t_2-1)} + \lambda e^{\phi(t_1 t_2-1)} - \lambda_1 - \lambda_2 - \lambda). \tag{99}$$

For this distribution Holgate obtains the simple (separable) recurrence relation given below

$$(x_1 + 1)P(x_1 + 1, x_2) = \lambda_1 \phi e^{-\phi} \sum_{i=0}^{x_1} (\phi^i/i!)P(x_1 - i, x_2) + \lambda\phi e^{-\phi} \sum_{i=0}^{x_1-1} (\phi^i/i!)P(x_1 - i, x_2 - 1 - i) \tag{100}$$

(if $x_2 - 1 - i < 0$, $P(x_1 - i, x_2 - 1 - i) = 0$),

but he suggests that the direct calculation from Type A probabilities may be equally simple.

The only methods so far available for fitting each of the three types of distributions are based on the use of the five sample moments of first and second order.

8. Compound Multinomial and Negative Multinomial Distributions

8.1 *Dirichlet (β–) Compound Multinomial Distribution*

Compound multivariate distributions are formed, in the same way as compound univariate distributions, by assigning distributions to some (or all) of the parameters of a multivariate distribution. As noted in Chapter 8, there are an unlimited number of possible combinations of initial and compounding distributions. We will describe only two which have attracted serious attention. (It may be noted that the negative multinomial itself can be regarded as a compound distribution (see Section (3.2.).

The first can be described symbolically as

$$\text{Multinomial}\,(N, \mathbf{p}_1, \mathbf{p}_2, \ldots, \mathbf{p}_k) \underset{\mathbf{p}_1, \ldots, \mathbf{p}_k}{\wedge} \text{Dirichlet}\,(\alpha_1, \alpha_2, \ldots, \alpha_k). \tag{101}$$

The distribution is defined by

$$(102)\qquad P(\mathbf{n}_1,\mathbf{n}_2,\ldots,\mathbf{n}_k) = \frac{N!}{\prod_{j=1}^{k} \mathbf{n}_j!} E\left[\prod_{j=1}^{k} \mathbf{p}_j^{n_j}\right]$$

$$= \frac{N!\Gamma\left(\sum_{j=1}^{k}\alpha_j\right)}{\Gamma\left(N+\sum_{j=1}^{k}\alpha_j\right)} \prod_{j=1}^{k} \frac{\Gamma(\mathbf{n}_j+\alpha_j)}{\Gamma(\alpha_j)}$$

$$= \frac{N!}{\left(\sum_{j=1}^{k}\alpha_j\right)^{[N]}} \prod_{j=1}^{k} \frac{\alpha_j^{[\mathbf{n}_j]}}{\mathbf{n}_j!} \qquad \left(\mathbf{n}_j \geq 0, \sum_{j=1}^{k} \mathbf{n}_j = N\right).$$

This simple formula is a natural generalization of the Binomial $(N,\mathbf{p}) \underset{\mathbf{p}}{\wedge}$ Beta distribution studied in Chapter 8, Section 2(x) and other places. In fact, the marginal distributions of (102) are of this form. The distribution may be called a *Dirichlet* (*or* $\beta-$) *compound multinomial* distribution.

Distribution (102) was derived in the way described above by Mosimann [52] and has also been studied by Ishii and Hayakawa [39]. It is called the *multivariate binomial-beta distribution* by the latter authors.

The marginal distribution of $\mathbf{n}_j$ is

$$\text{Binomial } (N,\mathbf{p}_j) \underset{\mathbf{p}_j}{\wedge} \text{Beta}\left(\alpha_j, \sum_{i=1}^{k}\alpha_i - \alpha_j\right)$$

for which

(103)

$$\Pr[\mathbf{n}_j = x]$$

$$= \binom{N}{x}\left[\beta\left(\alpha_j, \sum_{i=1}^{k}\alpha_i - \alpha_j\right)\right]^{-1} \int_0^1 p_j^{x+\alpha_j-1}(1-p_j)^{N-x+\sum_1^k\alpha_i-\alpha_j-1}\, dp_j$$

$$= \binom{N}{x} \frac{\Gamma\left(\sum_{i=1}^{k}\alpha_i\right)}{\Gamma(\alpha_j)\Gamma\left(\sum_{i=1}^{k}\alpha_i-\alpha_j\right)} \frac{\Gamma(x+\alpha_j)\Gamma\left(N-x+\sum_{i=1}^{k}\alpha_i-\alpha_j\right)}{\Gamma\left(N+\sum_{i=1}^{k}\alpha_i\right)}$$

$$= \binom{x+\alpha_j-1}{x}\binom{N-x+\sum_{i=1}^{k}\alpha_j-\alpha_i-1}{N-x}\Big/\binom{N+\sum_{i=1}^{k}\alpha_i-1}{N}.$$

This is (for integer values of α_j's) a negative hypergeometric distribution (Chapter 6, Section 8.3) with parameters $N+\sum_{i=1}^{k}\alpha_i - 1$, α_j-1, N. The distribution (102) is sometimes called a *negative multivariate hypergeometric* distribution (Cheng Ping [17]).

The joint distribution of the subset $\mathbf{n}_{a_1}, \mathbf{n}_{a_2}, \ldots, \mathbf{n}_{a_s}$ (and $N - \sum_{i=1}^{s} \mathbf{n}_{a_i}$) is also of form (102) with parameters $N, \alpha_{a_1}, \alpha_{a_2}, \ldots \alpha_{a_s}, \sum_{j=1}^{k} \alpha_j - \sum_{i=1}^{s} \alpha_{a_i}$. This can be established by the following argument. If $\mathbf{n}_1, \mathbf{n}_2, \ldots \mathbf{n}_k$ have the joint multinomial distribution (1), then $\mathbf{n}_{a_1}, \mathbf{n}_{a_2}, \ldots, \mathbf{n}_{a_s}$ and $N - \sum_{i=1}^{s} \mathbf{n}_{a_i}$ have a joint multinomial distribution with parameters $N, p_{a_1}, \ldots p_{a_s}, 1 - \sum_{i=1}^{s} p_{a_i}$. Also if $\mathbf{p}_1, \ldots \mathbf{p}_k$ have a joint Dirichlet distribution with parameters $\alpha_1, \alpha_2 \ldots, \alpha_k$ then $\mathbf{p}_{a_1}, \mathbf{p}_{a_2}, \ldots, \mathbf{p}_{a_s}, 1 - \sum_{i=1}^{s} \mathbf{p}_{a_i}$ have a similar distribution with parameters $\alpha_{a_1}, \alpha_{a_2}, \ldots, \alpha_{a_s}, \sum_{i=1}^{k} \alpha_j - \sum_{i=1}^{s} \alpha_{a_i}$.

A similar argument leads to the conclusion that the conditional distribution of the remaining $\mathbf{n}_j$'s, given $\mathbf{n}_{a_1}, \mathbf{n}_{a_2}, \ldots, \mathbf{n}_{a_s}$ is of form (102) with parameters $N - \sum_{i=1}^{s} n_{a_s}, \{\alpha_j, j \neq a_1, \ldots, a_s\}$.

The factorial moments of distribution (102) are:

$$\mu_{(r_1, r_2, \ldots, r_k)} = E\left[\prod_{j=1}^{k} \mathbf{n}_j^{(r_j)}\right] = N^{\left(\sum_1^k r_j\right)} \left(\prod_{j=1}^{k} \alpha_j^{[r_j]}\right) \Big/ \left(\sum_{j=1}^{k} \alpha_j\right)^{\left[\sum_1^k r_j\right]}. \tag{104}$$

In particular

$$E(\mathbf{n}_j) = Np'_j; \ \mathrm{var}(\mathbf{n}_j) = \left(N + \sum_{i=1}^{k} \alpha_i\right)\left(1 + \sum_{i=1}^{k} \alpha_j\right)^{-1} Np'_j(1 - p'_j) \tag{105}$$

$$\mathrm{cov}(\mathbf{n}_i, \mathbf{n}_j) = -\left(N + \sum_{i=1}^{k} \alpha_i\right)\left(1 + \sum_{i=1}^{k} \alpha_i\right)^{-1} Np'_i p'_j \tag{106}$$

and hence

$$\mathrm{corr}(\mathbf{n}_i, \mathbf{n}_j) = -\sqrt{p'_i p'_j (1 - p'_i)^{-1} (1 - p'_j)^{-1}} \tag{107}$$

where

$$p'_j = \alpha_j \left(\sum_{i=1}^{k} \alpha_i\right)^{-1}.$$

We observe that the variance-covariance matrix of distribution (102) is equal to

$$\frac{\left(N + \sum_1^k \alpha_i\right)}{1 + \sum_1^k \alpha_i} \times \text{(variance-covariance matrix of a multinomial distribution with parameters } N, p'_1, p'_2, \ldots, p'_k). \tag{108}$$

(Note that $\sum_{i=1}^{k} p'_i = 1$.)

The multiple regression of $\mathbf{n}_j$ on $\mathbf{n}_{a_1},\mathbf{n}_{a_2},\ldots\mathbf{n}_{a_s}$ $(j \neq a_1,a_2,\ldots a_s)$ is

$$(109)\qquad E[\mathbf{n}_j \mid \mathbf{n}_{a_1},\ldots,\mathbf{n}_{a_s}] = \left(N - \sum_{i=1}^{s} \mathbf{n}_{a_i}\right)\alpha_j\left(\sum_{j=1}^{k}\alpha_j - \sum_{i=1}^{s}\alpha_{a_i}\right)^{-1}.$$

The distribution (102) depends on the parameters $N, \alpha_1, \ldots, \alpha_k$. Mosimann [52] has discussed the problem of estimating the parameters α_j (N being supposed known) given independent sets of observed values $(n_{1t},n_{2t},\ldots,n_{kt})$ with $\sum_{i=1}^{k} n_{it} = N$ $(t = 1,2,\ldots m)$. If a method of moments is used then from (105) we obtain

$$(110)\qquad \bar{\mathbf{n}}_j = N\tilde{\alpha}_j\left(\sum_{i=1}^{k}\tilde{\alpha}_i\right)^{-1} \qquad (j = 1,2,\ldots k)$$

where

$$\bar{\mathbf{n}}_j = m^{-1}\sum_{t=1}^{m}\mathbf{n}_{jt}.$$

From the k equations (110) the *ratios* of the $\tilde{\alpha}_j$'s, one to another are determined.

In order to obtain an equation from which $\sum_{i=1}^{k}\tilde{\alpha}_i$ can be obtained we make use of (108), equating the determinant of the sample variance-covariance matrix (sample generalized variance) to $[(N + \sum_{i=1}^{k}\tilde{\alpha}_i)/(1 + \sum_{i=1}^{k}\tilde{\alpha}_i)]^{k-1}$ times the variance-covariance matrix of a multinomial distribution with parameters $N, \bar{\mathbf{n}}_1/N, \bar{\mathbf{n}}_2/N, \ldots, \bar{\mathbf{n}}_k/N$, (only $(k-1)$ of the variables being used, to avoid having singular matrices). From the resulting equation $\sum_{i=1}^{k}\tilde{\alpha}_i$ can be calculated, and the individual values $\tilde{\alpha}_1, \tilde{\alpha}_2, \ldots, \tilde{\alpha}_k$ then calculated from (110). Numerical examples, applied to counts of pollen, are given by Mosimann [52]. The accuracy of this method of estimation is not yet well established.

8.2 *Dirichlet* ($\beta-$) *Compound Negative Multinomial Distribution*

If, in place of the multinomial distribution, a negative multinomial distribution is compounded by a Dirichlet distribution, the resulting distribution might well be called a *Dirichlet (or* $\beta-$*) compound negative binomial distribution.* This is in fact the name given to distributions represented symbolically as

$$(111)\qquad \text{Negative Multinomial }(N,\mathbf{P}_1,\mathbf{P}_2,\ldots,\mathbf{P}_k) \bigwedge_{\{\mathbf{P}_j/\mathbf{Q}\},\mathbf{Q}^{-1}} \text{Dirichlet }(\alpha_1,\alpha_2,\ldots,\alpha_{k+1})$$

where $\mathbf{Q} = 1 + \sum_{j=1}^{k}\mathbf{P}_j$ (Mosimann [53]). (Note that since there is no exact relationship among the P_j's, it is necessary to have $(k + 1)$ parameters in the Dirichlet distribution (compare (101).)

The distribution is defined by the formula

(112)

$$P(\mathbf{n}_1,\mathbf{n}_2,\ldots\mathbf{n}_k) = \frac{\Gamma\left(N+\sum_{j=1}^{k}\mathbf{n}_j\right)\Gamma\left(\sum_{j=1}^{k+1}\alpha_j\right)\Gamma(N+\alpha_{k+1})}{\left(\prod_{j=1}^{k}\mathbf{n}_j!\right)\Gamma(N)\Gamma(\alpha_{k+1})\Gamma\left(N+\sum_{j=1}^{k+1}\alpha_j+\sum_{j=1}^{k}n_j\right)} \times \prod_{j=1}^{k}\frac{\Gamma(\mathbf{n}_j+\alpha_j)}{\Gamma(\alpha_j)}$$

$$= \frac{N^{\left[\sum_1^k \mathbf{n}_j\right]}}{\left(\sum_{j=1}^{k+1}\alpha_j\right)^{\left[N+\sum_1^k \mathbf{n}_j\right]}} \cdot \alpha_{k+1}^{[N]} \prod_{j=1}^{k}\frac{\alpha_j^{[\mathbf{n}_j]}}{\mathbf{n}_j!}.$$

If the α_j s are integers the marginal distributions are generalized hypergeometric distributions. The properties of distribution (112) parallel very closely those of (102). Thus the distribution of a subset $\mathbf{n}_{a_1}, \ldots \mathbf{n}_{a_s}$ is of form (112) with parameters $N, \alpha_{a_1}, \alpha_{a_2}, \ldots, \alpha_{a_s}, \alpha_{k+1}$; the conditional distribution of the remaining $\mathbf{n}_j$'s given $\mathbf{n}_{a_1}, \mathbf{n}_{a_2} \ldots \mathbf{n}_{a_s}$ is also of form (112), with parameters $N + \sum_{i=1}^{s} \mathbf{n}_{a_i}, \{\alpha_j; j \neq \alpha_i\} \alpha_{k+1} + \sum_{i=1}^{s} \alpha_{a_i}$ and so on.

Mosimann [53] gives a table of expected values, variances, covariances and correlations. (He also gives formulas for these quantities which can be used with a general compounding distribution applied to either multinomial or negative multinomial distributions.)

9. Multivariate Multinomial Distributions

This distribution is the joint distribution of a number of multinomial distributions. As such, it is particularly suitable for representing data which can be classified in multi-dimensional contingency tables.

Consider an m-way $h_1 \times h_2 \times \ldots \times h_m$ cross-classified contingency table, the number of factor levels for the jth classification factor being h_j. For example in taking measurements on individuals sampled from a human population, factors of classification might be height, eye color, hair color, age and sex. In this case $m = 5$. If eye color is assessed as belonging to one of eight groups (levels) then $h_2 = 8$, while $h_5 = 2$ corresponding to the 'levels' male and female. We will denote the probability of obtaining the combination of levels $a_1, a_2, \ldots a_m$ by $p_{a_1 a_2 \ldots a_m}$, and *assume this is constant for all individuals selected.* Denoting the number of individuals, among a sample of size N, with this combination of levels, by $n_{a_1 a_2 \ldots a_m}$, the marginal distribution of $\mathbf{n}_1^{(1)}$,

$\mathbf{n}_2^{(1)}, \ldots, \mathbf{n}_{h_1}^{(1)}$, where

$$\mathbf{n}_j^{(1)} = \sum_{a_2 \ldots a_m} \sum \mathbf{n}_{ja_2 \ldots a_m}$$

will be (*under our assumptions*) multinomial with parameters $N, p_1^{(1)}, \ldots, p_{h_1}^{(1)}$ where

$$p_j^{(1)} = \sum_{a_2 \ldots a_m} \sum p_{ja_2 \ldots a_m}.$$

We then naturally wish to consider the joint distribution of the m sets of variables

$$\mathbf{n}_1^{(i)}, \mathbf{n}_2^{(i)}, \ldots, \mathbf{n}_{h_i}^{(i)} \qquad (i = 1,2,\ldots m).$$

Each set has a multinomial distribution, the joint distribution is called a *multivariate multinomial* distribution.

The joint probability generating function of the $\mathbf{n}_{a_1 a_2 \ldots a_m}$ is

$$\left(\sum_{a_1} \sum_{a_2} \cdots \sum_{a_m} p_{a_1 a_2 \ldots a_m} t_{a_1 a_2 \ldots a_m}\right)^N \tag{113}$$

and the joint probability generating function of the $\mathbf{n}_j^{(i)}$'s is

$$\left(\sum_{a_1} \sum_{a_2} \cdots \sum_{a_m} p_{a_1 a_2 \ldots a_m} t_{a_1}^{(1)} t_{a_2}^{(2)} \cdots t_{a_m}^{(m)}\right)^N \tag{114}$$

in an obvious notation.

Wishart [82] has given general formulas for calculating cumulants and product cumulants (and hence moments and product moments) of the distribution. He devoted particular attention to the case $m = 2$, $h_1 = h_2 = 2$. This corresponds to 2×2 contingency tables. The corresponding distribution (a special form of multivariate multinomial) is called a *bivariate binomial* distribution. This has also been studied by Aitken and Gonin [2] and by Capobianco [15].

In the general case, the joint distribution of any two variables $\mathbf{n}_{a_i}^{(i)}$, $\mathbf{n}_{a_j}^{(j)}$, $(j \neq i)$ is a bivariate binomial distribution. This can be seen by noting that the remaining $m - 2$ factors can be amalgamated into a single 'factor', as can all the variates $\mathbf{n}_u^{(i)}(u \neq a_i)$ and $\mathbf{n}_u^{(j)}(u \neq a_j)$ giving a 2×2 table,

	$n_{a_j}^{(j)}$	$N - n_{a_j}^{(j)}$
$n_{a_i}^{(i)}$		
$N - n_{a_i}^{(i)}$		

(with only the marginal totals shown).

By considering the limiting form of a bivariate binomial distribution as $N \to \infty$, with $Np_1^{(1)} = \theta_1$, $Np_1^{(2)} = \theta_2$, Aitken [1] derived the bivariate Poisson distribution described earlier in this chapter.

Steyn [71] has constructed another distribution which he also calls a "multivariate multinomial distribution". He supposes that members of an infinite population fall into one of $(k + 1)$ cells. If an individual is chosen from the i-th cell it causes an event, E, to occur i times. The probability generating function for the number of times E occurs when N individuals are chosen is

$$(p_0 + p_1t + p_2t^2 + \cdots + p_kt^k)^N. \tag{115}$$

The corresponding distribution is called by Steyn [66] a 'univariate multinomial distribution'. It is, of course, different from the multinomial distribution discussed in Section 2, which is a multivariate distribution. Steyn [71] defines a 'multivariate multinomial distribution' by means of the probability generating function

$$\left(p_0 + \sum_i \sum_j p_{ij}t_i^j\right)^N \tag{116}$$

where

$$p_0 \geq 0, p_{ij} \geq 0;\ p_0 + \sum_i \sum_j p_{ij} = 1 \quad (i = 1, \ldots, k; j = 1, \ldots, n_i)$$

and N is a positive integer. Formula (116) is a natural generalization of (115).

10. Multivariate Negative Multinomial Distribution

The *multivariate negative multinomial* distribution is defined in analogous fashion to the multivariate multinomial distribution. Using $\mathbf{n}_{a_i \ldots a_m}$, $\mathbf{n}_{a_j}^{(j)}$ with the same meanings, the joint probability generating function of $\mathbf{n}_{a_1 \ldots a_m}$ is

$$p_{11\ldots1}^N \left(t_{11\ldots1} - \sum_{\substack{a_1=1 \\ (a_1,\ldots,a_m)\neq(1,\ldots,1)}}^{h_1} \sum_{a_2=1}^{h_2} \cdots \sum_{a_m=1}^{h_m} t_{a_1a_2\ldots a_m}p_{a_1a_2\ldots a_m}\right)^{-N} \tag{117}$$

The joint probability generating function of the marginal totals $\{\mathbf{n}_{a_j}^{(j)}\}$ is

$$p_{11\ldots1}^N \left(t_1^{(1)}t_1^{(2)} \ldots t_1^{(m)} - \sum_{\substack{a_1 \\ (a_1,\ldots,a_m)\neq(1,\ldots,1)}} \sum_{a_2} \cdots \sum_{a_m} t_{a_1}^{(1)}t_{a_2}^{(2)} \ldots t_{a_m}^{(m)}p_{a_1a_2\ldots a_m}\right)^{-N} \tag{118}$$

The marginal joint distribution of $\mathbf{n}_1^{(j)}, \mathbf{n}_2^{(j)}, \ldots \mathbf{n}_{h_j}^{(j)}$ is negative multinomial. Wishart [82] indicates how to obtain formulas for the joint cumulants of $\{\mathbf{n}_{a_j}^{(j)}\}$. Similarly to the multivariate multinomial distribution, the joint distribution of $\mathbf{n}_{a_i}^{(i)}$ and $\mathbf{n}_{a_j}^{(j)}$ (with $i \neq j$) is a bivariate negative binomial distribution.

Note that the multivariate negative binomial distribution (Section 3) corresponds to the case $m = 1$.

Steyn [71] defines a 'multivariate negative multinomial distribution' in a similar way to his 'multivariate multinomial distribution'.

11. Tallis' 'Generalized Multinomial Distribution'

Tallis [76], [77] has constructed a 'generalized multinomial distribution'*

*It should be realized that Tallis used the word 'multinomial' to describe the *univariate* distribution of each $\mathbf{x}_j$.

which is the joint distribution of n discrete variables $\mathbf{x}_1, \mathbf{x}_2, \ldots \mathbf{x}_n$, each having the same marginal distribution

$$\Pr[\mathbf{x} = j] = p_j \qquad (j = 0,1,2,\ldots,k)$$

and such that the correlation between any two different $\mathbf{x}$'s has a specified value ρ. This joint distribution is defined by the corresponding joint probability generating function

$$\rho\left\{\sum_{j=0}^{k} p_j \left(\prod_{i=1}^{n} t_i\right)^j\right\} + (1-\rho)\prod_{i=1}^{n}\left\{\sum_{j=1}^{k} p_j t_i^j\right\}. \tag{119}$$

Tallis [76] describes how the values of $p_0, p_1, p_2 \ldots p_k$ and ρ may be estimated, using N sets of observed values of $\mathbf{x}_1, \mathbf{x}_2, \ldots, \mathbf{x}_n$. In [76], n is allowed to be a random variable with the ratios $p_0 : p_1 : p_2 : \ldots$ fixed, and also, with n fixed, $(p_0, p_1, \ldots, p_k)$ is supposed to have a Dirichlet distribution. Some general results, with both n and $(p_0, p_1, \ldots, p_k)$ varying, are also obtained.

12. Series Expansions

Representation of distributions by series expansions is used less often for discrete than for continuous distributions, and we have devoted little attention to it in this volume. Since the theory for the multivariate case is not much more complex than for univariate distributions, we judged it convenient to summarize the salient features, here, for both classes of distributions. The following brief discussion can be extended by reference to Aitken and Gonin [2], Gonin [28], [29], van Heerden and Gonin [35], Campbell [14], and Krishnamoorthy, [46].

A discrete distribution defined by

$$\Pr[\mathbf{x} = j] = P_j \qquad (j = 0,1,\ldots)$$

(with $\sum_{j=0}^{\infty} P_j = 1$) has associated with it orthogonal polynomials $G_0(x)$, $G_1(x)$, $\ldots$, where $G_r(x)$ is a polynomial of order r in x and

$$\sum_{j=0}^{\infty} P_j G_r(j) G_s(j) = 0 \qquad \text{for all } r \neq s. \tag{120}$$

For the Poisson distribution, the $G(x)$'s are the *Charlier Type B polynomials* which will be discussed in Chapter 12 (*Continuous Univariate Distributions*). For the binomial distribution with

$$P_j = \binom{n}{j} p^j (1-p)^{n-j} \qquad (j = 0, 1, \ldots, n)$$

the $G_r(x)$'s are the *Krawtchouk polynomials* (see Chapter 1, Section 3)

(121)

$$G_r(x) = (1 + p\Delta)^{-n+r-1}x^{(r)}$$
$$= x^{(r)} - \binom{r}{1}(n - r + 1)px^{(r-1)} + \binom{r}{2}(n - r + 1)^{[2]}p^2x^{(r-2)} - \cdots$$
$$\cdots + (-1)^r(n - r + 1)^{[r]}p^r.$$

For the negative binomial with

$$P_j = \binom{N - 1 + j}{N - 1}(P + 1)^{-N}\left(\frac{P}{P + 1}\right)^j \qquad (j = 0, 1, 2, \ldots)$$

the orthogonal polynomials are

(122)

$$G_r(x) = (1 - P\Delta)^{N+r-1}x^{(r)}$$
$$= x^{(r)} - \binom{r}{1}(N + r - 1)Px^{(r-1)} + \binom{r}{2}(N + r - 1)^{(2)}P^2x^{(r-2)} - \cdots$$
$$\cdots + (-1)^r(N + r - 1)^{(r)}P^r.$$

Returning to the general case, consider the distribution

(123) $$P'_j = \Pr[\mathbf{x} = j] = \{a_0G_0(j) + a_1G_1(j) + \cdots + a_rG_r(j)\}P_j \qquad (j = 0,1,2,\ldots).$$

Note that from (120), $\sum_{j=0}^{\infty} P_jG_r(j) = 0$ for $r > 0$, and so $a_0G_0(j)$ must be equal to 1, since

$$\sum_{j=0}^{\infty} P'_j = 1.$$

The coefficients of the G_r's are determined (apart from a factor of proportionality) by the moments of the distribution of $\mathbf{x}$. In particular, $G_1(x)$ must be proportional to $\{x - E[\mathbf{x}]\}$. The usual convention is to make the coefficient of x^r to $G_r(x)$ equal to 1, so that, for example, $G_1(x) = x - E[\mathbf{x}]$.

Bivariate distributions can be constructed of form

(124) $$\Pr[\mathbf{x}_1 = j_1, \mathbf{x}_2 = j_2]$$
$$= [1 + a_1G_1(j_1)G_1(j_2) + a_2G_2(j_1)G_2(j_2) + \ldots$$
$$\ldots + a_rG_r(j_1)G_r(j_2)]P_{j_1}P_{j_2}.$$

This is a natural generalization of (123).

Since $\sum_{j=0}^{\infty} G_r(j)P_j = 0$, it is clear that the marginal distribution of $\mathbf{x}_t$ $(t = 1,2)$ is

$$\Pr[\mathbf{x}_t = j] = P_j.$$

Also,

$$E[\mathbf{x}_1\mathbf{x}_2] = E[\mathbf{x}_1]E[\mathbf{x}_2] + a_1E[\mathbf{x}_1G_1(\mathbf{x}_1)]E[\mathbf{x}_2G_2(\mathbf{x}_2)]$$

and if $G(x) = x - E[\mathbf{x}]$, then

$$a_1 = \operatorname{corr}(\mathbf{x}_1,\mathbf{x}_2)\,[\operatorname{var}(\mathbf{x}_1)\operatorname{var}(\mathbf{x}_2)]^{-\frac{1}{2}}. \tag{125}$$

Campbell [14], gave the special form

(126)

$$\Pr[\mathbf{x}_1 = j_1, \mathbf{x}_2 = j_2] = \left[e^{-\theta_1}\frac{\theta_1^{j_1}}{j_1!}\right]\left[e^{-\theta_2}\frac{\theta_2^{j_2}}{j_2!}\right]\left[1 + \frac{\bar{\theta}}{1!}G_1(j_1,\theta_1)G_1(j_2,\theta_2) + \frac{\bar{\theta}^2}{2!}G_2(j_1,\theta_1)G_2(j_2,\theta_2) + \cdots\right]$$

where $\bar{\theta} = \rho(\theta_1\theta_2)^{\frac{1}{2}}$ and the G's are Charlier Type B polynomials.

Gonin [29], gave

(127)

$$\Pr[\mathbf{x}_1 = j_1, \mathbf{x}_2 = j_2] = \binom{n}{j_1} p_1^{j_1}(1 - p_1)^{n-j_1}\binom{n}{j_2} p_2^{j_2}(1 - p_2)^{n-j_2} \times\left[1 + \binom{n}{1}\epsilon G_1(j_1,\theta_1)G_2(j_2,\theta_2) + \binom{n}{2}\epsilon^2 G_2(j_1,\theta_1)G_2(j_2,\theta_2) + \cdots\right]$$

where $\epsilon = \rho[p_1(1 - p_1)p_2(1 - p_2)]^{\frac{1}{2}}$ and the G's are Krawtchouk polynomials.

Similar formulas for joint distributions of m variables have been given by Krishnamoorthy [46]. His 'multivariate Poisson' distribution is defined by the following equation

(128)

$$\Pr\left[\bigcap_{j=1}^{m}(\mathbf{x}_j = k_j)\right] = \left[\prod_{j=1}^{m}\frac{e^{-\theta_j}\theta_j^{k_j}}{k_j!}\right]\left[\exp\left\{\sum_i\sum_j \theta_{ij}G(k_i)G(k_j) + \sum_i\sum_j\sum_l \theta_{ijl}G(k_i)G(k_j)G(k_l) + \cdots + \theta_{12\ldots m}\prod_{j=1}^{m}G(k_j)\right\}\right]$$

and his 'multivariate binomial' by

(129)

$$\Pr\left[\bigcap_{j=1}^{m}(\mathbf{x}_j = k_j)\right] = \left[\prod_{j=1}^{m}\binom{n}{k_j} p_j^{k_j}(1 - p_j)^{n-k_j}\right] \times\left[1 + \sum_i\sum_j d_{ij}\frac{G(k_i)}{Np_iq_i}\frac{G(k_j)}{Np_jq_j} + \cdots + d_{12\ldots m}\prod_{j=1}^{m}\left\{\frac{G(k_j)}{Np_jq_j}\right\}\right]^n$$

with

$$d_{ij\ldots l} = p_{ij\ldots l} - p_ip_j\cdots p_l$$

where in the expansions of (128) and (129), $[G(k_i)]^r$ is replaced by $G_r(k_i)$ and N^r by $N^{(r)}$. (Of course the G's are Charlier Type B polynomials in (128) and Krawtchouk polynomials in (129).)

The factorial moment generating function of (129) is

$$\left(1 + \sum_{i=1}^{m} p_i t_i + \sum_i \sum_j p_{ij} t_i t_j + \cdots + p_{12\ldots m} t_1 t_2 \ldots t_m\right)^n.$$

Distribution (128) is obtained from (129) as the limit as $n \to \infty$ and $p_{ij\ldots\ell} \to 0$ with $np_{ij\ldots\ell} = \theta_{ij\ldots\ell}$ for all $i, j \ldots, \ell$. (Compare with the probability generating function (113).)

REFERENCES

[1] Aitken, A. C. (1936). A further note on multivariate selection, *Proceedings of the Edinburgh Mathematical Society*, **5,** 37–40.

[2] Aitken, A. C. and Gonin, H. T. (1935). On fourfold sampling with and without replacement, *Proceedings of the Royal Society of Edinburgh*, **55,** 114–125.

[3] Arbous, A. G. and Kerrich, J. E. (1951). Accident statistics and the concept of accident proneness, *Biometrics*, **7,** 340–432.

[4] Arbous, A. G. and Sichel, H. S. (1954). New techniques for the analysis of absenteeism data, *Biometrika*, **41,** 79–90.

[5] Argentiero, P. D., Morris, R. A. and Tolson, R. H. (1967). *The X^2 statistics and the goodness of fit test*, Report X-551-67-519, Goddard Space Flight Center, Greenbelt, Maryland.

[6] Asano, C. (1965). On estimating multinomial probabilities by pooling incomplete samples, *Annals of the Institute of Statistical Mathematics, Tokyo*, **17,** 1–13.

[7] Banerjee, D. P. (1959). On some theorems on Poisson distribution, *Proceedings of the National Academy of Science, India, Section A*, **28,** 30–33.

[8] Bates, Grace E. and Neyman, J. (1952). Contributions to the theory of accident proneness, *University of California, Publications in Statistics,* **1,** 215–253.

[9] Batschelet, E. (1960). Über eine Kontingenztafel mit fehlenden Daten, *Biometrische Zeitschrift*, **2,** 236–243.

[10] Bennett, B. M. and Nakamura, E. (1968). Percentage points of the range from a symmetric multinomial distribution, *Biometrika*, **55,** 377–379.

[11] Bennett, R. W. (1962). Sizes of the χ^2 test in the multinomial distribution, *Australian Journal of Statistics*, **4,** 86–88.

[12] Bhat, B. R. and Kulkarni, N. V. (1966). On efficient multinomial estimation, *Journal of the Royal Statistical Society, Series B*, **28,** 45–52.

[13] Camp, B. H. (1929). The multinomial solid and the chi test, *Transactions of the American Mathematical Society*, **31,** 133–144.

[14] Campbell, J. T. (1934). The Poisson correlation function, *Proceedings of the Edinburgh Mathematical Society*, **4,** 18–26.

[15] Capobianco, M. F. (1964). *On the bivariate binomial distribution and related problems*, Ph.D. Thesis, Polytechnic Institute of Brooklyn, Department of Mathematics.

[16] Chatfield, C., Ehrenberg, A. S. C. and Goodhardt, G. J. (1966). Progress on a simplified model of stationary purchasing behaviour, *Journal of the Royal Statistical Society, Series A*, **129,** 317–367.

[17] Cheng Ping (1964). Minimax estimates of parameters of distributions belonging to the exponential family, *Acta Mathematica Sinica*, **5,** 277–299.

[18] Cochran, W. G. (1954). Some methods for strengthening the common χ^2 tests, *Biometrics*, **10,** 417–451.

[19] Dwass, M. and Teicher, H. (1957). On infinitely divisible random vectors, *Annals of Mathematical Statistics*, **28,** 461–470.

[20] Edwards, Carol B. (1962). *Multivariate and multiple Poisson distributions*, Ph.D. Thesis, Iowa State University.

[21] Feller, W. (1957). *An Introduction to Probability Theory and its Applications*, New York: John Wiley & Sons, Inc.

[22] Finucan, H. M. (1964). The mode of a multinomial distribution. *Biometrika*, **51,** 513–517.

[23] Fisz, M. (1964). The limiting distributions of the multinomial distribution, *Studia Mathematica*, **14,** 272–275.

[24] Gart, J. J. (1966). Alternative analyses of contingency tables, *Journal of the Royal Statistical Society*, *Series B*, **28,** 164–179.

[25] Geppert, M. P. (1961). Erwartungstreue plausibelste Schätzen aus dreieckig gestützten Kontingenztafeln, *Biometrische Zeitscrift*, **3,** 55–67.

[26] Girshick, M. A., Mosteller, F. and Savage, L. J. (1946). Unbiased estimates for certain binomial sampling problems with applications, *Annals of Mathematical Statistics*, **17,** 13–23.

[27] Gold, R. (1962). *On comparing multinomial probabilities*, School of Aerospace Medicine, U.S.A.F. Medical Division, Brooks Air Force Base, Texas. Report No. 62–81.

[28] Gonin, H. T. (1961). The use of orthogonal polynomials of the positive and negative binomial functions in curve fitting by Aitken's method, *Biometrika*, **48,** 115–123.

[29] Gonin, H. T. (1966). Poisson and binomial frequency surfaces, *Biometrika*, **53,** 617–619.

[30] Goodman, L. A. (1965). On simultaneous confidence intervals for multinomial proportions, *Technometrics*, **7,** 247–254.

[31] Greenwood, M. and Yule, G. U. (1920). An inquiry into the nature of frequency distributions of multiple happenings, *Journal of the Royal Statistical Society*, *Series A*, **83,** 255–279.

[32] Guldberg, S. (1935). Recurrences formulae for the semi-invariants of some discontinuous frequency functions of *n* variables, *Skandinavisk Aktuarietidskrift*, **18,** 270–278.

[33] Gupta, S. S. and Nagel, K. (1966). *On Selection and Ranking Procedures and Order Statistics from the Multinomial Distribution*, Purdue University, Mimeo Series No. 77.

[34] Gurian, J. M., Cornfield, J. and Mosimann, J. F. (1964). Comparisons of power for some exact multinomial significance tests, *Psychometrika*, **29,** 409–419.

[35] Heerden, D. F. I. van and Gonin, H. T. (1966). The orthogonal polynomials of power series probability distributions and their uses, *Biometrika*, **53,** 121–128.

[36] Hoel, P. G. (1938). On the chi-square distribution for small samples, *Annals of Mathematical Statistics*, **9,** 158–165.

[37] Holgate, P. (1964). Estimation for the bivariate Poisson distribution, *Biometrika*, **51,** 241–245.

[38] Holgate, P. (1966). Bivariate generalizations of Neyman's Type A distribution, *Biometrika*, **53,** 241–244.

[39] Ishii, G. and Hayakawa, R. (1960). On the compound binomial distribution, *Annals of the Institute of Statistical Mathematics, Tokyo*, **12,** 69–80. (Errata, **12,** 208)

[40] Johnson, N. L. (1960). An approximation to the multinomial distribution, some properties and applications, *Biometrika*, **47**, 93–102.

[41] Johnson, N. L. and Tetley, H. (1950). *Statistics: An Intermediate Text Book*, Vol. II, London: Cambridge University Press. (Chapter 12).

[42] Johnson, N. L. and Young, D. H. (1960). Some applications of two approximations to the multinomial distribution, *Biometrika*, **47**, 463–469.

[43] Kesten, H. and Morse, N. (1959). A property of the multinomial distribution, *Annals of Mathematical Statistics*, **30**, 120–127.

[44] Khatri, C. G. and Mitra, S. K. (1968). *Some identities and approximations concerning positive and negative multinomial distributions*, Technical Report No. 1/68, Indian Statistical Institute, Calcutta.

[45] Kozelka, R. M. (1956). Approximate upper percentage points for extreme values in multinomial sampling, *Annals of Mathematical Statistics*, **27**, 507–512.

[46] Krishnamoorthy, A. S. (1951). Multivariate binomial and Poisson distributions, *Sankhyā*, **11**, 117–124.

[47] Lancaster, H. O. (1961). Significance tests in discrete distributions, *Journal of the American Statistical Association*, **56**, 223–234. (Errata, **57**, 919)

[48] Lancaster, H. O. and Brown, T. A. I. (1965). Sizes of the χ^2 test in the symmetrical multinomials, *Australian Journal of Statistics*, **7**, 40–44.

[49] Lewontin, R. C. and Prout, T. (1956). Estimation of the number of different classes in a population, *Biometrics*, **12**, 211–223.

[50] Lundberg, O. (1940). *On Random Processes and Their Application to Sickness and Accident Statistics*, Uppsala: Almquist and Wiksell.

[51] Mallows, C. L. (1968). An inequality involving multinomial probabilities, *Biometrika*, **55**, 422–424.

[52] Mosimann, J. E. (1962). On the compound multinomial distribution, the multivariate β-distribution and correlations among proportions, *Biometrika*, **49**, 65–82.

[53] Mosimann, J. E. (1963). On the compound negative multinomial distribution and correlations among inversely sampled pollen counts, *Biometrika*, **50**, 47–54.

[54] Muhamedhanova, R. and Suleimanova, M. (1961). Unbiased estimates for selection problems in the case of multinomial distributions, *Proceedings of the Institute of Mathematics, AN UzSSR*, **22**, 121–129. (In Russian)

[55] Neyman, J. (1963). Certain chance mechanisms involving discrete distributions (Inaugural Address), *Proceedings of the International Symposium on Discrete Distributions, Montreal*, 4–14.

[56] Olbrich, E. (1965). Geometrical interpretation of the multinomial distribution and some conclusions, *Biometrische Zeitschrift*, **7**, 96–101. (In German)

[57] Olkin, I. and Sobel, M. (1965). Integral expressions for tail probabilities of the multinomial and negative multinomial distributions, *Biometrika*, **52**, 167–179.

[58] Patil, G. P. and Bildikar, Sheela (1967). Multivariate logarithmic series distribution as a probability model in population and community ecology and some of its statistical properties, *Journal of the American Statistical Association*, **62**, 655–674.

[59] Pearson, K. (1900). On the criterion that a given system of deviations from the probable in the case of a correlated system of variables is such that it can reasonably be supposed to have arisen from random sampling, *Philosophical Magazine, 5th Series*, **50,** 157–175.

[60] Quesenberry, C. P. and Hurst, D. C. (1964). Large sample simultaneous confidence intervals for multinomial proportions, *Technometrics*, **6,** 191–195.

[61] Rao, R. (1957). Maximum likelihood estimation for the multinomial distribution, *Sankhyā*, **18,** 139–148.

[62] Rjauba, B. (1958). Asymptotic laws for the trinomial distribution, *Vilnus Universitat, Mosklo Akademijos Darbai, Series B*, **25,** 17–22. (In Russian)

[63] Rüst, H. (1965). Die Momente der Testgrösse des χ^2-Tests, *Zeitschrift für Wahrscheinlichkeitstheorie und verwandte Gebiete*, **4,** 222–231.

[64] Shanawany, M. R. El (1936). An illustration of the accuracy of the χ^2 approximation, *Biometrika*, **28,** 179–187.

[65] Sibuya, M., Yoshimura, I. and Shimizu, R. (1964). Negative multinomial distribution, *Annals of the Institute of Statistical Mathematics, Tokyo*, **16,** 409–426.

[66] Steyn, H. S. (1951). On discrete multivariate probability functions, *Proceedings Koninklijke Nederlandse Akademie van Wetenschappen, Series A*, **54,** 23–30.

[67] Steyn, H. S. (1955). On discrete multivariate probability functions of hypergeometric type, *Proceedings Koninklijke Nederlandse Akademie van Wetenschappen, Series A*, **58,** 588–595.

[68] Steyn, H. S. (1956). On the univariate series $F(t) \equiv F(a;b_1,b_2...b_k;c;t,t^2,...t^k)$ and its applications in probability theory, *Proceedings Koninklijke Nederlandse Akademie van Wetenschappen, Series A*, **59,** 190–197.

[69] Steyn, H. S. (1957). On regression properties of discrete systems of probability functions, *Proceedings Koninklijke Nederlandse Akademie van Wetenschappen, Series A*, **60,** 119–127.

[70] Steyn, H. S. (1959). On χ^2-tests for contingency tables of negative multinomial types, *Statistica Neerlandica*, **13,** 433–444.

[71] Steyn, H. S. (1963). On approximations for the distributions obtained from multiple events, *Proceedings Koninklijke Nederlandse Akademie van Wetenschappen, Series A*, **66,** 85–96.

[72] Steyn, H. S. and Wiid, A. J. B. (1958). On eightfold probability functions, *Proceedings Koninklijke Nederlandse Akademie van Wetenschappen, Series A*, **61,** 129–138.

[73] Stoka, M. I. (1966). Asupra functiei de repartitie a unei repartitii multinomiale, *Studii si Cercetari Matematice*, **18,** 1281–1285. (In Roumanian)

[74] Studer, H. (1966). Prüfung der exakten χ^2 Verteilung durch die stetige χ^2 Verteilung, *Metrika*, **11,** 55–78.

[75] Sutherland, I. and Whitwell, G. P. B. (1948). Studies in occupational morbidity (2), *British Industrial Medicine*, **5,** 77.

[76] Tallis, G. M. (1962). The use of a generalized multinomial distribution in the estimation of correlation in discrete data, *Journal of the Royal Statistical Society, Series B*, **24,** 530–534.

[77] Tallis, G. M. (1964). Further models for estimating correlation in discrete data, *Journal of the Royal Statistical Society, Series B*, **26,** 82–85.

[78] Teicher, H. (1954). On the multivariate Poisson distribution, *Skandinavisk Aktuarietidskrift*, **37,** 1–9.

[79] Vora, S. A. (1950). *Bounds on the distribution of chi-square*, Ph.D. Thesis. University of North Carolina.

[80] Wise, M. E. (1963). Multinomial probabilities and the χ^2 and X^2 distributions, *Biometrika*, **50,** 145–154.

[81] Wise, M. E. (1964). A complete multinomial distribution compared with the X^2 approximation and an improvement to it, *Biometrika*, **51,** 277–281.

[82] Wishart, J. (1949). Cumulants of multivariate multinomial distributions, *Biometrika*, **36,** 47–58.

[83] Wolfowitz, J. (1947). The efficiency of sequential estimates and Wald's equation for sequential processes, *Annals of Mathematical Statistics*, **18,** 215–230.

[84] Young, D. H. (1961). Quota fulfilment using unrestricted random sampling, *Biometrika*, **48,** 333–342.

[85] Young, D. H. (1962). Two alternatives to the standard χ^2 test of the hypothesis of equal cell frequencies, *Biometrika*, **49,** 107–116.

[86] Young, D. H. (1967). A note on the first two moments of the mean deviation of the symmetrical multinomial distribution, *Biometrika*, **54,** 312–314.

Acknowledgements

Figures

P. 39	Adapted from Ord, J. K. (1967). Graphical methods for a class of discrete distributions, *Journal of the Royal Statistical Society*, *Series A*, **130,** 232–238.
P. 42	Adapted from Katz, L. (1963). Unified treatment of a broad class of discrete probability distributions, *Proceedings of the International Symposium on discrete distributions*, *Montreal*, 175–182.
P. 60	Adapted from Pearson, E. S. and Hartley, H. O. (Ed.) (1958). *Biometrika Tables*, 1, London: Cambridge University Press.
P. 71 & 72	Adapted from Larson, H. R. (1966). A nomograph of the cumulative binomial distribution, *Industrial Quality Control*, **23,** 270–278. Figure 3 is reproduced by permission of the author and Western Electric Company.
P. 133	Adapted from Anscombe, F. J. (1950). Sampling theory of the negative binomial and logarithmic series distributions, *Biometrika*, **37,** 358–382.
P. 222 & 223	Adapted from Martin, D. C. and Katti, S. K. (1962). Approximations to the Neyman Type A distribution for practical problems, *Biometrics*, **18,** 354–364.
P. 225	(left and right) Adapted from Katti, S. K. and Gurland, J. (1962). Efficiency of certain methods of estimation for the negative binomial and the Neyman Type A distributions, *Biometrika*, **49,** 215–226.
P. 268 & 269	Adapted from Borenius, G. (1953). On the statistical distribution of mine explosions, *Skandinavisk Aktuarietidskrift*, **36,** 151–157.
P. 272	Adapted from Oliver, R. M. (1961). A traffic counting distribution, *Operations Research*, **9,** 802–810.

Tables

P. 38	Adapted from Ord, J. K. (1967). Graphical methods for a class of discrete distributions, *Journal of the Royal Statistical Society*, *Series A*, **130,** 232–238.
P. 67	Adapted from Dunin-Barkovsky, I. V. and Smirnov, N. V. (1955). *Theory of Probability and Mathematical Statistics in Engineering*, Moscow: GITTL.

P. 75 Adapted from Shah, S. M. (1966). On estimating the parameter of a doubly truncated binomial distribution, *Journal of the American Statistical Association*, **61,** 259–263.

P. 76 Adapted from Gart, J. J. (1968). A simple nearly efficient alternative to the simple sib method in the complete ascertainment case, *Annals of Human Genetics, London*, **31,** 283–291.

P. 97 Adapted from Mantel, N. (1962). (Appendix C to Haenzel, W., Loveland, D. B., and Sorken, M. B.) Lung cancer mortality as related to residence and smoking histories. I. White males, *Journal of the National Cancer Institution*, **28,** 947–997.

P. 100 & 101 Adapted from Eisenhart, C. and Zelen, M. (1958). Elements of Probability, Chapter 12, *Handbook of Physics*, New York, N.Y.: McGraw-Hill, Inc.

P. 129 & 130 Adapted from Bartko, J. J. (1966). Approximating the negative binomial, *Biometrics*, **8,** 340–342.

P. 147 Adapted from Chapman, D. G. (1951). Some properties of the hypergeometric distribution with applications to zoological sample censuses, *University of California Publications in Statistics*, **1,** 131–159.

P. 151 Adapted from Bol'shev, L. N. (1964). Distributions related to the hypergeometric distribution, *Teoriya Veroyatnostei i ee Primeneniya*, **9,** 687–692. (In Russian)

P. 154 155, & 156 Adapted from Ross, A. S. C. (1950). Philological probability problems, *Journal of the Royal Statistical Society, Series B*, **12,** 19–41.

P. 161 Adapted from Kemp, C. D. and Kemp, A. W. (1956). Generalized hypergeometric distributions, *Journal of the Royal Statistical Society, Series B*, **18,** 202–211.

P. 171 Adapted from Blackman, G. E. (1935). A study of statistical methods of the distribution species in a grassland association, *Annals of Botany* (*New Series*), **49,** 749–777.

P. 172 Adapted from Rowe, J. A. (1942). Mosquito light trap catches from ten American cities, 1940, *Iowa State College Journal of Science*, **16,** 487–518.

P. 175 Adapted from Patil, G. P. (1962). Some methods of estimation for the logarithmic series distribution, *Biometrics*, **18,** 68–75.

P. 193 Adapted from Katti, S. K. and Gurland, J. (1962). Some methods of estimation for the Poisson binomial distribution, *Biometrics*, **18,** 42–51.

P. 208 Adapted from Yoneda, K. (1962). Estimations in some modified Poisson distributions, *Yokohama Mathematical Journal*, **10,** 73–96.

P. 211 Adapted from Thomas, Marjorie (1949). A generalization of Poisson's binomial limit for use in ecology, *Biometrika*, **36,** 18–25.

P. 242 Adapted from Walther, A. (1926). Anschauliches zür Riemannschen Zetafunktion, *Acta Mathematica*, **48,** 393–400.

P. 243 & 244 Adapted from Moore, P. G. (1956). The geometric, logarithmic and discrete Pareto forms of series, *Journal of the Institute of Actuaries*, **82,** 130–136.

P. 260 Adapted from Foster, F. G. and Stuart, A. (1954). Distribution-free tests in time-series based on the breaking of records, *Journal of the Royal Statistical Society, Series B*, **16,** 1–13.

Index

This Index is intended as an auxiliary to the detailed Table of Contents. It contains references only to distributions, and only to those occurrences not easily identifiable from the Table of Contents. For example, there are no references to 'Binomial' for any pages in Chapter 3, *which is devoted to the Binomial distribution.*

ABCDEFGHIJ–M–7876543210/69

Applied Probability and Statistics (Continued)

DUNN · Basic Statistics: A Primer for the Biomedical Sciences, *Second Edition*

DUNN and CLARK · Applied Statistics: Analysis of Variance and Regression

ELANDT-JOHNSON · Probability Models and Statistical Methods in Genetics

ELANDT-JOHNSON and JOHNSON · Survival Models and Data Analysis

FLEISS · Statistical Methods for Rates and Proportions

GALAMBOS · The Asymptotic Theory of Extreme Order Statistics

GIBBONS, OLKIN, and SOBEL · Selecting and Ordering Populations: A New Statistical Methodology

GNANADESIKAN · Methods for Statistical Data Analysis of Multivariate Observations

GOLDBERGER · Econometric Theory

GOLDSTEIN and DILLON · Discrete Discriminant Analysis

GROSS and CLARK · Survival Distributions: Reliability Applications in the Biomedical Sciences

GROSS and HARRIS · Fundamentals of Queueing Theory

GUPTA and PANCHAPAKESAN · Multiple Decision Procedures: Theory and Methodology of Selecting and Ranking Populations

GUTTMAN, WILKS, and HUNTER · Introductory Engineering Statistics, *Second Edition*

HAHN and SHAPIRO · Statistical Models in Engineering

HALD · Statistical Tables and Formulas

HALD · Statistical Theory with Engineering Applications

HARTIGAN · Clustering Algorithms

HILDEBRAND, LAING, and ROSENTHAL · Prediction Analysis of Cross Classifications

HOEL · Elementary Statistics, *Fourth Edition*

HOLLANDER and WOLFE · Nonparametric Statistical Methods

HUANG · Regression and Econometric Methods

JAGERS · Branching Processes with Biological Applications

JESSEN · Statistical Survey Techniques

JOHNSON and KOTZ · Distributions in Statistics
Discrete Distributions
Continuous Univariate Distributions–1
Continuous Univariate Distributions–2
Continuous Multivariate Distributions

JOHNSON and KOTZ · Urn Models and Their Application: An Approach to Modern Discrete Probability Theory

JOHNSON and LEONE · Statistics and Experimental Design in Engineering and the Physical Sciences, Volumes I and II, *Second Edition*

KALBFLEISCH and PRENTICE · The Statistical Analysis of Failure Time Data

KEENEY and RAIFFA · Decisions with Multiple Objectives

LANCASTER · An Introduction to Medical Statistics

LEAMER · Specification Searches: Ad Hoc Inference with Nonexperimental Data

McNEIL · Interactive Data Analysis

MANN, SCHAFER, and SINGPURWALLA · Methods for Statistical Analysis of Reliability and Life Data

MEYER · Data Analysis for Scientists and Engineers